Microcontinuum
Field Theories

Springer
New York
Berlin
Heidelberg
Barcelona
Hong Kong
London
Milan
Paris
Singapore
Tokyo

A. Cemal Eringen

Microcontinuum Field Theories

I. Foundations and Solids

With 46 Figures

 Springer

A. Cemal Eringen
Emeritus Professor, Princeton University
15 Red Tail Drive
Littleton, CO 80126-5001
USA

Library of Congress Cataloging-in-Publication Data
Eringen, A. Cemal.
    Microcontinuum field theories : foundations and solids / A. Cemal
  Eringen
        p.   cm.
    Includes bibliographical references.
    ISBN 0-387-98620-0 (alk. paper)
      1. Unified field theories.   2. Elasticity.   3. Fluid dynamics.
    4. Electromagnetic theory.
    QC173.7.E75   1998
    530.14'2 – dc21   98-30560

Printed on acid-free paper.

Production managed by Allan Abrams; manufacturing supervised by Jeffrey Taub.
Typeset by The Bartlett Press, Inc., Marietta, GA.
Printed and bound by Maple-Vail Book Manufacturing Group, York, PA.
Printed in the United States of America.

9 8 7 6 5 4 3 2 1

ISBN 0-387-98620-0 Springer Verlag New York Berlin Heidelberg   SPIN 10691374

Preface

This two-volume work is devoted to the development of the *microcontinuum field theories* for material bodies that possess inner structures that can deform and interact with mechanical and elecromagnetic (E-M) fields. The terminology *microcontinuum field theories* is used here to denote theories of *micromorphic, microstretch,* and *micropolar* (3M) continua (solids, fluids, E-M) that were introduced by the author. These theories constitute logical and systematic extensions of the classical field theories (elasticity, fluid dynamics, E-M theory) covering a much broader scope of important physical phenomena, beyond those of classical field theories. The significance of these theories has been demonstrated amply by the thousands of papers published during the last three decades, covering the applications of these theories to a variety of different fields of physics and engineering. Yet there is much that remains to be done, both on mathematical grounds and with applications. Two centuries of continuing research in classical field theories can provide a dramatic demonstration of the scope of the future research work left in these promising fields.

The main purpose of this book is to present the micromorphic, microstretch, and micropolar (3M) field theories (which will be referred to also as 3M continua) on a unified foundation and, by selected solutions, to direct attention to the new physical phenomena they predict. Based on this foundation, continuum theories are also given for more complex substances (e.g., liquid crystals, blood, bubbly fluids, see Volume II).

Presently, there exists no published treatise on the subject covered in these volumes, except some partial expositions and limited reviews, that are scattered throughout the literature. We believe this book, with its unified and precise approach to the development of nonlinear and linear theories, and with its breadth through applications to diverse active fields of physics should be valuable for graduate study and research. A little observation will show that the methodology is easily adaptable to the development of parallel continuum field theories. In this sense, it may be considered a theory of continuum theories!

Chapter 1 presents kinematics of micromorphic, microstretch, and micropolar (3M) continua on a unified basis. Chapter 2 develops balance laws and thermodynamics. In Chapter 3, we give the general constitutive theory and obtain exact constitutive equations for nonlocal, memory-dependent elastic solids and viscous fluids and their local forms. Chapter 4 brings the E-M interactions into formulation. These results are then used in the following chapters to obtain linear theories. Chapters 1–3 are essential to the remainder of the book, which is devoted to a variety of applications. Volume I explores applications of the elastic 3M media, and Volume II of the fluid media.

Volume II begins with the presentation of basic equations of micropolar viscous fluids. Solutions are presented for selected problems, to demonstrate new predictions offered beyond the classical Navier–Stokes theory. Formulations continue to theories of liquid crystal, E-M interactions, suspension theory, and several other topics (as outlined in the Preface to Volume II). Space limitations have made the inclusion of many other fields impossible. Interesting among them are the 3M mixture theories, composites, porous media, and their interactions with E-M fields.

Microcontinuum field theories are nonlocal in character, with limited nonlocality. Parallel development of nonlocal field theories requires separate volumes. Presently, this is only in the planning stage.

I express my thanks to Greg Stiehl, a serious graduate student, for searching and copying some references for me. Professor Mark Lusk was kind to locate an undergraduate student, James Casey, who typed the text. My love is forever for my daughter, Meva Eringen, who read the nontechnical parts of the manuscript and edited them.

Littleton, Colorado   A. Cemal Eringen
                                                                September 1998

Contents

Introduction

Microcontinuum field theories constitute extensions of the classical field theories concerned with the deformations, motions, and electromagnetic (E-M) interactions of material media, as continua, in microscopic space and short time scales. In terms of a physical picture, a material body may be envisioned as a collection of a large number of *deformable particles* (subcontinua or microcontinua) that contribute to the macroscopic behavior of the body. This picture is similar to those frequently used in statistical mechanics and transport theory, namely, a small volume element containing a large number of *small particles* for which statistical laws are valid.

A question arises: How can we reconcile the concept of the deformable particle (which implies finite size) with the continuum hypothesis? Mathematically, material particles are posited to be geometrical points that possess physical and mathematical properties, e.g., mass, charge, deformable directors. The field equations constructed with this model are expected then to unveil many new and wider classes of physical phenomena that fall outside classical field theories.

The concept of microcontinuum naturally brings length and time scales into field theories. The response of the body is influenced heavily with the ratio of the characteristic length λ (associated with the external stimuli) to the internal characteristic length l. When $\lambda/l \gg 1$, the classical field theories give reliable predictions since, in this case, a large number of particles act collaboratively. However, when $\lambda/l \approx 1$, the response of constituent subcontinua (particles) becomes important, so that the axiom of locality underlying classical field theories fails. For example, classical elasticity predicts two branches of nondispersive acoustical waves whose short wavelength behavior departs drastically from experimental observations. Moreover, there exist several other observed branches (so-called optical branches) of the dispersion curves, in the high-frequency range, that are not predicted at all by the classical theory of elasticity (cf. Sections 5.11, 6.3, 7.4). This is because the ratio of the internal to external characteristic times $\tau/T = \Omega/\omega (\Omega \equiv$ external frequency, $\omega =$ internal frequency) is missing in classical elasticity, but is present in microelasticity. These are indicative of the fact that microcontinuum theories are *nonlocal* in character, with limited nonlocality.

With microcontinuum field theories, in essence, we have a fundamental departure from the classical ideas of local action. The physics of lattice dynamics clearly points to necessity for this departure. However, we need not point only to the atomic scale phenomena to justify the *raison d'être* of these theories. Nature abounds with many substances which clearly point to the necessity for the incorporation of micromotions into mechanics. Suspensions, blood flow, liquid crystals, porous media, polymeric substances, solids with microcracks, dislocations and disclinations, turbulent fluids with vortices, bubbly fluids, slurries, and composites are but a few examples which require consideration of the motions of their microconstituents, e.g., blood cells, suspended particles, fibers, grains, swarms of liquid crystals, etc.

The departure from classical (local) theories begins with polar theories. In these theories, the material points are considered to possess orientations. A material point carrying three deformable directors (*micromorphic continuum*) introduces nine extra degrees of freedom over the classical theory. When the directors are constrained to have only breathing-type microdeformations, then we have *microstretch continuum*, and the extra degrees of freedom are reduced to four: three microrotations and one microstretch. In *micropolar continuum*, a point is endowed with three rigid directors only. A material point is then equipped with the degrees of freedom for rigid rotations only, in addition to the classical translational degrees of freedom.

By associating tensors of various order with the material points, higher-order theories can be constructed. However, this process soon ceases to be mathematically tractable. Micromorphic, microstretch, and micropolar (3M) theories of the first grade (three directors only), introduced by the present author, are the only ones that have had wide attraction to research workers during the last three decades. Consequently, this two-volume work is primarily devoted to formulations and diverse applications of these theories.

It appears that the polar nature of crystalline solids was recognized by Voigt [1887]. In fact, he gave equations of equilibrium for such crystals (including moment equilibrium) and explored crystal properties. Later E. and F. Cosserat [1909], in a remarkable memoir, developed a theory of elasticity, by means of a variational principle which they called "l'Action Euclidienne." With the Cosserats, a concept, "triedre," was introduced to elasticity. The Cosserats obtained equations for the balance of momentum for the dynamical case. But they did not give a specific microinertia, nor a conservation law for the microinertia tensor, which are crucial to the construction of constitutive equations and dynamical problems in solids and fluent media, e.g., liquid crystals, suspension, etc. The rigid directors used by the Cosserats, and later by others to represent rigid rotations, did not have metrical significance. Consequently, difficulties are encountered in the treatment of material symmetry regulations in constitutive equations. For example, none of the work employing rigid directors discusses isotropy, e.g., E. and F. Cosserat did not give any specific constitutive equations.

The Cosserats book remained dormant for over half a century. In the 1960s the subject matter was reopened, independently, by several authors. Grad [1952] obtained some conservation laws by means of statistical mechanics. Günther [1958] and Schaëfer [1967] recapitulated the Cosserats elasticity and remarked on its connection to dislocations. At this time, also popular, was a theory of indeterminate couple stress which is mostly abandoned now. In this theory, the axisymmetric part of the stress tensor is redundant and it remains indeterminate. Some of these early theories are discussed in various review articles, cf. Eringen [1967b], Ariman et al. [1973, 1974].

These theories, as well as the Cosserats elasticity, have contact with the nonlinear theory of microelasticity published by Eringen and Şuhubi [1964]. In the same year, Eringen [1964a][1] published his theory of microfluids, uncovering a new balance law: the *law of conservation of microinertia*. This law was missing from all other previous work. Without the balance law of the microinertia tensor, basic field equations are incomplete and the evolution of the constitution of the body with motion cannot be determined.

Material particles of bodies, generally, are of arbitrary shapes. The microinertia is a tensor (not necessarily scalar as in the spherical particles). The evolution of the inertia tensor with motion determines the anisotropic character of the body at any time. Thus, for example, mechanics of granular and porous solids, those of composites with their chopped fibers oriented in the same direction and deformations and motions of crystalline solids, cannot be addressed. Material properties of liquid crystals (e.g., viscosities, elasticities) depend significantly on the orientations and shapes of their molecules (cf. Volume II). Moreover, the phase transitions are the result of the change of the microinertia tensor with temperature.

A complete and *well-posed* linear theory of micropolar elasticity was introduced by Eringen [1966a].[2] This was followed by an exact nonlinear

---

[1] See also Eringen [1964b], which includes theories of micromorphic fluids, anisotropic fluids, and micro-viscoelasticity.

[2] A technical report containing the full text of this paper was distributed widely as an ONR report in 1964. Part of this report with additional materials on micropolar fluids, micropolar viscoelasticity, nonlocal micropolar elasticity, and solutions of channel

theory, Eringen [1968a, 1970a]. For a comprehensive *exposé*, including microphic theories of solids, fluids, and nonlocal theories, see Eringen and Kafadar [1976]. For the relativistic aspect, see Kafadar and Eringen [1971].

Theories of micropolar fluids, Eringen [1966b], and microstretch fluids, Eringen [1969a], and microstretch elastic solids, Eringen [1971b], complete the cycle of basic theories of microcontinua. While the extension of microcontinuum theories to other fields continued and continues even today, the mechanical foundation was complete by 1971. Among other developments, I cite: memory-dependent media (solids, fluids, polymers), mixtures, diffusion, E-M interactions, microplasticity, dislocations, disclinations, suspensions, blood flow, porous media, nonlocal polar theories, etc. Some of these topics are the subject of Volume II and, for the nonlocal theories, another volume is planned.

The published work in microcontinuum mechanics is so large that it is not possible to do justice to all contributors by mere mention of their names. The majority of these contributions consists of applications of the theory, i.e., solutions of the equations of micropolar, microstretch, and micromorphic (3M) solids and fluids of Eringen. Contributions dealing with theoretical problems (e.g., existence, uniqueness, asymptotic behavior, etc.) are not rare. In appropriate places, references are made to some of the contributors of the topic under discussion.

Partial reviews of various sectors of these fields exist. Eringen [1968a] gave an extensive account on micropolar elasticity. Ariman et al. [1973, 1974] published a comprehensive review of microcontinuum fluid mechanics. Nowacki's [1970] lecture notes contain solutions of various problems in micropolar elasticity. Eringen [1970a] presented mathematical foundations of micropolar thermoelasticity. Eringen and Kafadar [1976] presented microcontinuum fluids and solids (all 3M continua). This book also contains accounts on the nonlocal continuum theory. Bedford and Drumheller [1983] reviewed theories of immiscible and structured fluids. Petrosyan [1984] (in Russian) explored solutions to some problems in micropolar fluids. In a book, Stokes [1984] discussed various theories and some applications of the fluids with microstructure. For a thorough discussion of the theory of micropolar fluids (especially mathematical aspects, such as existence, well-posedness), see Lukaszewicz [1998][3]

The present two volumes rely heavily on the works of Eringen and his co-workers in regard to the formulations of the three different microcontinuum theories (we also call them "3M continua"): micromorphic, microstretch, and micropolar (3M) theories. Over the years, these theories have undergone deep scrutiny, and we have learned more fundamental and better

---

flow was presented at the Midwestern Mechanics Conference at Wisconsin University, in August, 1965, and published in the Proceedings of this conference, Eringen [1965].

[3]I am indebted to Dr. Lukaszewicz for sending me a copy of Petrosyan's book with the translation of its Table of Contents, and a copy of his forthcoming book prepublication.

approaches to their formulations. It is now possible to present all three fields on a unified and firm foundation. Most of the formulations are new and have not been published in the present form. In some places we give a different and simpler approach to the published accounts.

The main purpose of these books is the development of these theories. Example solutions are provided to demonstrate the power and potential of theories and some mathematical techniques are given for solving problems. By no means can any claim be made in regard to completeness of the coverage.

Chapter 1 is concerned with the kinematics of the 3M continua. Strain measures, deformation-rate tensors, and relative strain tensors, essential to fluid media, are introduced. Compatibility conditions, material time-rates of tensors, principles of objectivity, and the definitions of mass, inertia, momenta, and kinetic energy essential to dynamics are presented. In Chapter 2 we present the concept of stress tensors and give balance laws. The second law of thermodynamics and the concept of dissipation, fundamental to the construction of the constitutive theory, are the subject of this chapter. Chapter 3 introduces fundamental ideas for the construction of constitutive equations. Elastic solids, viscous fluids, memory-dependent solids, and fluids are discussed. In Chapter 4, the topic of Chapter 3 is extended to the constructions of the constitutive equations of 3M continua interacted by E-M fields. Nonlinear constitutive equations are obtained for elastic solids, viscous fluids, and memory-dependent solids and fluids. These results are used in succeeding chapters to obtain linear constitutive equations for 3M continua interacted by E-M fields, e.g., magnetohydrodynamics, piezoelectricity of micropolar solids, liquid crystals, etc.

Chapters 1 to 3 are essential to the rest of both volumes. With Chapter 5 we begin to undertake a deeper study for the simplest of the 3M continua: micropolar elasticity. By and large, this topic is the most developed of the three fields. Here, we present the solutions of many problems, discuss new physical phenomena predicted by the theory, and establish contact with lattice dynamics, waves, dislocations, and disclinations. Also, a new plate theory is presented. This chapter may be of interest to those who are working with applications and solving problems in micropolar elasticity.

Chapter 6 presents fundamental ideas and predictions of the theory of microstretch elasticity. The uniqueness theorem is given. Fundamental solutions are presented, surface waves are discussed, and the relationship to crystal lattice dynamics is provided.

Chapter 7 is an *exposé* of micromorphic elasticity. By far, this is the least-developed field, and yet it is the most promising research area. Here we obtain linear constitutive equations of micromorphic elasticity. Micromorphic theory, in special cases, gives microstretch and micropolar elasticities. Passages to these theories are established. Material stability restrictions are obtained. Solutions for plane harmonic waves are given, displaying the intimate connection with lattice dynamics. Finally, in Chapter 8, we discuss the subject of E-M interactions with micropolar media. Linear constitutive equations are constructed for anisotropic and isotropic media. Field equations are given and material stability regulations are obtained. Piezoelectricity and magnetoelasticity are discussed briefly.

Volume II of this work should provide even stronger motivation for research in these fields. Fluid media, with promises for rational formulation of turbulence, liquid crystals, and dense suspensions should provide challenges to research workers. E-M interactions with these materials possess other exciting mysteries to be discovered by future engineers, scientists, and mathematicians.

To them, I offer the logo of
the Larousse dictionary:

Je séme a tout vent

1
Kinematics

Arguments against new ideas generally pass through distinct stages from:
"It's not true" to
"Well it may be true, but not important" to
"It's true and it's important, but it's not new–we knew it all along."

From *The Artful Universe*
by John D. Barron

1.0 Scope

This chapter is concerned with the kinematics of deformations and motions
of micromorphic, microstretch, and micropolar (3M) continua. In Section
1.1 we present a physical picture to show how microdeformation of bodies
gives rise to extra degrees of freedom necessary to characterize microstructural continua. To this end, examples are chosen from lattice dynamics of
crystalline solids, liquid crystals, suspensions, animal blood, and composites. Definitions are then given for the microcontinua based on this picture.
In Section 1.2, motion and micromotions are defined. Micromorphic continua are then characterized by the macromotions of a particle and a set
of deformable directors that represent the mathematical model of a deformable particle. The rotation is the subject of Section 1.3, where finite
macrorotation of directors is introduced and the fundamental theorem of
rotation is given. In Section 1.4 we define microstretch and micropolar
continua and show that they arise as special *subcontinua*, by means of constraints placed on micromorphic continua. Possible applications of these
continua are mentioned briefly. Section 1.5 is devoted to strain measures
of micromorphic, microstretch, and micropolar continua. To eliminate repetitions, we shall also call them *3M continua.*

In Section 1.6, we introduce relative motions and relative strain measures
by considering a spatial reference configuration $\mathbf{x}$ at time t and by referring
to motion at a later time $t' > t$, relative to their configuration at time t.

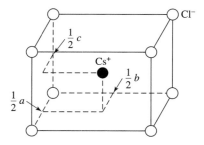

FIGURE 1.1.1. Cesium Chloride

The relative strain measures, based on the relative motion, are fundamental
to fluent media.

Section 1.7 deals with the compatibility conditions of 3M continua. Given
the strain tensors of simply connected 3M continua, what equations must
they satisfy in order to represent a single-valued displacement field? These
are given by theorems (compatibilty conditions). Violation of the compatibility conditions indicates the presence of defects (dislocations, disclinations, etc.) in the body.

Section 1.8 is devoted to material time-rates of tensors. Velocity, acceleration, microgyration, and spin inertia are discussed. Time-rates of strain
measures are calculated and deformation-rate tensors are introduced for
3M continua. A representation theorem is presented for the microgyration
tensor.

In Section 1.9 we introduce the concept of *objectivity*, alternatively known
as *material-frame indifference*. Objective tensors are candidates for the
constitutive equations.

The closing section of Chapter 1, Section 10, is devoted to the concepts of
mass, microinertia, momenta, and kinetic energy. The laws of conservation
of mass and microinertia follow from statistical moments of the micromass.
These concepts play central roles in the dynamical equations of microcontinua.

1.1   Physical Picture

In the atomic scale, crystalline solids possess primitive cells in the form of
geometrical figures (lattice structures) like cubes, hexagons, etc. For example, cesium chloride (CsCl) has a cubic structure with Cs^+ located at the
center of the cube and Cl^- at eight corners of the cube (Figure 1.1.1). In
this case, Cs^+ is the central location of the cell, with eight discrete vectors
$(\frac{1}{2}, \frac{1}{2}, \frac{1}{2})$ marking the positions of the eight Cl^-. There exist many other
crystalline solids with different ions occupying different positions in their
primitive lattices. This is also true for more complex structures consist-

ing of molecules. Phonon dispersion experiments display several branches confirming the presence of the lattice structure. For a few examples, we mention experiments by Pope for germanium, and by Iengar et al. for magnesium in Vallis [1965, pp. 147 and pp. 230]. For silicon see W. Cochran, and for CuZn, Brockhouse in Stevenson [1966, pp. 68 and 137]; for diamond, see Warren et al. [1965], and for NaI, Woods et al. [1960]. See also Section 5.13.

There exist fluids with oriented molecules. For example, liquid crystals possess dipolar elements in the form of short bars and platelets. Nematic and smectic liquid crystals possess layers of bars imbedded in liquid substances. The classical example, p-azoxyanisole (PAA) is composed of rigid bars of length approximately 20 Å and width 5 Å imbedded in a liquid. A number of synthetic polypeptides, in appropriate solvents, have rodlike formations with rod length of the order of 300 Å and width 20 Å. In cholesteric liquid crystals, the rod orientations from one plane to the next, along the common normal to the planes, varies slightly, forming a helical structure. These oriented fluids are excellent examples of the micropolar media (see Volume II).

Animal blood carrying deformable platelets, clouds with smoke, slurries, suspensions, bubbly fluids, concrete, granular solids, composite materials are other examples of the media with microstructures.

In all these examples, we notice that the primitive elements of the media are *stable elements* (e.g., primitive lattice, bar or plate elements of liquid crystals, blood cells, chopped fibers in a composite). These stable elements are considered deformable, but not destructible. For brevity, we shall call these stable elements *particles*.

Definition. A microcontinuum is a continuous collection of deformable point particles.

Physically, the particles are point particles, i.e., they are infinitesimal in size. They do not violate continuity of matter, and yet, they are deformable. Clearly, the deformability of the material point places microcontinuum theories beyond the scope of the classical continuum theory. The questions arises then:

How can we represent the intrinsic deformation of a point particle?

This question is settled by replacing the deformable particle with a geometrical point P and some vectors attached to P that denote the orientations and intrinsic deformations of the material points of P. This is compatible with the classical picture where a material point in a continuum is endowed with physical properties such as mass density, electric field, stress tensor, etc. Here, the vectors assigned to P represent, in addition, the degrees of freedom arising from the deformations of the material points of the particle. Accordingly, a particle P is identifed by its position vector (or its coordinates X_K) $K = 1, 2, 3$, in the reference state B and vectors at-

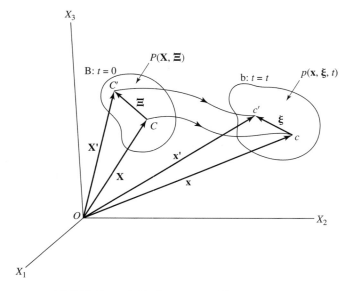

FIGURE 1.1.2. Deformation of Microelement

tached to P, representing the inner structure of P by Ξ_α, $\alpha = 1, 2, \ldots, N$.
Both $\boldsymbol{X}$ and Ξ_α have their own motions

$$\boldsymbol{X} \xrightarrow{t} \boldsymbol{x}, \qquad \Xi_\alpha \xrightarrow{\boldsymbol{X},t} \boldsymbol{\xi}_\alpha, \qquad \alpha = 1, 2, \ldots, N. \qquad (1.1.1)$$

Such a medium may be called microcontinuum of *grade N*. Presently, no
general theory of this magnitude exists. This book is concerned with the
case of grade 1 ($\alpha = 1$).

From a physical (and practical) point of view, a visual benefit will re-
sult if we consider Ξ and $\boldsymbol{\xi}$ as the relative positions of the material points
(sometimes called *microelements*). These are contained in the particle P
(sometimes called the *macroelement*), respectively, in the reference (*material*) and *spatial* configurations (i.e., in the undeformed and deformed
states) (Figure 1.1.2)

$$\boldsymbol{x}' = \boldsymbol{x}(\boldsymbol{X},t) + \boldsymbol{\xi}(\boldsymbol{X},\Xi,t). \qquad (1.1.2)$$

In fact, the subject of micromorphic continuum involves a special form of
(1.1.2), namely,

$$\boldsymbol{x}' = \boldsymbol{x}(\boldsymbol{X},t) + \boldsymbol{\chi}_K(\boldsymbol{X},t)\Xi_K \qquad (1.1.3)$$

(see Section 1.2). Thus, in this sense, for example, the cesium chloride
is represented with a mathematical model of a point P which carries de-
formable vectors, $\boldsymbol{\chi}_K$. Here, $\boldsymbol{x}(\boldsymbol{X},t)$ accounts for the motion of the Cs$^+$
ion located at the centroid and $\boldsymbol{\chi}_K$ accounts for the inner motions of the
Cl$^-$ ions located at Ξ with respect to the centroid of the cesium cell (i.e.,

Cs$^+$). Indeed, this representation is fairly universal. Clearly, it is applicable to liquid crystals, blood cells, suspension, and many other media.

1.2   Motions and Deformations

A physical body B is considered to be a collection of a set of material particles $\{P\}$. The body is imbedded in a three-dimensional Euclidean space E^3, at all times. The set B is considered to be a subset of the *universal set* U, consisting of B and its *complement* B', the set of elements that are not in B. Both B and B' may contain subsets. In the natural state B, B' is not interacted with B.

A material point $P(\boldsymbol{X}, \boldsymbol{\Xi}) \in B$ is characterized by its centroid C and vector $\boldsymbol{\Xi}$ attached to C. The point C is identified by its rectangular coordinates X_1, X_2, X_3 in a coordinate network $X_K, K = 1, 2, 3$, and the vector $\boldsymbol{\Xi}$ by its components Ξ_1, Ξ_2, Ξ_3 (in short Ξ_K) in the coordinate frame X_K.

Deformation carries $P(\boldsymbol{X}, \boldsymbol{\Xi})$ to $p(\boldsymbol{x}, \boldsymbol{\xi})$ in a spatial frame of reference b, so that, $X_K \longrightarrow x_k$, $\Xi_K \longrightarrow \xi_k$ $(K = 1, 2, 3; k = 1, 2, 3)$, Figure 1.1.2. In general, these mappings are expressed by

$$\boldsymbol{X} \longrightarrow \boldsymbol{x} = \hat{\boldsymbol{x}}(\boldsymbol{X}, t) \qquad \text{or} \qquad x_k = \hat{x}_k(X_K, t), \qquad (1.2.1)$$

$$\boldsymbol{\Xi} \longrightarrow \boldsymbol{\xi} = \hat{\boldsymbol{\xi}}(\boldsymbol{X}, \boldsymbol{\Xi}, t) \qquad \text{or} \qquad \xi_k = \hat{\xi}_k(X_K, \Xi_K, t). \qquad (1.2.2)$$

The mapping (1.2.1) is called the *macromotion* (or simply, the motion) and (1.2.2) the *micromotion*.

Material particles are considered to be of very small size (infinitesimally small) as compared to macroscopic scales of the body. Consequently, a linear approximation in $\boldsymbol{\Xi}$ is permissible for the micromotion (1.2.2) replacing it by

$$\xi_k = \chi_{kK}(\boldsymbol{X}, t)\Xi_K, \qquad (1.2.3)$$

where, and henceforth, the summation convention on repeated indices is understood.

Note that $\boldsymbol{X}$ is taken to be the centroid of P, i.e., $\boldsymbol{\xi}(\boldsymbol{X}, \boldsymbol{0}, t) = \boldsymbol{0}$.

Definition 1 (Micromorphic Continuum[1]).   A material body is called a micromorphic continuum of grade one (or simply micromorphic continuum) if its motions are described by (1.2.1) and (1.2.3) which possess continuous partial derivatives with respect to X_K and t, and they are invertible uniquely, i.e.,

$$X_K = \hat{X}_k(\boldsymbol{x}, t), \qquad k = 1, 2, 3, \qquad (1.2.4)$$

$$\Xi_K = \mathfrak{X}_{Kk}(\boldsymbol{x}, t)\xi_k, \qquad K = 1, 2, 3, \quad k = 1, 2, 3. \qquad (1.2.5)$$

---

[1]The origin of the theory goes back to two papers by Eringen and Şuhubi [1964]. The theory was completed by Eringen [1964a] with the introduction of the microinertia conservation law. The present terminology was introduced by Eringen [1964b].

The two-point tensors χ_{kK} and $\mathfrak{X}_{Kk}$ are called *microdeformation* and *inverse microdeformation* tensors, respectively (alternatively, *deformable directors*). The mathematical idealization (1.2.3) is valid from the continuum viewpoint, only when the particles are considered to be infinitesimally small, so that the continuity of matter is not violated.

The existence of solutions (1.2.4) and (1.2.5) or (1.2.1) and (1.2.3) requires that the following implicit function theorem of calculus is valid:

Theorem 1 (Implicit Functions).   If, for a fixed t, the function $\hat{x}_k(\boldsymbol{X}, t)$ is continuous and possesses continuous first-order partial derivatives with respect to X_K in a neighborhood $|\boldsymbol{X}' - \boldsymbol{X}| < \Delta$ of a point C and if the Jacobians

$$J \equiv \det\left(\frac{\partial x_k}{\partial X_k}\right), \qquad j \equiv \det \chi_{kK}, \qquad (1.2.6)$$

do not vanish there, then unique inverses of (1.2.1) and (1.2.3) in the forms (1.2.4) and (1.2.5) exist in a neighborhood $|\boldsymbol{x}' - \boldsymbol{x}| < \delta$ of a point c, at time t, and (1.2.4) possess continuous first-order partial derivatives with respect to x_k.

In order to retain the right-hand screw orientations of the frames-of-references, we assume

$$J \equiv \det\left(\frac{\partial x_k}{\partial X_k}\right) > 0, \qquad j \equiv \det \chi_{kK} = 1/\det \mathfrak{X}_{Kk} > 0. \qquad (1.2.7)$$

The existence of unique inverses (1.2.4) and (1.2.5) expresses the physical assumption of continuity, indestructibility, and impenetrability of matter. No region of positive finite volume is deformed into one of zero or infinite volume. Every region goes into a region, every surface into a surface, and every curve into a curve. The three independent directors $\mathfrak{X}_K$ go to the *independent directors* χ_k, Figure 1.2.1,

$$\chi_K = \chi_{kK}(\boldsymbol{X}, t)\boldsymbol{i}_k, \qquad \mathfrak{X}_k = \mathfrak{X}_{Kk}(\boldsymbol{x}, t)\boldsymbol{I}_K, \qquad (1.2.8)$$

where $\boldsymbol{I}_K$ and $\boldsymbol{i}_k$ denote, respectively, Cartesian unit vectors in the material and spatial frames of references B and b.

A material point in the body is now considered to possess *three deformable directors*, which represent the degrees of freedom arising from microdeformations of the physical particle. Thus, a micromorphic continuum is none other than a classical continuum endowed with extra degrees of freedom represented by the deformable directors χ_K and $\mathfrak{X}_k$.

For some purposes, it is useful to introduce the second-order tensors

$$\chi_{kl} = \chi_{kK}\delta_{Kl}, \qquad \mathfrak{X}_{KL} = \mathfrak{X}_{Kk}\delta_{kL}, \qquad (1.2.9)$$

where δ_{Kl} and δ_{kL} are called *shifters*, defined by

$$\boldsymbol{I}_K \cdot \boldsymbol{i}_l = \delta_{Kl}, \qquad \boldsymbol{i}_k \cdot \boldsymbol{I}_L = \delta_{kL}, \qquad (1.2.10)$$

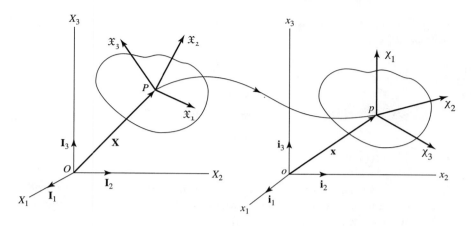

FIGURE 1.2.1. Deformable Directors

which are none other than the cosine directors of the unit vectors of $\boldsymbol{B}$ and $\boldsymbol{b}$.

Gradients of deformation satisfy

$$x_{k,K}X_{K,l} = \delta_{kl}, \qquad x_{k,K}X_{L,k} = \delta_{KL}. \tag{1.2.11}$$

Also, by substitution from (1.2.5) into (1.2.3), we have

$$\chi_{kK}\mathfrak{X}_{Kl} = \delta_{kl}, \qquad \chi_{kK}\mathfrak{X}_{Lk} = \delta_{KL}, \tag{1.2.12}$$

where, and henceforth, we drop the carat (^) distinguishing the mapping function from its image, and use a comma to denote partial derivatives

$$x_{k,K} = \frac{\partial x_k}{\partial X_K}, \qquad X_{K,l} = \frac{\partial X_K}{\partial x_l}.$$

The linear equations (1.2.11) and (1.2.12) can be solved for $X_{K,k}$ and $\mathfrak{X}_{Kk}$:

$$X_{K,k} = \frac{1}{2J}\epsilon_{KLM}\epsilon_{klm}x_{l,L}x_{m,M}, \tag{1.2.13}$$

$$\mathfrak{X}_{Kk} = \frac{1}{2j}\epsilon_{KLM}\epsilon_{klm}\chi_{lL}\chi_{mM}, \tag{1.2.14}$$

where ϵ_{klm} and ϵ_{KLM} are the permutation symbols and

$$J \equiv \det x_{k,K} = \frac{1}{6}\epsilon_{KLM}\epsilon_{klm}x_{k,K}x_{l,L}x_{m,M},$$
$$j \equiv \det \chi_{kK} = \frac{1}{6}\epsilon_{KLM}\epsilon_{klm}\chi_{kK}\chi_{lL}\chi_{mM}. \tag{1.2.15}$$

By differentiation, we obtain the following useful results:

$$\frac{\partial J}{\partial x_{k,K}} = \text{cofactor } x_{k,K} = JX_{K,k}, \qquad \frac{\partial j}{\partial \chi_{kK}} = j\mathfrak{X}_{Kk}, \tag{1.2.16}$$

$$\left(JX_{K,k}\right)_{,K} = 0, \qquad \left(J^{-1}x_{k,K}\right)_{,k} = 0. \tag{1.2.17}$$

1. Kinematics

The following identities are useful in these and in other algebraic manipulations:

$$\epsilon_{ijk}\epsilon_{mnp} = \begin{vmatrix} \delta_{im} & \delta_{in} & \delta_{ip} \\ \delta_{jm} & \delta_{jn} & \delta_{jp} \\ \delta_{km} & \delta_{kn} & \delta_{kp} \end{vmatrix}, \qquad \epsilon_{ijk}\epsilon_{inp} = \delta_{jn}\delta_{kp} - \delta_{kn}\delta_{jp},$$

$$\epsilon_{ijk}\epsilon_{ijp} = 2\delta_{kp}, \qquad\qquad \epsilon_{ijk}\epsilon_{ijk} = 3! = 6. \qquad (1.2.18)$$

1.3 Rotation

According to a theorem of Cauchy, a matrix $\boldsymbol{F}$ may be decomposed as products of two matrices, one of which is orthogonal and the other a symmetric matrix (cf. Eringen [1980, p. 46]), i.e.,

$$\boldsymbol{F} = \boldsymbol{R}\boldsymbol{U} = \boldsymbol{V}\boldsymbol{R} \qquad (1.3.1)$$

with

$$\boldsymbol{U}^2 = \boldsymbol{F}^T\boldsymbol{F}, \qquad \boldsymbol{V}^2 = \boldsymbol{F}\boldsymbol{F}^T \qquad (1.3.2)$$

where a superscript T denotes transpose. If we take $F_{kK} = x_{k,K}$, then $\boldsymbol{R}$ represents a classical macrorotation tensor. If we take $F_{kK} = \chi_{kK}$, then $\boldsymbol{R}$ represents the microrotation tensor. $\boldsymbol{U}$ and $\boldsymbol{V}$ are called *right* and *left stretch tensors* for macro- and microdeformations. In the case of microdeformation, the above equations read:

$$\boldsymbol{\chi} = \boldsymbol{r}\boldsymbol{u} = \boldsymbol{v}\boldsymbol{r} \qquad (1.3.3)$$

and

$$\boldsymbol{u}^2 = \boldsymbol{\chi}^T\boldsymbol{\chi}, \qquad \boldsymbol{v}^2 = \boldsymbol{\chi}\boldsymbol{\chi}^T. \qquad (1.3.4)$$

Besides microstretch tensors $\boldsymbol{u}$ and $\boldsymbol{v}$, there exist microstretch tensors arising from the gradients of $\boldsymbol{\chi}$. From (1.3.3), we have

$$\boldsymbol{\chi}_{,K} = \boldsymbol{r}_{,K}\boldsymbol{u} + \boldsymbol{r}\boldsymbol{u}_{,K} = \boldsymbol{v}_{,K}\boldsymbol{r} + \boldsymbol{v}\boldsymbol{r}_{,K}. \qquad (1.3.5)$$

The case of $\boldsymbol{u}^2 = \boldsymbol{v}^2 = \boldsymbol{1}$ is significant, because in this case $\boldsymbol{\chi}^T = \boldsymbol{\mathfrak{X}}$ and $\boldsymbol{\chi}\boldsymbol{\mathfrak{X}} = \boldsymbol{1}$. This case is known as *micropolar*. In this case, (1.3.5) gives

$$\boldsymbol{\chi}_{,K} = \boldsymbol{r}_{,K}\boldsymbol{1} = \boldsymbol{1}\boldsymbol{r}_{,K} \qquad (1.3.6)$$

so that the microdeformation gradient is equal to the microrotation gradient and through (1.3.3) we have

$$\boldsymbol{\chi} = \boldsymbol{r}, \qquad \boldsymbol{u}^2 = \boldsymbol{v}^2 = \boldsymbol{1}. \qquad (1.3.7)$$

It is now clear that for the micropolar continua, $\boldsymbol{\chi}$ is none other than the *microrotation* tensor.

From these considerations, it follows that, in a micromorphic continuum, deformation of a particle P is composed of (Figure 1.3.1):

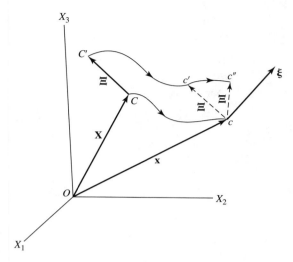

FIGURE 1.3.1. Microdeformation: $\vec{CC'} = \Xi$ at natural state, $\vec{cc'}$ translation of Ξ to c, $\vec{cc''}$ microrotation of Ξ at c, and ξ stretch of Ξ at c and final position.

(i) Classical macrodeformation consisting of a translation that carries C to a spatial place c, a macrorotation at c, and a macrostretch at c;

(ii) microdeformation that carries directors at C with the translation from C to c;

(iii) microrotation of directors at c;

(iv) microstretch of directors at c which may accompany a further microrotation of directors.

These compositions are product compositions so that, in general, the measures of intermediate states will depend on the order; the result, however, being the same, whichever order may have been followed.

Finite Rotation of Rigid Directors

For micropolar continuum, directors are rigid. Here it is possible to represent the motion of directors as a rigid body rotation with respect to an axis.

Theorem (Finite Microrotation Tensor). The finite microrotation tensor χ_{kl} is characterized by

$$\chi_{kl} = \cos\phi\,\delta_{kl} - \sin\phi\,\epsilon_{klm}n_m + (1 - \cos\phi)n_k n_l, \qquad (1.3.8)$$

$$\phi = (\phi_k\phi_k)^{1/2}, \qquad n_k = \frac{\phi_k}{\phi}, \qquad (1.3.9)$$

$$\chi_{kl} = \chi_{kL}\delta_{Ll}. \qquad (1.3.10)$$

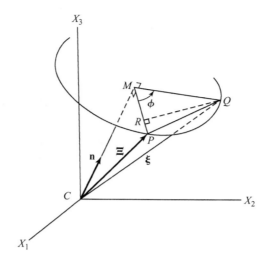

FIGURE 1.3.2. Finite Microrotation

Here, δ_{Ll} are the shifters (see (1.2.10)). A geometrical proof of (1.3.8) is as follows. Consider a centroidal vector $\vec{CP} = \boldsymbol{\Xi}$ at the undeformed particle P, Figure 1.3.2. After rotation about an axis $\vec{CM}$, the point of P moves to Q,

$$\boldsymbol{\xi} = \boldsymbol{\Xi} + \vec{PQ}, \qquad \vec{PQ} = \vec{PR} + \vec{RQ},$$

where $\vec{QR}$ is perpendicular to $\vec{PM}$. But

$$\vec{PR} = -(1 - \cos\phi)\left(\boldsymbol{\Xi} - \vec{CM}\right), \qquad \vec{RQ} = \boldsymbol{n} \times \boldsymbol{\Xi}\sin\phi,$$
$$\vec{CM} = (\boldsymbol{n} \cdot \boldsymbol{\Xi})\,\boldsymbol{n}.$$

Combining, I have

$$\boldsymbol{\xi} = \cos\phi\boldsymbol{\Xi} + \boldsymbol{n} \times \boldsymbol{\Xi}\sin\phi + (1 - \cos\phi)(\boldsymbol{n} \cdot \boldsymbol{\Xi})\,\boldsymbol{n}. \qquad (1.3.11)$$

Upon using (1.2.3) and (1.3.10), we get (1.3.8).

If we denote any rotation about a unit vector $\boldsymbol{n}$ by $r(\boldsymbol{n}, \phi)$, then the rotation given by (1.3.8) is multivalued, since

$$r(\boldsymbol{n}, \phi + 2k\pi) = r(-\boldsymbol{n}, -\phi + 2k\pi), \qquad (1.3.12)$$

where k is an integer. Although the multiplicity of rotation can be reduced by taking (see Synge [1960, p. 17])

$$\boldsymbol{\phi} = \boldsymbol{n}\sin(\phi/2), \qquad \boldsymbol{\phi} \cdot \boldsymbol{\phi} = \sin^2(\phi/2). \qquad (1.3.13)$$

In order to consider the reflection of $\boldsymbol{n}$ (i.e., $-\boldsymbol{n}$) in the plane of rotation, we retain the form (1.3.8).

If the directors undergo successive rotations $\chi^{(1)}, \chi^{(2)}, \ldots$, then the final rotation is given by

$$\xi_k^{(n)} = \chi_{kl_{n-1}}^{(n)} \chi_{k_{n-1}k_{n-2}}^{(n-1)} \cdots \xi_{k_1l}^{(1)} \Xi_L \delta_{Ll} \tag{1.3.14}$$

(cf. Jeffreys and Jeffreys [1950, p. 96]). For small rotations, (1.3.8) gives

$$\chi_{kl} \simeq \delta_{kl} - \epsilon_{klm}\phi_m, \tag{1.3.15}$$

indicating that the rotation tensor can be represented by a vector ϕ_k.

From (1.3.8) it also follows that, for $\phi = 0$, $\chi_{kl} = \delta_{kl}$. Hence:

Corollary. The necessary and sufficient conditions that the micropolar directors remain parallel to their original directions, are the vanishing ϕ.

Here, we take cognizance of the difference between the macrorotation tensor of classical continuum mechanics and the microrotation. The macrorotation tensor due to the change of line elements, in classical continuum mechanics, is given by

$$R_{kK} = x_{k,L} \overset{-1/2}{C}_{LK} \tag{1.3.16}$$

(see Eringen [1980, p. 46]), where C_{LK} is the classical deformation tensor defined by

$$C_{LK} = x_{k,L}x_{k,K}. \tag{1.3.17}$$

In the linear theory, (1.3.16) reduces to

$$R_{km} - \delta_{km} \equiv \tilde{R}_{km} = \tfrac{1}{2}\left(u_{k,m} - u_{m,k}\right), \tag{1.3.18}$$

where u_k is the displacement vector.

The difference between macrorotation tensors $\boldsymbol{R}$, given by (1.3.16), and microrotation tensor $\boldsymbol{\chi}$, given by (1.3.8), is now fully clarified. For small deformations, we have the vector representations (1.3.18) and (1.3.15). The fact that this difference is responsible for the exta stress fields in a micropolar body, indicates the importance of this distinction.

1.4   Microstretch and Micropolar Continua

Definition 1 (Microstretch Continuum).   A micromorphic continuum is called microstretch,[2] if its directors are related by

$$\mathfrak{X}_{Kk} = \frac{1}{j^2}\chi_{kK}. \tag{1.4.1}$$

---

[2]The theory microstretch continua was introduced by Eringen [1969a, 1971b], originally with the terminology "Micropolar Fluid with Stretch" and "Micropolar Elastic Solids with Stretch." The present terminology was coined by Eringen in later publications.

From orthogonality relations (1.2.12), it follows that

$$j = \det \chi_{kK} = 1/\det \mathfrak{X}_{Kk}. \tag{1.4.2}$$

Consequently, directors χ_{kK} of microstretch continua satisfy

$$\chi_{kK}\chi_{lK} = j^2\delta_{kl}, \qquad \chi_{kK}\chi_{kL} = j^2\delta_{KL}. \tag{1.4.3}$$

Since $j = \det \chi_{kK}$ represents the microvolume change with microdeformations, we see that:

Theorem 1. A microstretch continuum is a micromorphic continuum that is constrained to undergo microrotation and microstretch (expansion and contraction) without microshearing (breathing microrotations).

Definition 2 (Micropolar Continuum). A micromorphic continuum is called micropolar[3] if its directors are orthonormal, *i.e.*,

$$\chi_{kK}\chi_{lK} = \delta_{kl}, \qquad \mathfrak{X}_{Kk}\mathfrak{X}_{Lk} = \delta_{KL}. \tag{1.4.4}$$

By multiplying the first one of these by $\mathfrak{X}_{Ll}$ and using (1.2.12) we find that

$$\chi_{kK} = \mathfrak{X}_{Kk}. \tag{1.4.5}$$

Consequently, for micropolar continua, (1.2.12) takes the form

$$\chi_{kK}\chi_{lK} = \delta_{kl}, \qquad \chi_{kK}\chi_{kL} = \delta_{KL}. \tag{1.4.6}$$

It follows that

$$j = \det \chi_{kK} = 1. \tag{1.4.7}$$

Equation (1.4.6) *expresses that the directors of micropolar continuum are rigid.* Consequently, micromotion is a rigid-body rotation.

Classification of microcontinua is shown in the **following sketch** (Figure 1.4.1).

Possible Substances that Can Be Modeled by 3M Continua

(i) Micromorphic Continua

Polymers with flexible molecules, liquid crystals with side chains, animal blood with deformable cells, suspensions with deformable elements, turbulent fluids with flexible vortices.

---

[3]The Eringen and Şuhubi paper [1964] contains a preliminary version of the micropolar theory under the terminology "Linear theory of couple stress." The complete linear theory was given by Eringen [1966a]. Nonlinear theories are found in later expositions, e.g., Eringen [1970a], Kafadar and Eringen [1971]. Cosserats' elasticity is related in principle, but not in full content, e.g., it lacks the law of conservation of microinertia and several other aspects.

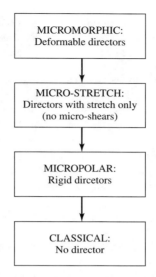

FIGURE 1.4.1. Sub-classes of Micromorphic Continua

(ii) Microstretch Continua

Animal lungs, bubbly fluids, slurries, polluted air, and fluids, springy suspension, mixtures with breathing elements, porous media, lattices with base, biological fluids: insect, small animal, and fish colonies that live in the ground, air, and sea.

(iii) Micropolar Continua

Liquid crystals with rigid molecules, rigid suspensions, animal blood with rigid cells, chopped fiber composites, bones, magnetic fluids, clouds with dusts, concrete with sand, muddy fluids.

In the remainder of this book, we formulate and discuss these three continuum theories as applied to various classes of solids and fluids.

1.5   Strain Measures

A. Micromorphic Continua

Essential to the deformation of micromorphic continuua are deformation gradients

$$x_{k,K}, \qquad X_{K,k}, \tag{1.5.1}$$

and the microdeformations and their gradients

$$\chi_{kK}, \qquad \mathfrak{X}_{Kk},$$

$$\chi_{kK,L}, \qquad \mathfrak{X}_{Kk,l}. \tag{1.5.2}$$

When $x_{k,K}$ is given, $X_{K,k}$ is determined by (1.2.13). Similarly, $\mathfrak{X}_{Kk}$ is given by (1.2.14) in terms of χ_{kK}. By differentiating (1.2.14) and using orthogonality relations (1.2.11) and (1.2.12), we also obtain

$$\chi_{kK,L} = -\mathfrak{X}_{Ml,m}x_{m,L}\chi_{kM}\chi_{lK},$$
$$\mathfrak{X}_{Kk,l} = -\chi_{mL,M}X_{M,l}\mathfrak{X}_{Km}\mathfrak{X}_{Lk}. \tag{1.5.3}$$

These results indicate that we need only one from each pair (1.5.1) and (1.5.2) to form the strain measures. Any other set will not be independent from such a set of strain measures.

Eringen and Şuhubi [1964] have constructed several sets of strain tensors. One such set of strain measures is given by

$$C_{KL} \equiv x_{k,K}x_{k,L}, \qquad \overset{*}{\mathfrak{C}}_{KL} \equiv x_{k,K}\chi_{kL}, \qquad \overset{*}{\Gamma}_{KLM} \equiv x_{k,K}\chi_{kL,M}. \tag{1.5.4}$$

This set is form-invariant under the rigid motions of the spatial frame of reference (i.e., it is objective) and it determines uniquely the motion and micromotion to within the same rigid motion, subject to compatibility conditions (see Section 1.7). The justification of this set as legitimate strain measures for micromorphic continua can also be made by calculating the square of the deformed arc length. From (1.1.3), we have

$$d\boldsymbol{x}' = \boldsymbol{x}_{,K}dX_K + \boldsymbol{\chi}_K\,d\Xi_K + \boldsymbol{\chi}_{K,L}\Xi_K\,dX_L. \tag{1.5.5}$$

By scalar multiplication, this gives, for the arc length,

$$(ds')^2 = d\boldsymbol{x}' \cdot d\boldsymbol{x}' = [\boldsymbol{x}_{,K} \cdot \boldsymbol{x}_{,L} + 2(\boldsymbol{x}_{,K} \cdot \boldsymbol{\chi}_{,L} + \boldsymbol{x}_{,L} \cdot \boldsymbol{\chi}_{M,K})\Xi_M$$
$$+ \boldsymbol{\chi}_{M,K} \cdot \boldsymbol{\chi}_{N,L}\Xi_M\Xi_N]dX_K\,dX_L \tag{1.5.6}$$
$$+ 2(\mathbf{x}_{,K} \cdot \boldsymbol{\chi}_L + \boldsymbol{\chi}_L \cdot \boldsymbol{\chi}_{M,K}\Xi_M)dX_K d\Xi_L + \boldsymbol{\chi}_K \cdot \boldsymbol{\chi}_L d\Xi_K d\Xi_L.$$

From (1.5.4), we have

$$\overset{-1}{C}_{KL} = X_{K,k}X_{L,k}, \qquad \chi_{kL} = \overset{*}{\mathfrak{C}}_{KL}\,X_{K,k}, \qquad \chi_{kL,M} = \overset{*}{\Gamma}_{KLM}\,X_{K,k}. \tag{1.5.7}$$

Using these and (1.5.4) in (1.5.6) we obtain

$$(ds')^2 = (C_{KL} + 2\Xi_M\,\overset{*}{\Gamma}_{KML} + \Xi_M\Xi_N\,\overset{*}{\Gamma}_{PML}\overset{*}{\Gamma}_{RNK}\overset{-1}{C}_{PR})\,dX_K\,dX_L$$
$$+ 2(\overset{*}{\mathfrak{C}}_{KL} + \Xi_M\,\overset{*}{\mathfrak{C}}_{NL}\overset{*}{\Gamma}_{RMK}\overset{-1}{C}_{NR})\,d\Xi_K\,dX_L$$
$$+ \overset{*}{\mathfrak{C}}_{MK}\overset{*}{\mathfrak{C}}_{NL}\overset{-1}{C}_{MN}\,d\Xi_K\,d\Xi_L. \tag{1.5.8}$$

This shows that the square of the deformed arc length is expressed in terms of the strain measures (1.5.4).

The square of the arc length in the undeformed body is given by

$$(dS')^2 = dX_K\,dX_K + 2\,dX_K\,d\Xi_K + d\Xi_K\,d\Xi_K. \tag{1.5.9}$$

An examination of (1.5.7) shows that when

$$C_{KL} = \overset{*}{\mathfrak{C}}_{KL} = \delta_{KL}, \qquad \overset{*}{\Gamma}_{KLM} = 0, \qquad (1.5.10)$$

then $(ds')^2 = (dS')^2$. This then justifies the proposition that (1.5.4) represents an admissible set of deformation tensors for the micromorphic continua.

Another set of strain measures that lead to somewhat simpler results, particularly in the constitutive equations, is

$$\mathfrak{C}_{KL} \equiv x_{k,K} \mathfrak{X}_{Lk}, \qquad \mathcal{C}_{KL} \equiv \chi_{kK} \chi_{kL} = \mathcal{C}_{LK}, \qquad \Gamma_{KLM} \equiv \mathfrak{X}_{Kk} \chi_{kL,M}. \tag{1.5.11}$$

Here $\mathfrak{C}_{KL}$ is called the *deformation tensor*, $\mathcal{C}_{KL}$ the *microdeformation tensor*, and Γ_{KLM} the *wryness tensor*. The relations of these to (1.5.4) are

$$C_{KL} = \mathfrak{C}_{KR} \mathcal{C}_{RS} \mathfrak{C}_{LS} = C_{LK}, \qquad \overset{*}{\mathfrak{C}}_{KL} = \mathfrak{C}_{KR} \mathcal{C}_{RL},$$

$$\Gamma_{KLM} = \mathfrak{C}_{KR} \mathcal{C}_{RS} \Gamma_{SLM}. \tag{1.5.12}$$

The inverse deformation tensors are given by

$$\overset{-1}{\mathfrak{C}}_{KL} = \chi_{kK} X_{L,k}, \qquad \overset{-1}{\mathcal{C}}_{KL} = \mathfrak{X}_{Kk} \mathfrak{X}_{Lk}, \tag{1.5.13}$$

which satisfy

$$\overset{-1}{\mathfrak{C}}_{KL} \mathfrak{C}_{LM} = \delta_{KM}, \qquad \overset{-1}{\mathcal{C}}_{KL} \mathcal{C}_{LM} = \delta_{KM}.$$

For the theory of micromorphic elasticity, the following strain tensors are useful, especially in the linear theory,

$$\mathfrak{E}_{KL} \equiv \mathfrak{C}_{KL} - \delta_{KL}, \qquad \mathcal{E}_{KL} = \mathcal{C}_{KL} - \delta_{KL}. \tag{1.5.14}$$

These tensors vanish when a body assumes the natural (undeformed) state.

For isotropic bodies, strain tensors expressed in the spatial frame become useful. These are defined by

$$\mathfrak{c}_{kl} = X_{K,k} \chi_{lK} = \mathfrak{X}_{Kk} \chi_{lL} \overset{-1}{\mathfrak{C}}_{KL},$$

$$\varsigma_{kl} = \mathfrak{X}_{Kk} \mathfrak{X}_{Kl} = \chi_{kK} \overset{-1}{\mathcal{C}}_{KL} \mathfrak{X}_{Ll},$$

$$\gamma_{klm} = \chi_{kK,M} \mathfrak{X}_{Kl} X_{M,m} = \Gamma_{KLM} \chi_{kK} \mathfrak{X}_{Ll} X_{M,m}. \tag{1.5.15}$$

The inverses of $\mathfrak{c}_{kl}$ and ς_{kl} are given by

$$\overset{-1}{\mathfrak{c}}_{kl} = \mathfrak{X}_{Kk} x_{l,K}, \qquad \overset{-1}{\varsigma}_{kl} = \chi_{kK} \chi_{lK}. \tag{1.5.16}$$

B. Microstretch Continua

Upon using (1.4.1), micromorphic deformation tensors reduce to

$$\mathfrak{C}_{KL} = j^{-2}x_{k,K}\chi_{kL}, \qquad \mathcal{C}_{KL} = j^2\delta_{KL}, \qquad \Gamma_{KLM} = j^{-2}\chi_{kK}\chi_{kL,M}.$$
$$(1.5.17)$$

While these tensors are possible candidates for the strain measures of microstretch media, they are coupled and they do not constitute an independent set. In fact, the number of independent components of the strain tensors of a microstretch continuum is much less than 42. An independent set is obtained by setting

$$\chi_{kK} = j\overline{\chi}_{kK}, \qquad \mathfrak{X}_{Kk} = \frac{1}{j}\overline{\mathfrak{X}}_{Kk}, \qquad (1.5.18)$$

where $\overline{\chi}_{kK}$ and $\overline{\chi}_{Kk}$ are subject to

$$\overline{\chi}_{kK}\overline{\mathfrak{X}}_{Kl} = \delta_{kl}, \qquad \overline{\chi}_{kK}\overline{\mathfrak{X}}_{Lk} = \delta_{KL}. \qquad (1.5.19)$$

The deformation tensors of microstretch continuum are then defined by

$$\overline{\mathfrak{C}}_{KL} = x_{k,K}\overline{\chi}_{kL}, \qquad\qquad \mathcal{C}_{KL} = j^2\delta_{KL},$$
$$\Gamma_{KL} \equiv \tfrac{1}{2}\epsilon_{KMN}\overline{\chi}_{kM,L}\overline{\chi}_{kN}, \qquad \Gamma_K = j^{-1}j_{,K}. \qquad (1.5.20)$$

These are related to the reduced micromorphic deformation tensors (1.5.17) by

$$\mathfrak{C}_{KL} = j^{-1}\overline{\mathfrak{C}}_{KL}, \qquad \mathcal{C}_{KL} = j^2\delta_{KL}, \qquad \Gamma_{KLM} = \Gamma_M\delta_{KL} - \epsilon_{KLR}\Gamma_{RM}.$$
$$(1.5.21)$$

We note that (1.5.20) is an independent set which has 21 independent components. $\overline{\mathfrak{C}}_{KL}$ and Γ_{KL} are also deformation tensors of the micropolar media.

The spatial deformation tensors of the microstretch media are defined by

$$\overline{\mathfrak{c}}_{kl} = X_{K,k}\overline{\chi}_{lK}, \qquad\qquad \overline{c}_{kl} = \delta_{kl},$$
$$\gamma_{kl} = \tfrac{1}{2}\epsilon_{kmn}\overline{\chi}_{mK}\overline{\chi}_{nK,l}, \qquad \gamma_m = \frac{j_{,m}}{j}. \qquad (1.5.22)$$

These are related to the reduced spatial deformation tensors of the micromorphic continuum by

$$\mathfrak{c}_{kl} = j\overline{\mathfrak{c}}_{kl}, \qquad \varsigma_{kl} = j^{-2}\delta_{kl},$$
$$\gamma_{klm} = \gamma_m\delta_{kl} - \epsilon_{klr}\gamma_{rm}. \qquad (1.5.23)$$

C. Micropolar Continua

The deformation tensors of micropolar continuum follow from those of microstretch continuum by setting $j = 1$.

$$\overline{\mathfrak{C}}_{KL} \equiv x_{k,K}\overline{\chi}_{kL}, \qquad \Gamma_{KL} \equiv \tfrac{1}{2}\epsilon_{KMN}\overline{\chi}_{kM,L}\overline{\chi}_{kN}. \qquad (1.5.24)$$

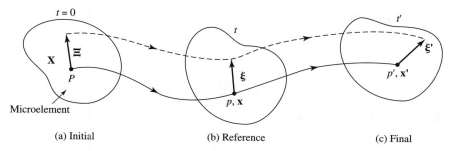

FIGURE 1.6.1. Relative Motion

We have named $\overline{\mathfrak{C}}_{KL}$, defined by (1.5.24), as *Cosserats' Deformation Tensor*, in honor of the Cosserats,[4] and Γ_{KL} as the *Wryness tensor*, Eringen [1970a, p. 24].

The spatial deformation tensors[5] $\overline{\mathfrak{c}}_{kl}$ and γ_{kl} of the micropolar media are given by

$$\overline{\mathfrak{c}}_{kl} = X_{K,k}\overline{\chi}_{lk}, \qquad \gamma_{kl} = \tfrac{1}{2}\epsilon_{kmn}\overline{\chi}_{mK}\overline{\chi}_{nK,l}. \tag{1.5.25}$$

1.6   Relative Motion, Relative Strain Measures

Strain measures introduced in Section 1.5 are based on the absolute motion of the body with respect to a fixed reference frame. For some media (e.g., fluids), it is more convenient to employ the spatial frame-of-reference at time t as a reference configuration and to record changes that occur at a later time $t' > t$, from this configuration, Figure 1.6.1. For the mapping from the initial configuration at time $t = 0$, to the reference configuration at time t, as usual, we have

$$\mathbf{x} = \mathbf{x}(\mathbf{X}, t) \qquad \longleftrightarrow \qquad \mathbf{X} = \mathbf{X}(\mathbf{x}, t), \tag{1.6.1}$$

$$\boldsymbol{\chi} = \boldsymbol{\chi}(\mathbf{X}, t). \tag{1.6.2}$$

In terms of the physical picture of the microelement, (1.6.2) expresses the micromotion of $\boldsymbol{\Xi}$ to $\boldsymbol{\xi}$, i.e.,

$$\xi_k = \chi_{kK}(\mathbf{X}, t)\Xi_K. \tag{1.6.3}$$

---

[4]We could not locate a deformation tensor defined by the Cosserats among the maze of long formulas in their book [1909].

[5]$\overline{\mathfrak{c}}_{kl}$ defined here is not the $\mathfrak{c}_{kl} = x_{k,K}\chi_{lK}$ given by Eringen [1970a]. But it agrees with that given by Eringen and Şuhubi [1964] and Eringen and Kafadar [1976, p. 8]. Of course, other equally acceptable strain measures can be defined. Similarly, Eringen and Kafadar used $\overset{*}{\gamma}_{kl} = \chi_{kK}\chi_{lL}\Gamma_{KL}$ in place of γ_{kl} defined here. However, the present γ_{kl} and Γ_{KL} possess nice symmetries and relate to the master notions γ_{klm} and Γ_{KLM}.

The mapping from the initial configuration to the final configuration is given by

$$\mathbf{x}' = \mathbf{x}(\mathbf{X}, t'), \qquad (1.6.4)$$

$$\xi'_k = \chi_{kK}(\mathbf{X}, t')\Xi_K. \qquad (1.6.5)$$

If we use the inverse motions

$$\mathbf{X} = \mathbf{X}(\mathbf{x}, t), \qquad \Xi_K = \mathfrak{X}_{Kk}(\mathbf{x}, t)\xi_k, \qquad (1.6.6)$$

in (1.6.4) and (1.6.5), we obtain

$$\mathbf{x}' = \mathbf{x}(\mathbf{X}(\mathbf{x}, t), t') \equiv \mathbf{x}_{(t)}(\mathbf{x}, t'), \qquad (1.6.7)$$

$$\xi'_k = \chi_{(t)kl}(\mathbf{x}, t')\xi_l, \qquad (1.6.8)$$

where

$$\chi_{(t)kl}(\mathbf{x}, t') = \chi_{kK}(\mathbf{X}, t')\mathfrak{X}_{Kl}(\mathbf{x}, t), \qquad (1.6.9)$$

and, consequently,

$$\mathfrak{X}_{(t)kl}(\mathbf{x}, t') = \mathfrak{X}_{Kl}(t')\chi_{kK}. \qquad (1.6.10)$$

Note that

$$\chi_{(t)kl}(t) = \mathfrak{X}_{(t)kl}(t) = \delta_{kl}.$$

Hence, the relative motion of a point $\mathbf{x}$, in the reference configuration, is given by (1.6.7) and that of a director by (1.6.9).

We assume that $\mathbf{x}_{(t)}$, given by (1.6.7), is invertible, i.e.,

$$\mathbf{x} = \mathbf{x}(\mathbf{x}', t'). \qquad (1.6.11)$$

It is then required that the Jacobian of the relative deformation be positive:

$$J_{(t)}(t') = \det\left(\frac{\partial x'_k}{\partial x_l}\right) = J(t')/J(t) > 0, \quad J(t') = \det\left(\frac{\partial x'_k}{dX_K}\right). \qquad (1.6.12)$$

Note that $J_{(t)}(t) = 1$.

The relative deformation gradient is defined by

$$x'_{k,l}(t') = F_{(t)kl} = \frac{\partial x'_k}{\partial x_l}, \qquad \mathbf{F}_{(t)}(t') = \nabla\mathbf{x}', \qquad (1.6.13)$$

which satisfy

$$x'_{k,l}(t) = \delta_{kl} \qquad \text{or} \qquad \mathbf{F}_{(t)}(t) = \mathbf{1}. \qquad (1.6.14)$$

A. Micromorphic Continua

The relative strain measures of micromorphic continua are defined similar to (1.5.11):

$$\mathfrak{C}_{(t)kl}(t') \equiv x_{(t)m,k}\mathfrak{X}_{(t)lm}, \qquad \mathcal{C}_{(t)kl}(t') \equiv \chi_{(t)mk}\chi_{(t)ml},$$
$$\Gamma_{(t)klm}(t') = \mathfrak{X}_{(t)kp}\chi_{(t)pl,m}. \tag{1.6.15}$$

These are related to $\mathfrak{C}_{KL}, \mathcal{C}_{KL}$ and Γ_{KLM} by

$$\mathfrak{C}_{(t)kl}(t') = X_{K,k}(t)\chi_{lL}(t)\mathfrak{C}_{KL}(t'),$$
$$\mathcal{C}_{(t)kl}(t') = \mathfrak{X}_{Kk}(t)\mathfrak{X}_{Ll}(t)\mathcal{C}_{KL}(t'),$$
$$\Gamma_{(t)klm}(t') = \chi_{kK}(t)\mathfrak{X}_{Ll}(t)X_{M,m}(t)[\Gamma_{KLM}(t') - \Gamma_{KLM}(t)]. \tag{1.6.16}$$

B. Microstretch Continua

The relative deformation tensors of microstretch continua are defined similar to (1.5.20), i.e.,

$$\bar{\mathfrak{C}}_{(t)kl}(t') = x_{(t)m,k}\bar{\chi}_{(t)ml},$$
$$\bar{\mathcal{C}}_{(t)kl}(t') = j^2_{(t)}(t')\delta_{kl},$$
$$\Gamma_{(t)kl}(t') = \tfrac{1}{2}\epsilon_{kmn}\bar{\chi}_{(t)pm,l}\bar{\chi}_{(t)pm},$$
$$\Gamma_{(t)k}(t') = \frac{j_{(t)}(t')_{,k}}{j_{(t)}(t')}. \tag{1.6.17}$$

By using (1.6.9) and (1.6.10), it can be shown that these are related to $\mathfrak{C}_{KL}, \Gamma_{KL}$, and Γ_K by

$$\bar{\mathfrak{C}}_{(t)kl}(t') = X_{K,k}\bar{\chi}_{lL}\bar{\mathfrak{C}}_{KL}(t'),$$
$$\Gamma_{(t)kl}(t') = j^{-2}\bar{\chi}_{kK}X_{L,l}[\Gamma_{KL}(t') - \Gamma_{KL}],$$
$$\Gamma_{(t)k}(t') = [\Gamma_K(t') - \Gamma_K(t)]X_{K,k}. \tag{1.6.18}$$

C. Micropolar Continua

The relative deformation tensors of micropolar continuum are identical to those of the microstretch continuum, excluding $\bar{\mathcal{C}}_{(t)kl}(t')$ and $\Gamma_{(t)k}(t')$,

$$\bar{\mathfrak{C}}_{(t)kl}(t') = X_{K,k}(t)\bar{\chi}_{lL}(t)\bar{\mathfrak{C}}_{KL}(t'),$$
$$\Gamma_{(t)kl}(t') = [\Gamma_{KL}(t') - \Gamma_{KL}(t)]\bar{\chi}_{kK}(t)X_{L,l}(t). \tag{1.6.19}$$

1.7 Compatibility Conditions

Given a set of deformation tensors $\mathfrak{C}_{KL}, \mathcal{C}_{KL}$, and Γ_{KLM} we have 42 first-order partial differential equations (1.5.11) to determine 12 unknowns x_k

and χ_{kK}. For the existence of single-valued displacement and director fields corresponding to the deformation tensors, certain relations must be satisfied among the deformation tensors. These are the integrability conditions for the partial differential equations (1.5.11). They are known as the *compatibility conditions*[6].

A. Micromorphic Continua

Theorem 1 (Compatibility Conditions).   For a simply-connected micromorphic body, the necessary and sufficient conditions for the integrability of the system of partial differential equations (1.5.11) are:

$$\epsilon_{KPQ}(\mathfrak{C}_{PL,Q} + \mathfrak{C}_{PR}\Gamma_{LRQ}) = 0,$$
$$\epsilon_{KPQ}(\Gamma_{LMP,Q} + \Gamma_{LRQ}\Gamma_{RMP}) = 0,$$
$$\mathfrak{C}_{KL,M} - (\Gamma_{PKM}\mathfrak{C}_{LP} + \Gamma_{PLM}\mathfrak{C}_{KP}) = 0. \qquad (1.7.1)$$

PROOF.   For a simply-connected body, the necessary and sufficient conditions that $x_{k,K} = 0$ and $\chi_{kK,L} = 0$ be integrable are

$$x_{k,PQ} = x_{k,QP}, \qquad \chi_{kK,PQ} = \chi_{kK,QP}.$$

By use of the orthogonality relations (1.2.11) and (1.2.12), from (1.5.11) we have

$$x_{k,P} = \mathfrak{C}_{PR}\chi_{kR}, \qquad \chi_{kL,Q} = \Gamma_{KLQ}\chi_{kK}. \qquad (1.7.2)$$

Using $(1.7.2)_1$, we form

$$x_{k,PQ}\mathfrak{X}_{Lk} = \mathfrak{C}_{PL,Q} + \mathfrak{C}_{PR}\Gamma_{LRQ}.$$

From this, there follows $(1.7.1)_1$.

Next, we consider $(1.7.2)_2$, and form

$$\mathfrak{X}_{Lk}\chi_{kM,PQ} = \Gamma_{LMP,Q} + \Gamma_{RMP}\Gamma_{LRQ}.$$

From this, $(1.7.1)_2$ follows.

Finally, using $(1.5.11)_2$ we calculate

$$\mathfrak{C}_{KL,M} = \chi_{kK,M}\chi_{kL} + \chi_{kK}\chi_{kL,M}.$$

Substituting for $\chi_{kK,Q}$ from $(1.7.2)_2$, we obtain

$$\mathfrak{C}_{KL,M} = \Gamma_{PKM}\mathfrak{C}_{PL} + \Gamma_{PLM}\mathfrak{C}_{KP},$$

which is identical to $(1.7.1)_3$.   □

---

[6]Compatibility conditions were obtained by Eringen [1967c, 1968a, 1969b, 1970a]. See also Kafadar and Eringen [1971]. The linear form of compatibility equations of micropolar media was also discussed by Sandru [1966].

B. Microstretch Continua

Theorem 2. For a simply-connected microstretch continuum, the necessary and sufficient conditions for the integrability of equations (1.5.20) are

$$\epsilon_{KPQ}\left(\bar{\mathfrak{C}}_{PL,Q} + \epsilon_{MLN}\Gamma_{NQ}\bar{\mathfrak{C}}_{PM}\right) = 0,$$

$$\epsilon_{KPQ}\left(\Gamma_{LQ,P} + \frac{1}{2}\epsilon_{LMN}\Gamma_{MP}\Gamma_{NQ}\right) = 0. \tag{1.7.3}$$

The proof of these follows by substituting (1.5.21) into (1.7.1).

C. Micropolar Continua

The compatibility conditions of micropolar continua are identical to (1.7.3):

$$\epsilon_{KPQ}\left(\bar{\mathfrak{C}}_{PL,Q} + \epsilon_{MLN}\Gamma_{NQ}\bar{\mathfrak{C}}_{PM}\right) = 0,$$

$$\epsilon_{KPQ}\left(\Gamma_{LQ,P} + \frac{1}{2}\epsilon_{LMN}\Gamma_{MP}\Gamma_{NQ}\right) = 0. \tag{1.7.4}$$

Compatibility conditions for the spatial strain measures are obtained similarly.

Micromorphic continua:

$$\epsilon_{kpq}(\mathfrak{c}_{pl,q} - \mathfrak{c}_{pr}\gamma_{lrq}) = 0,$$
$$\epsilon_{kpq}(\gamma_{lmp,q} + \gamma_{lrq}\gamma_{rmp}) = 0,$$
$$\varsigma_{kl,m} + \gamma_{pkm}\mathfrak{c}_{pl} + \gamma_{plm}\varsigma_{kp} = 0. \tag{1.7.5}$$

Microstretch continua:

$$\epsilon_{kpq}(\bar{\mathfrak{c}}_{pl,q} + \epsilon_{lmn}\mathfrak{c}_{pm}\gamma_{nq}) = 0,$$
$$\epsilon_{kpq}(2\gamma_{lp,q} + \epsilon_{lmn}\gamma_{np}\gamma_{mq}) = 0. \tag{1.7.6}$$

Micropolar continua:

$$\epsilon_{kpq}(\bar{\mathfrak{c}}_{pl,q} + \epsilon_{lmn}\bar{\mathfrak{c}}_{pm}\gamma_{nq}) = 0,$$
$$\epsilon_{kpq}(2\gamma_{lp,q} + \epsilon_{lmn}\gamma_{np}\gamma_{mq}) = 0. \tag{1.7.7}$$

D. Linear Compatibility Conditions

These are obtained by linearizing (1.7.5) to (1.7.7).

Micromorphic:

$$\epsilon_{kpq}(\epsilon_{pl,q} + \gamma_{lpq}) = 0,$$
$$\epsilon_{kpq}\gamma_{lmp,q} = 0,$$
$$-2e_{kl,m} + \gamma_{klm} + \gamma_{lkm} = 0, \tag{1.7.8}$$

where ϵ_{kl}, e_{kl}, and γ_{klm} are the linear strain tensors defined by

$$\epsilon_{kl} = \delta_{kl} - \mathfrak{c}_{kl} = u_{l,k} - \phi_{lk},$$
$$2e_{kl} = \delta_{kl} - \varsigma_{kl} = \phi_{kl} + \phi_{lk},$$
$$\gamma_{klm} = \phi_{kl,m}. \tag{1.7.9}$$

Microstretch:

$$-\epsilon_{kpq}\epsilon_{pl,q} + \gamma_{pp}\delta_{kl} - \gamma_{kl} = 0,$$
$$\epsilon_{kpq}\gamma_{lq,p} = 0, \tag{1.7.10}$$

where

$$\epsilon_{kl} = u_{l,k} + \epsilon_{lkm}\phi_m, \qquad \gamma_{kl} = \phi_{k,l}. \tag{1.7.11}$$

Micropolar: These are identical to those of microstretch (1.7.10).

It must be remarked that when the compatibility conditions are not satisfied, the body contains defects. For example, when $(1.7.4)_1$ is violated, the body contains *dislocations*. When $(1.7.4)_2$ is violated, it contains *disclinations*. These defects are sources of plastic deformations and fracture. Dislocations are the result of the displacement discontinuities, while disclinations arise from rotational misfits. Thus, the micropolar continuum theory legitimately brings the study of the disclination theory within the scope of the continuum theory, which is not within the scope of the classical field theories. The micromorphic continuum theory suggests other types of defects beyond dislocations and disclinations. Preliminary discussions of these are contained in Eringen and Claus Jr. [1969, 1970, 1971]. Some discussions of disclinations that occur in liquid crystals are to be found in the works of Vertogen and de Jeu [1988] de Gennes and Prost [1995], and Eringen [1994].

Dislocation theory is a classical field of study. Here we will mention just a few classical books on the subject: Hirth and Lothe [1968], Nabarro [1967], Teodosiu [1982]. Later in Section 5.26 and in Volume II, we shall return to this subject.

1.8   Material Time-Rate of Tensors

Definition 1. The material time-rate of a function $f(\mathbf{x}, \mathbf{\Xi}, t)$ is defined as

$$\dot{f} = \frac{df}{dt} = \left.\frac{\partial f}{\partial t}\right|_{\mathbf{X},\mathbf{\Xi}}, \tag{1.8.1}$$

where the subscript $\mathbf{X}, \mathbf{\Xi}$ accompanying a vertical bar denotes that $\mathbf{X}$ and $\mathbf{\Xi}$ are held constants in the differentiation. If $f = f(\mathbf{X}, t)$, then

$$\dot{f} = \frac{\partial f(\mathbf{X}, t)}{\partial t}. \tag{1.8.2}$$

If $f = f(\mathbf{x}, t)$, then

$$\dot{f} = \frac{\partial f}{\partial t} + \frac{\partial f}{\partial x_k} \dot{x}_k. \qquad (1.8.3)$$

For example, the velocity vector $\mathbf{v}$ and acceleration vector $\mathbf{a}$ are defined by

$$\mathbf{v} = \frac{\partial \mathbf{x}(\mathbf{X}, t)}{\partial t} = \dot{\mathbf{x}},$$

$$\mathbf{a} = \dot{\mathbf{v}} = \frac{\partial^2 \mathbf{x}(\mathbf{X}, t)}{\partial t^2}. \qquad (1.8.4)$$

If we substitute $\mathbf{X} = \mathbf{X}(\mathbf{x}, t)$ into (1.8.4), we obtain

$$\mathbf{v}(\mathbf{X}(\mathbf{x}, t), t) \equiv \hat{\mathbf{v}}(\mathbf{x}, t). \qquad (1.8.5)$$

In this way $\hat{\mathbf{v}}$ represents the velocity of a material point at $\mathbf{x}$. However, $\hat{\mathbf{v}}(\mathbf{x}, t)$ does not indicate which material point is at the spatial place $\mathbf{x}$, at time t. Thus, $\hat{\mathbf{v}}$ is a velocity field in the space occupied by the body at time t. This is the *Eulerian point of view*, as contrasted to *Lagrangian representation* (1.8.4) in which the identity of the material point at $\mathbf{X}$ is known. When $\hat{\mathbf{v}}$ is given we can also determine the identity of the point. But this requires solving a set of first-order equations

$$\frac{d\mathbf{x}}{dt} = \hat{\mathbf{v}}(\mathbf{x}, t) \qquad (1.8.6)$$

subject to the initial conditions

$$\mathbf{x}(\mathbf{X}, 0) = \mathbf{x}_0(\mathbf{X}). \qquad (1.8.7)$$

The acceleration field $\mathbf{a}(\mathbf{x}, t)$ is obtained by

$$\mathbf{a} = \dot{\hat{\mathbf{v}}} = \frac{\partial \hat{\mathbf{v}}}{\partial t} + \hat{\mathbf{v}}_{,k} \hat{v}_k.$$

Henceforth we drop the hat ($\hat{\ }$), for brevity and express it as

$$\mathbf{a} = \frac{\partial \mathbf{v}}{\partial t} + \mathbf{v}_{,k} v_k, \qquad (1.8.8)$$

which, in component notation, reads

$$a_k = \frac{\partial v_k}{\partial t} + v_{k,l} v_l \equiv \frac{D v_k}{Dt}. \qquad (1.8.9)$$

The symbol D/Dt is universal and is called the *material derivative*.

The following two lemmas are useful in the calculations of material rates of various tensors:

Lemma 1. The material derivative of $x_{k,K}$ is given by

$$\frac{D}{Dt}(x_{k,K}) = v_{k,l}x_{l,K} \qquad \text{or} \qquad \frac{D}{Dt}(dx_k) = v_{k,l}\,dx_l. \qquad (1.8.10)$$

The proof of (1.8.10) follows from the observation that the operators D/Dt and $\partial/\partial X_K$ commute. Thus, we form

$$\frac{D}{Dt}(d\mathbf{x}) = \frac{D}{Dt}(\mathbf{x}_{,K}\,dX_K) = \mathbf{v}_{,K}\,dX_K = v_{k,l}x_{l,K}\mathbf{i}_k\,dX_K = v_{k,l}\,dx_l\mathbf{i}_k,$$

where the coefficient of $\mathbf{i}_k$ is (1.8.10) and, thus, the proof of the lemma.

Corollary. The material time rate of $X_{K,k}$ is given by

$$\frac{D}{Dt}(X_{K,k}) = -X_{K,l}v_{l,k}. \qquad (1.8.11)$$

This follows by differentiating

$$X_{K,k}x_{k,L} = \delta_{KL}$$

and using $(1.8.10)_1$ and this expression once again to solve for $D(X_{K,k})/Dt$.

A. Micromorphic Continuua

Definition 2. Microgyration tensor ν_{kl} is defined by

$$\nu_{kl} = \dot{\chi}_{kK}\mathfrak{X}_{Kl}. \qquad (1.8.12)$$

Lemma 2. The material derivatives of χ_{kK} and $\mathfrak{X}_{Kk}$ are given by

$$\dot{\chi}_{kK} = \nu_{kl}\chi_{lK}, \qquad \dot{\mathfrak{X}}_{Kk} = -\nu_{lk}\mathfrak{X}_{Kl}. \qquad (1.8.13)$$

The first of these follows from (1.8.12) upon multiplication by χ_{lL}. To show the second, we calculate the time rate of the orthogonality relation (1.2.12)

$$\dot{\chi}_{kK}\mathfrak{X}_{Lk} + \chi_{kK}\dot{\mathfrak{X}}_{Lk} = 0.$$

Upon multiplication by $\mathfrak{X}_{Km}$ and using (1.2.12), we obtain $(1.8.13)_2$. The material derivative of $\boldsymbol{\xi}$ is given by

$$\dot{\xi}_k = \nu_{kl}\xi_l. \qquad (1.8.14)$$

Theorem 1. The material derivative of the microdisplacement gradient $\chi_{kK,L}$ is given by

$$\frac{D}{Dt}(\chi_{kK,L}) \equiv \dot{\chi}_{kK,L} = \nu_{kl}\chi_{lK,L} + \nu_{kl,m}\chi_{lK}x_{m,L}. \qquad (1.8.15)$$

To show this, we use (1.8.12) and note that D/Dt commutes with $\partial/\partial X_L$. A corollary of this theorem is

$$\frac{D}{Dt}(d\xi_k) = \nu_{km}\,d\xi_m + \nu_{kl,m}\xi_l\,dx_m. \qquad (1.8.16)$$

This is obtained by taking the material derivative of

$$d\xi_k = \chi_{kK,L}\Xi_K\,dX_L + \chi_{kK}\,d\Xi_K \qquad (1.8.17)$$

and using (1.8.15) and (1.8.17).

Theorem 2. The material time derivatives of strain measures of micromorphic continua are given by

$$\dot{\mathcal{C}}_{KL} = a_{kl}x_{k,K}\mathfrak{X}_{Ll}, \qquad \dot{\mathcal{C}}_{KL} = 2c_{kl}\chi_{kK}\chi_{lL},$$
$$\dot{\Gamma}_{KLM} = b_{klm}\mathfrak{X}_{Kk}\chi_{lL}x_{m,M}, \qquad (1.8.18)$$

where **a**, **b**, and **c** are called deformation-rate tensors. They are defined by

$$a_{kl} \equiv \nu_{l,k} - \nu_{lk}, \qquad 2c_{kl} \equiv \nu_{kl} + \nu_{lk}, \qquad b_{klm} \equiv \nu_{kl,m}. \qquad (1.8.19)$$

Theorem 3. The material derivative of the square of the arc length, in the deformed micromorphic continuum, is given by

$$\frac{D}{Dt}(ds'^2) = [v_{k,l} + v_{l,k} + (\nu_{kr,l} + \nu_{lr,k})\xi_r]\,dx_k\,dx_l$$
$$+ 2(v_{l,k} + \nu_{kl} + \nu_{lr,k}\xi_r)\,dx_k\,d\xi_l + (\nu_{kl} + \nu_{lk})\,d\xi_k\,d\xi_l. \qquad (1.8.20)$$

To show this, we calculate the material time derivative of

$$(ds')^2 = (dx_k + d\xi_k)(dx_k + d\xi_k) \qquad (1.8.21)$$

and use (1.8.10) and (1.8.16).

Theorem 4. The necessary and sufficient conditions for a micromorphic body to undergo rigid motion are

$$v_{l,k} - \nu_{lk} = 0, \qquad \nu_{kl} + \nu_{lk} = 0, \qquad \nu_{kl,m} = 0. \qquad (1.8.22)$$

These generalize the classical Killing's theorem.

PROOF.   Clearly when (1.8.22) is valid, then from (1.8.20) it follows that $D(ds'^2)/Dt = 0$. Conversely, $D(ds'^2)/Dt$ vanishes for arbitrary dx_k, $d\xi_k$, and ξ_k, if (1.8.22) is fulfilled.   □

The symmetric and antisymmetric parts of (1.8.22)$_1$ upon using (1.8.22)$_2$ give

$$v_{(k,l)} \equiv 0, \qquad \nu_{kl} - v_{[k,l]} = 0, \qquad (1.8.23)$$

where, as usual, parentheses enclosing indices indicates the symmetric part of the tensor and brackets, the antisymmetric part, i.e.,

$$v_{(k,l)} = \tfrac{1}{2}(v_{k,l} + v_{l,k}), \qquad v_{[k,l]} \equiv \tfrac{1}{2}(v_{k,l} - v_{l,k}). \qquad (1.8.24)$$

The general solution of (1.8.22) is

$$x_k = R_{kK}(t)X_K + b_k(t), \qquad \chi_{kK} = R_{kK}(t). \qquad (1.8.25)$$

Here R_{kK} is an arbitrary, time-dependent, orthogonal tensor and $b_k(t)$ is an arbitrary time-dependent vector. Clearly (1.8.25) represents an arbitrary time-dependent rigid motion.

The material time rates of the relative deformation tensors (1.6.15) are useful for the formulation of the constitutive equations of micromorphic viscous fluids. To this end we define *micromorphic deformation-rate tensors* of grade n by

$$\mathbf{a}_n = \left.\frac{D^n}{Dt'^n}\mathfrak{C}_{(t)}(t')\right|_{t'=t}, \qquad \mathbf{b}_n = \left.\frac{D^n}{Dt'^n}\boldsymbol{\Gamma}_{(t)}(t')\right|_{t'=t},$$

$$\mathbf{c}_n = \left.\frac{1}{2}\frac{D^n}{Dt'^n}\mathfrak{C}_{(t)}(t')\right|_{t'=t}. \qquad (1.8.26)$$

For $n = 1, 2$, these have specific forms:

$$a_{1kl} = v_{l,k} - \nu_{lk}, \qquad b_{1klm} = \nu_{kl,m}, \qquad c_{1kl} = \nu_{(kl)},$$

$$a_{2kl} = \dot{v}_{l,k} - 2v_{r,k}\nu_{lr} - \dot{\nu}_{lk} + \nu_{rk}\nu_{lr}, \qquad \frac{Dj}{Dt} = jc_{kk},$$

$$b_{2klm} = \dot{\nu}_{kl,m} + \nu_{kr,m}\nu_{rl} - \nu_{kr}\nu_{rl,m},$$

$$c_{2kl} = \dot{\nu}_{(kl)} + \nu_{(kr)}\nu_{rl} + \nu_{rk}\nu_{(rl)}. \qquad (1.8.27)$$

Note that $\mathbf{a}_1 = \mathbf{a}, \mathbf{b}_1 = \mathbf{b}$, and $\mathbf{c}_1 = \mathbf{c}$, are the same as in (1.8.19).

B. Microstretch Continua

From (1.5.19), by differentiation, we have

$$\dot{\bar{\chi}}_{kK} = \bar{\nu}_{kl}\bar{\chi}_{lK}, \qquad \dot{\bar{\mathfrak{X}}}_{Kk} = -\bar{\mathfrak{X}}_{Kl}\bar{\nu}_{lk}, \qquad (1.8.28)$$

where

$$\bar{\nu}_{kl} \equiv \dot{\bar{\chi}}_{kK}\bar{\chi}_{lK} = -\bar{\nu}_{lk}, \qquad 3\nu \equiv \frac{1}{j}\frac{Dj}{Dt}. \qquad (1.8.29)$$

Alternatively, introducing *microrotation rate* vector ν_k by

$$\nu_k = -\tfrac{1}{2}\epsilon_{klm}\bar{\nu}_{lm}, \qquad \bar{\nu}_{kl} = -\epsilon_{klm}\nu_m,$$

we have

$$\dot{\bar{\chi}}_{kK} = -\epsilon_{klm}\nu_m\bar{\chi}_{lK}. \qquad (1.8.30)$$

The time-rate of the directors of micromorphic media for the microstretch case is related to these by

$$\dot{\chi}_{kK} = \frac{D}{Dt}(j\bar{\chi}_{kK}) = (\nu\delta_{kl} - \epsilon_{klm}\nu_m)\chi_{lK}, \qquad (1.8.31)$$

which establishes the relation

$$\nu_{kl} = \nu\delta_{kl} - \epsilon_{klm}\nu_m = \nu\delta_{kl} + \bar{\nu}_{kl}. \qquad (1.8.32)$$

From (1.8.31) it is clear that a micromorphic continuum, that is constrained to undergo a uniform microstretch (a breathing motion) represented by ν, and rigid microrotation, represented by ν_k, is a microstretch continuum.

Theorem 5. The material time-rates of the deformation tensors of microstretch continuum are given by

$$\dot{\mathfrak{C}}_{KL} = (v_{l,k} + \epsilon_{lkm}\nu_m)x_{k,K}\bar{\chi}_{lL},$$
$$\dot{\Gamma}_{KL} = j^{-2}\nu_{k,l}x_{l,L}\bar{\chi}_{kK},$$
$$\dot{\Gamma}_K = \nu_{,K}, \qquad \dot{\mathfrak{C}}_{KL} = 6j^2\nu\delta_{KL}. \qquad (1.8.33)$$

These are readily deduced by calculating the time-rates of (1.5.20) and using (1.8.10) and (1.8.30). An alternative simpler approach is through taking the material time-rate (1.5.21) and using (1.8.18).

In the special case of microstretch medium, the deformation-rate tensors of micromorphic continuum take the special form

$$a_{kl} = v_{l,k} + \bar{\nu}_{kl} - \nu\delta_{kl}, \qquad c_{kl} = \nu\delta_{kl},$$
$$b_{klm} = \bar{\nu}_{kl,m} + \nu_{,m}\delta_{kl}. \qquad (1.8.34)$$

C. Micropolar Continua

In the case of micropolar continuum, $j = 1$. Consequently, the time-rates of strain tensors are identical to (1.8.33) excluding $\dot{\Gamma}_K$, i.e.,

$$\dot{\mathfrak{C}}_{KL} = (v_{l,k} + \epsilon_{lkm}\nu_m)x_{k,K}\bar{\chi}_{lL},$$
$$\dot{\Gamma}_{KL} \equiv \nu_{k,l}x_{l,L}\bar{\chi}_{kK}. \qquad (1.8.35)$$

In *summary*: The deformation-rate tensors of all three continua are:

Micromorphic Continua

$$a_{kl} = v_{l,k} - \nu_{lk}, \qquad b_{klm} = \nu_{kl,m}, \qquad c_{kl} = \nu_{(kl)}. \qquad (1.8.36)$$

Microstretch Continua

$$a_{kl} = v_{l,k} + \epsilon_{lkm}\nu_m, \qquad b_{kl} = \nu_{k,l}, \qquad c_{kl} = \nu\delta_{kl}. \qquad (1.8.37)$$

Micropolar Continua

$$a_{kl} = v_{l,k} + \epsilon_{lkm}\nu_m, \qquad b_{kl} = \nu_{k,l}. \tag{1.8.38}$$

Deformation-rate tensors are fundamental to the construction of constitutive equations of viscous fluids of 3M continua.

Theorem 6. The microgyration vector, $\boldsymbol{\nu}$, is related to the material time rate of the rotation vector $\boldsymbol{\phi}$ by, (Eringen [1970a], Eringen and Kafadar [1976])

$$\boldsymbol{\nu} = \dot{\phi}\mathbf{n} + \sin\phi\,\dot{\mathbf{n}} + (1 - \cos\phi)\mathbf{n} \times \dot{\mathbf{n}}, \tag{1.8.39}$$

where

$$\phi = (\boldsymbol{\phi}\cdot\boldsymbol{\phi})^{1/2}, \qquad \mathbf{n} = \boldsymbol{\phi}/\phi. \tag{1.8.40}$$

Alternatively,

$$\nu_k = \Lambda_{kl}\dot{\phi}_l, \tag{1.8.41}$$

where

$$\Lambda_{kl} \equiv \frac{\sin\phi}{\phi}\delta_{kl} - \frac{1-\cos\phi}{\phi^2}\epsilon_{klm}\phi_m + (1 - \frac{\sin\phi}{\phi})\frac{1}{\phi^2}\phi_k\phi_l. \tag{1.8.42}$$

This is proved by taking the material time-rate of (1.3.11). The inverse of (1.8.41) exists and is given by

$$\dot{\phi}_{lk} = \Lambda_{lk}^{-1}\nu_k, \tag{1.8.43}$$

where

$$\Lambda_{lk}^{-1} = \tfrac{\phi}{2}\cot\tfrac{\phi}{2}\delta_{kl} + \tfrac{1}{2}\epsilon_{klm}\phi_m + \left(1 - \tfrac{\phi}{2}\cot\tfrac{\phi}{2}\right)\tfrac{1}{\phi^2}\phi_k\phi_l. \tag{1.8.44}$$

For very small angles ϕ, the linear approximations for $\boldsymbol{\Lambda}$ and $\boldsymbol{\Lambda}^{-1}$ are

$$\Lambda_{kl} \simeq \delta_{kl} - \tfrac{1}{2}\epsilon_{klm}\phi_m, \qquad \Lambda_{lk}^{-1} \simeq \delta_{kl} + \tfrac{1}{2}\epsilon_{klm}\phi_m. \tag{1.8.45}$$

It then follows that, for the linear theory,

$$\nu_k \simeq \dot{\phi}_k. \tag{1.8.46}$$

1.9 Objective Tensors

It is intuitively clear that the material properties do not depend on the coordinate frame selected. The measurements made by an observer, whether he is in motion or not, should be the same. If this viewpoint is accepted, then the measurements made in one frame-of-reference are sufficient to determine the material properties in all other frames which are in rigid motion with respect to one another. In the formulation of the response functions, it is desirable to employ quantities that are not dependent on the motions of the observer. Such quantities are called *objective* or *material frame-indifferent*. For example, the velocity of an automobile will appear different to two observers, one of whom is resting, the other riding in another car. Therefore, the velocity vector is not objective. Similarly, the acceleration is not an objective vector. The distance between two points and angles between two directions are independent of the rigid motions of the frame-of-reference (the observer). Hence, they are objective quantities.

Definition 1. Two motions $x_k(\mathbf{X}, t)$ and $\tilde{x}_k(\mathbf{X}, \tilde{t})$ are called objectively equivalent if and only if

$$\tilde{x}_k(\mathbf{X}, \tilde{t}) = Q_{kl}(t)x_l(\mathbf{X}, t) + b_k(t), \qquad \tilde{t} = t - a, \qquad (1.9.1)$$

where a is a constant time shift, $\mathbf{b}(t)$ is a time-dependent translation, and $\{\mathbf{Q}(t)\}$ are time-dependent full orthogonal transformations, i.e.,

$$Q_{kl}Q_{ml} = Q_{lk}Q_{lm} = \delta_{km}, \qquad \det Q_{kl} = \pm 1, \qquad (1.9.2)$$

Q_{kl} consists of all rigid rotations $\det \mathbf{Q} = +1$, and inversions ($\det \mathbf{Q} = -1$).

Two objectively equivalent motions differ only in relative frame and time. For a fixed frame and time, the two motions can be made to coincide by superposing an arbitrary rigid motion to one and shifting the origin of time.

Definition 2 (Objectivity).   Any tensorial quantity that obeys the tensor transformation law under (1.9.1) is said to be objective or material frame-indifferent.

For example, a vector a_k and a tensor t_{kl} are objective if they obey the transformation laws

$$\tilde{a}_k(\mathbf{X}, \tilde{t}) = Q_{kl}(t)a_l(\mathbf{X}, t),$$
$$\tilde{t}_{kl}(\mathbf{X}, \tilde{t}) = Q_{km}(t)Q_{lm}(t)t_{mn}(\mathbf{X}, t). \qquad (1.9.3)$$

Vectors and tensors that do not depend on time are objective. For time-dependent quantities this is not always the case. Consider, for instance, the velocity vector $\mathbf{v} = \dot{\mathbf{x}}$. From (1.9.1), we have

$$\frac{D\tilde{x}_k}{Dt} = Q_{kl}\dot{x}_l + \dot{Q}_{kl}x_l + \dot{b}_k$$

or

$$\tilde{v}_k = Q_{kl}v_l + \dot{Q}_{kl}x_l + \dot{b}_k. \qquad (1.9.4)$$

This is not in the form (1.9.3)$_1$.

Theorem. The deformation-rate tensor $a_{kl} = v_{l,k} - \nu_{lk}$ is objective.

From (1.9.4), we have

$$\tilde{v}_{k,l} = Q_{km}v_{m,n}\frac{\partial x_n}{\partial \tilde{x}_l} + \dot{Q}_{km}\frac{\partial x_m}{\partial \tilde{x}_l}.$$

Using (1.9.1), we calculate $\partial x_n/\partial \tilde{x}_l = Q_{ln}$, so that

$$\tilde{v}_{l,k} = Q_{km}Q_{ln}v_{n,m} + Q_{km}\dot{Q}_{lm}. \qquad (1.9.5)$$

The transformation law for the directors is

$$\tilde{\chi}_{kK} = Q_{km}\chi_{mK}, \qquad \tilde{\mathfrak{X}}_{Kk} = Q_{km}\mathfrak{X}_{Km}. \qquad (1.9.6)$$

From (1.8.12), we have

$$\tilde{\nu}_{kl} = \dot{\tilde{\chi}}_{kK}\tilde{\mathfrak{X}}_{Kl}.$$

Using (1.9.6), this is expressed as

$$\tilde{\nu}_{kl} = Q_{km}Q_{ln}\nu_{mn} + \dot{Q}_{km}Q_{lm}. \qquad (1.9.7)$$

Adding this to (1.9.5), we obtain

$$\tilde{v}_{l,k} + \tilde{\nu}_{kl} = Q_{km}Q_{ln}(v_{n,m} + \nu_{mn}), \qquad (1.9.8)$$

since

$$Q_{km}\dot{Q}_{lm} + \dot{Q}_{km}Q_{lm} = 0.$$

From (1.9.7), we also have

$$\tilde{\nu}_{(kl)} = Q_{km}Q_{ln}\nu_{(mn)}, \qquad (1.9.9)$$

which shows that $\nu_{(kl)}$ is objective. Upon substituting

$$\tilde{\nu}_{kl} = -\tilde{\nu}_{lk} + 2\tilde{\nu}_{(kl)}, \qquad \nu_{mn} = -\nu_{nm} + 2\nu_{(mn)}$$

into (1.9.8) and using (1.9.9), we obtain

$$\tilde{v}_{l,k} - \tilde{\nu}_{lk} = Q_{km}Q_{ln}(v_{n,m} - \nu_{nm}), \qquad (1.9.10)$$

which shows that $a_{kl} = v_{l,k} - \nu_{lk}$ is objective, which proves the assertion.

From (1.9.7), it is clear that ν_{kl} is *not* objective. However, $c_{kl} \equiv \nu_{(kl)}$ is objective. Since Q_{kl} does not depend on $\mathbf{x}$, from (1.9.7) it also follows that $b_{klm} = \nu_{kl,m}$ is also objective.

Consequently, we have shown that all three deformation rate-tensors a_{kl}, b_{klm}, and c_{kl} given by (1.8.19) are objective tensors. Hence, they are proper candidates for the descriptions of dynamical properties of materials, e.g., viscous fluids.

The above proof, through the expressions (1.8.34), shows that the deformation-rate tensors of microstretch and micropolar continua are also objective tensors.

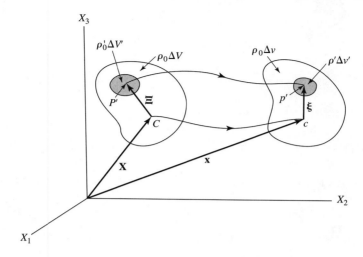

FIGURE 1.10.1. Micro-volume Elements

1.10   Mass, Inertia, Momenta, Kinetic Energy

A. Micromorphic Continua

The concepts of mass and inertia require a finite volume of the material body. Before we consider a limiting process of volume densities, we consider a particle P having volume element ΔV, in the reference state and its image p, in the spatial frame at time t (Figure 1.10.1). The total mass of these particles is the sum of the masses of microelements, i.e.,

$$\rho_0 \Delta V = \int_{\Delta V} \rho_0' \, dV', \qquad \rho \Delta v = \int_{\Delta v} \rho' \, dv', \qquad (1.10.1)$$

where primed quantities refer to microelements of P and p. The relative position vectors Ξ and ξ of the microelements are taken with respect to centroids C and c of ΔV and dv, so that

$$\int_{\Delta V} \rho_0' \Xi \, dV' = 0, \qquad \int_{\Delta v} \rho' \xi \, dv' = 0. \qquad (1.10.2)$$

However, the second moments of $\rho_0' \, dV'$ and $\rho' \, dv'$ do not vanish, and they are given by

$$\rho_0 I_{KL} \Delta V = \int_{\Delta V} \rho_0' \Xi_K \Xi_L \, dV', \qquad \rho i_{kl} \Delta v = \int_{\Delta v} \rho' \xi_k \xi_l \, dv'. \qquad (1.10.3)$$

We assume that the mass of the microelement is conserved during the motion

$$\rho_0' \, dV' = \rho' \, dv'. \qquad (1.10.4)$$

This implies that, in the limit as $\Delta V \to 0$ and $\Delta v \to 0$

$$\rho_0 \, dV = \rho \, dv. \tag{1.10.5}$$

This is the *law of conservation of mass* which remains valid for all 3M continua.

In the limit, (1.10.3) defines *microinertia tensors*

$$\rho_0 I_{KL} \, dV = \int_{dV} \rho_0' \Xi_K \Xi_L \, dV', \qquad \rho i_{kl} \, dv = \int_{dv} \rho' \xi_k \xi_l \, dv'. \tag{1.10.6}$$

Upon using (1.10.5) and (1.2.3), we obtain the *law of conservation of microinertia*[7]

$$i_{kl} = I_{KL} \chi_{kK} \chi_{lL}, \qquad I_{KL} = i_{kl} \mathfrak{X}_{Kk} \mathfrak{X}_{Ll}. \tag{1.10.7}$$

These results may also be obtained by defining probability densities

$$P'(\mathbf{X}, \boldsymbol{\Xi}) \equiv \rho_0'(\mathbf{X}, \boldsymbol{\Xi}) / \int \rho_0'(\mathbf{X}, \boldsymbol{\Xi}) \, dV'(\boldsymbol{\Xi}),$$

$$p'(\mathbf{x}, \boldsymbol{\xi}, t) = \rho'(\mathbf{x}, \boldsymbol{\xi}) / \int \rho'(\mathbf{x}, \boldsymbol{\xi}, t) \, dv'(\boldsymbol{\xi}), \tag{1.10.8}$$

subject to the conservation law

$$P' \, dV' = p' \, dv'. \tag{1.10.9}$$

Then I_{KL} and i_{kl} are defined by (in the limits)

$$I_{KL} = \int_{\Delta V} P' \Xi_K \Xi_L \, dV' = \langle \Xi_K \Xi_L \rangle \tag{1.10.10}$$

$$i_{kl} = \int_{\Delta v} p' \xi_k \xi_l \, dv' = \langle \xi_k \xi_l \rangle \tag{1.10.11}$$

Definition 1. The momentum density per unit volume, at a point of the deformed body, is defined by

$$\mathbf{P} = \rho \mathbf{v}, \qquad \mathbf{v} \equiv \dot{\mathbf{x}}. \tag{1.10.12}$$

Definition 2. Kinetic energy per unit mass is defined by

$$K \equiv \tfrac{1}{2} \langle (\dot{\mathbf{x}} + \dot{\boldsymbol{\xi}}) \cdot (\dot{\mathbf{x}} + \dot{\boldsymbol{\xi}}) \rangle. \tag{1.10.13}$$

Employing (1.2.3) this leads to

$$K = \tfrac{1}{2} \mathbf{v} \cdot \mathbf{v} + \tfrac{1}{2} i_{kl} \nu_{mk} \nu_{ml}. \tag{1.10.14}$$

---

[7]Introduced by Eringen [1964a].

Definition 3. The spin inertia per unit mass is defined by

$$\sigma_{kl} \equiv \langle \ddot{\xi}_k \xi_l. \rangle. \tag{1.10.15}$$

By means of (1.2.3) and (1.8.14) this gives[8]

$$\sigma_{kl} = i_{ml}(\dot{\nu}_{km} + \nu_{kn}\nu_{nm}). \tag{1.10.16}$$

It is a simple exercise to show that

$$\sigma_{kl}\nu_{kl} = \frac{1}{2}\frac{D}{Dt}(i_{kl}\nu_{mk}\nu_{ml}). \tag{1.10.17}$$

B. Microstretch Continua

From (1.8.32), we have

$$\nu_{kl} = \nu\delta_{kl} - \epsilon_{klm}\nu_m. \tag{1.10.18}$$

we decompose σ_{kl} for the microstretch continua as

$$\sigma_{kl} = \frac{\sigma}{3}\delta_{kl} - \tfrac{1}{2}\epsilon_{klm}\sigma_m, \tag{1.10.19}$$

which introduces microstretch scalar inertia σ and microstretch rotatory inertia σ_k. It is useful also to decompose i_{kl} and I_{KL} as

$$i_{kl} = \frac{1}{2}j_0\delta_{kl} - j_{kl}, \qquad j_{kl} = i_0\delta_{kl} - i_{kl}, \qquad j_0 \equiv j_{kk},$$

$$I_{KL} = \frac{1}{2}J_0\delta_{KL} - J_{KL}, \qquad J_{KL} = I_0\delta_{KL} - I_{KL}. \tag{1.10.20}$$

Substituting i_{kl} and j_{kl} from these into (1.10.14) and (1.10.16), we obtain

$$K = \tfrac{1}{2}\mathbf{v} \cdot \mathbf{v} + \tfrac{1}{4}j_0\nu^2 + \tfrac{1}{2}j_{kl}\nu_k\nu_l, \tag{1.10.21}$$

$$\sigma_k = j_{kl}\dot{\nu}_l + 2\nu j_{kl}\nu_l + \epsilon_{klm}j_{mn}\nu_l\nu_n,$$

$$\sigma = \tfrac{1}{2}j_0(\dot{\nu} + \nu^2) - j_{kl}\nu_k\nu_l. \tag{1.10.22}$$

C. Micropolar Continua

In micropolar continua, $\nu = 0$, $\sigma = 0$, and we have

$$K = \tfrac{1}{2}\mathbf{v} \cdot \mathbf{v} + \tfrac{1}{2}j_{kl}\nu_k\nu_l, \tag{1.10.23}$$

$$\sigma_k = j_{kl}\dot{\nu}_l + \epsilon_{klm}j_{mn}\nu_l\nu_n = \frac{D}{Dt}(j_{kl}\nu_l). \tag{1.10.24}$$

---

[8]To avoid a possible misinterpretation of $\dot{\sigma}_{kl}$ as the time-rate, I abandon my previous notation, and use σ_{kl} to represent the spin inertia.

Chapter 1 Problems

1.1. Prove that (1.2.13) and (1.2.14) are the unique solutions of $(1.2.11)_1$ and $(1.2.12)_1$, provided J and j do not vanish.

1.2. Express the rotation about a fixed point O, that carries an orthonormal triad $\mathbf{I}_k$ into $\mathbf{i}_k$ $(K,\ k = 1,\ 2,\ 3)$ in terms of Eulerian angles.

1.3. By examining deformations of a parallelepiped, discuss the physical meaning of strain and rotation tensors. What type of distortions are represented by the components of γ_{klm}?

1.4. Find the expression of $\dot{\boldsymbol{\xi}}$ in terms of Eulerian angles.

1.5. Calculate material time-rates:

   (a) $\frac{D}{Dt}\overset{-1}{\mathfrak{C}}_{KL}$,

   (b) $\frac{D}{Dt}\mathfrak{c}_{kl}$,

   (c) $\frac{D}{Dt}\gamma_{klm}$,

   (d) $\frac{D}{Dt}\overset{-1}{\mathfrak{c}}_{kl}$.

1.6. Obtain the compatibility conditions for the linear strain tensors ϵ_{kl} and γ_{klm}.

1.7. Obtain the expressions of the deformation-rate tensors a_{3kl}, b_{3klm}, and c_{3kl}.

1.8. Prove the validity of the expression (1.8.43).

1.9. Calculate the material time-rates of $\mathfrak{c}_{kl}$ and γ_{klm} and examine them for objectivity.

1.10. A $(*)$-derivative of a vector $\boldsymbol{\nu}$ is defined by

$$\overset{*}{\nu}_k = \dot{\nu}_k + \nu_l v_{k,l}.$$

Show that $\overset{*}{\nu}_k$ is objective. Construct a $(*)$-derivative for a tensor t_{kl}.

1.11. Show the validity of expressions (1.10.17) and (1.10.22).

1.12. (Short-Term Paper) Study the literature and establish connections for the incompatibility of strain measures with dislocations, disclinations, and other incompatible distortions.

1.13. (Short-Term Paper) Establish connections of micromorphic strain measures with the curvature and torsion of a curved space.

2
Stress

2.0 Scope

This chapter is devoted to the concept of stress, the balance laws, and
the general theory of constitutive equations. In Section 2.1, we present
the global balance law of energy for micromorphic continua. This law is
then specialized to microstretch and micropolar continua. The nonlocal
forms of the balance laws of 3M continua are obtained by subjecting the
global laws to Galilean invariance. A theorem shows that to each Galilean
symmetry group of the energy balance law there corresponds a balance
law of microcontinuum mechanics. Hence, we have the law of balance of
momentum, balance of momentum moments, and energy. The local laws
are obtained by means of the postulate of localization. This process leads
to all balance laws and jump conditions at discontinuity surface σ which
may be sweeping the body at its own velocity. The jump conditions at
the boundary surface of the body give the boundary conditions. Later,
in Section 5.10, we show how the conservation laws are deduced from a
theorem of Noether.

In Section 2.3, the second law of thermodynamics is postulated. The
Clausius–Duhem inequality and the dissipation inequality are obtained by
combining the entropy inequality with the energy balance equation. These
are essential to the development of the constitutive equations and their restrictions. The solution of the dissipation inequality leads to development
of the irreversible (dissipative) parts of the constitutive equations. This is
presented in Section 2.4. The Onsager postulate is announced and *mem-*

ory functionals, essential to the development of constitutive equations of memory-dependent materials, are discussed.

2.1   Balance of Energy

A. Micromorphic Continua

A micropolar body will deform when subjected to external loads. These loads may be of a variety of origins: mechanical, thermal, electrical, chemical, etc. They can be divided into two categories: (a) body loads; and (b) surface loads (or contact loads). In category (a) we mention gravitational and electromagnetic forces and couples. In category (b) are contact forces and couples arising from external bodies that are in contact with the surface of the body.

The applied loads cause internal stress in the body. By isolating a part of the body from other parts, we can replace the effect of the rest of the body on the part isolated, as surface forces and couples. In this way the internal loads may be considered as resulting from the effect of one part of the body on another part through the surface of contact. Internal loads are the result of the change of intermolecular attractions with the application of the external loads or the deformation of the body. According to Newton's third law of motion, the mutual attractions of a pair of particles balance each other, so that the resultant internal force and couple vanish.

It is well known that, in a micropolar body, the contact force gives rise to stress tensor t_{kl} and couple stress tensor m_{kl}. The body loads are the body force f_k per unit mass and body couple per unit mass l_k.

In a micromorphic body, a third-order tensor m_{klm} replaces m_{kl} and a second-order tensor l_{kl} replaces l_k. These quantities, in fact, are the result of surface and volume means, defined by[1]

$$m_{klm} = \langle t'_{kl}\xi_m \rangle_2, \qquad l_{kl} = \langle f'_k \xi_l \rangle, \qquad (2.1.1)$$

as originally introduced by Eringen and Şuhubi [1964]. Here $\langle \rangle_2$ represents the surface mean and $\langle \rangle$ denotes the volume average. The primed quantities refer to the microelement contained in a particle. We shall, however, leave these quantities undefined, except that they are restricted by a law of balance of energy, postulated for the entire body as follows:

Principle of Energy Balance: *The time-rate of the sum of the kinetic and internal energies is equal to the work done by all loads acting on the body per unit time.*

---

[1]Previously for m_{klm}, we also used λ_{klm} and t_{klm}. Eringen and Şuhubi [1964], Eringen [1964a], [1968b], Eringen and Kafadar [1976, Section 3.6].

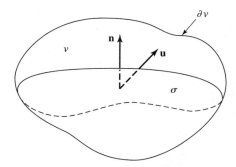

FIGURE 2.1.1. Discontinuity surface σ sweeping body with its own velocity $\boldsymbol{u}$

For a micromorphic body, this is expressed as

$$\frac{d}{dt} \int_{\mathcal{V}-\sigma} \rho(\epsilon + K)\, dv = \int_{\partial\mathcal{V}-\sigma} (t_{kl}v_l + m_{klm}\nu_{lm} + q_k)\, da_k$$
$$+ \int_{\mathcal{V}-\sigma} \rho(f_k v_k + l_{kl}\nu_{kl} + h)\, dv. \qquad (2.1.2)$$

Here, the left-hand side is the time-rate of the total internal and kinetic energies, ϵ represents the internal energy per unit mass. On the right-hand side, under the surface integral, the three terms, respectively, represent the *stress energy*, the *energy* of *stress moments*, and the *heat energy per unit time*. On the right-hand side, under the volume integral, the three terms denote, respectively, *the energy of the body force, the energy of body moments*, and *heat input per unit time*. The volume and surface integrals exclude the line and surface intersections of a discontinuity surface σ which may be sweeping the body with its own velocity $\mathbf{u}$, Figure 2.1.1. This is denoted by

$$\mathcal{V} - \sigma \equiv \mathcal{V} - \mathcal{V} \cap \sigma, \qquad \partial\mathcal{V} - \sigma \equiv \partial\mathcal{V} - \partial\mathcal{V} \cap \sigma. \qquad (2.1.3)$$

Note that the internal energy density ϵ is postulated to exist. With the existence of ϵ, the total energy is balanced. The kinetic energy per unit mass is given by (1.10.14). Initially, the expression (2.1.2) was derived by Eringen and Şuhubi [1964][2] through the integration of the mean energy in a particle. Here we take (2.1.2) as a postulate.

The energy balance law differs from the classical counterpart only in the terms involving m_{klm} and l_{kl}. The origin of these terms may be clarified by a physical picture. Consider a macrosurface element Δa on the surface of the body with exterior unit normal $\mathbf{n}$ (Figure 2.1.2). The work per unit time of a stress vector $\mathbf{t}'_k$ acting at a microsurface element da'_k, with unit normal $\mathbf{n}'$, upon integration over Δa gives the energy due to tractions on

---

[2]For another derivation, see Eringen and Kafadar [1976, Section 2.4].

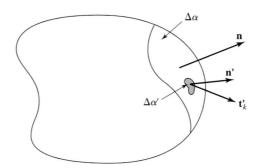

FIGURE 2.1.2. Traction at Microsurface Element $\Delta a'$

Δa. For the stress vector we have, classically, $\mathbf{t}'_k = t'_{kl}\mathbf{i}_l$, where $\mathbf{i}_l$ are the Cartesian unit vectors and t'_{kl} is the microstress tensor. The energy of $\mathbf{t}'_k$ is then given by

$$\int_{\Delta a} t'_{kl}\left(v_l + \dot{\xi}_l\right) da'_k = (t_{kl}v_l + m_{klm}\nu_{lm})\,\Delta a_k, \qquad (2.1.4)$$

where we used (1.8.14) and defined the *stress moment tensor* m_{klm} by the limit (as $\Delta a \to 0$) of

$$m_{klm}\,\Delta a_k = \int_{\Delta a} t'_{kl}\xi_m\, da'_k, \qquad (2.1.5)$$

in accordance with (2.1.1). The presence of the term involving $l_{kl}\nu_{kl}$ in (2.1.2) is similarly obtained by the volume average

$$\int_{\Delta v} \rho' f'_k(v_k + \dot{\xi}_k)\, dv' = (\rho f_k v_k + \rho l_{kl}\nu_{kl})\Delta v, \qquad (2.1.6)$$

where l_{kl} is defined in the limit as $\Delta v \to 0$

$$\rho l_{kl}\,\Delta v \equiv \int_{\Delta v} \rho' f'_k \xi_l\, dv'. \qquad (2.1.7)$$

B. Microstretch Continua

For microstretch continua, we decompose m_{klm} and l_{kl} as

$$m_{klm} = \tfrac{1}{3}m_k\delta_{lm} - \tfrac{1}{2}\epsilon_{lmr}m_{kr},$$
$$l_{kl} = \tfrac{1}{3}l\delta_{kl} - \tfrac{1}{2}\epsilon_{klr}l_r. \qquad (2.1.8)$$

With these, and using (1.10.18), (2.1.2) takes the form

$$\frac{d}{dt}\int_{\mathcal{V}-\sigma} \rho(\epsilon + K)\, dv = \int_{\partial\mathcal{V}-\sigma} (t_{kl}v_l + m_{kl}\nu_l + m_k\nu + q_k)\, da_k$$
$$+ \int_{\mathcal{V}-\sigma} \rho(f_k v_k + l\nu + l_k\nu_k + h)\, dv, \qquad (2.1.9)$$

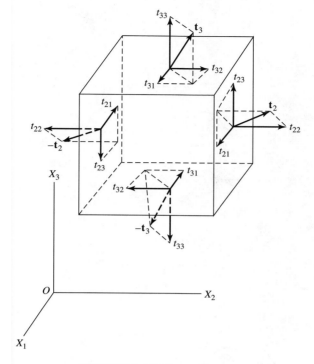

FIGURE 2.1.3. Stress Tensor

where K is given by (1.10.21).

Here m_k is the *microstretch vector* and m_{kl} is the *couple stress tensor*, l is the body *microstretch force density* and l_k the *body couple density*.

C. Micropolar Continua

In this case, $\nu = 0$, $m_k = 0$, and $l = 0$. The energy balance law reads

$$\frac{d}{dt} \int_{\mathcal{V}-\sigma} \rho(\epsilon + K)\, dv = \int_{\partial\mathcal{V}-\sigma} (t_{kl}v_l + m_{kl}\nu_l + q_k)\, da_k$$

$$+ \int_{\mathcal{V}-\sigma} \rho(f_k v_k + l_k \nu_k + h)\, dv. \quad (2.1.10)$$

The kinetic energy is given by (1.10.23).

The positive directions of stress tensor t_{kl} are shown in Figure 2.1.3 and those of the couple stress tensor m_{kl} are shown in Figures 2.1.4 and 2.1.5.

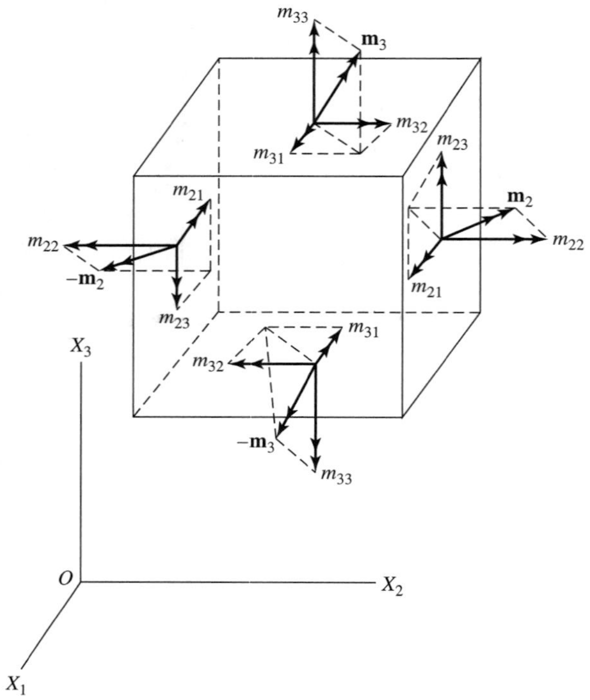

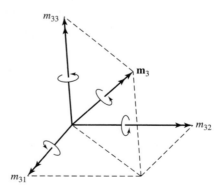

FIGURE 2.1.4. Couple Stress Tensor

FIGURE 2.1.5. Directions of Couple Stress

2.2 Balance Laws

A. Balance Laws of Micromorphic Continua[3]

For calculations of the time-rate of integrals and conversion of surface integrals to volume integrals, we need two formulas which are known as the Transport and Green–Gauss theorems:

Transport theorem:

$$\frac{d}{dt} \int_{\mathcal{V}-\sigma} \Upsilon \, dv = \int_{\mathcal{V}-\sigma} \left(\dot{\Upsilon} + \Upsilon \nabla \cdot \mathbf{v} \right) dv + \int_{\sigma} [\![\Upsilon (\mathbf{v} - \mathbf{u})]\!] \cdot d\mathbf{a}. \qquad (2.2.1)$$

Green–Gauss theorem:

$$\int_{\partial \mathcal{V}-\sigma} \mathbf{A} \cdot \mathbf{n} \, da = \int_{\mathcal{V}-\sigma} \nabla \cdot \mathbf{A} \, dv + \int_{\sigma} [\![\mathbf{A}]\!] \cdot \mathbf{n} \, da, \qquad (2.2.2)$$

where Υ is any tensor and $\mathbf{A}$ is any vector. Both are continuously differentiable in star-shaped domains. The double brackets $[\![\quad]\!]$ denote the jump of its enclosure across the discontinuity surface σ which may be sweeping the body with its own velocity $\mathbf{u}$. For a proof of these theorems, cf. Eringen [1967a, 1980].

Two of the local balance laws, conservation of mass and microinertia were already obtained in Section 1.10,

$$\rho \, dv = \rho_0 \, dV, \qquad (2.2.3)$$

$$i_{kl} = I_{KL} \chi_{kK} \chi_{lL}, \qquad I_{KL} = i_{kl} \mathfrak{X}_{Kk} \mathfrak{X}_{Ll}. \qquad (2.2.4)$$

For the entire body, these conservation laws are replaced by

$$\frac{d}{dt} \int_{\mathcal{V}-\sigma} \rho \, dv = 0, \qquad (2.2.5)$$

$$\frac{d}{dt} \int_{\mathcal{V}-\sigma} \rho i_{kl} \mathfrak{X}_{Kk} \mathfrak{X}_{Ll} \, dv = 0. \qquad (2.2.6)$$

These are called *Global Conservation Laws* (or integral conservation laws) of mass and microinertia.

By means of the transport theorem, (2.2.1) and (1.8.13), these are converted to

$$\int_{\mathcal{V}-\sigma} \hat{\rho} \, dv + \int_{\sigma} [\![\hat{R}_k]\!] n_k \, da = 0, \qquad (2.2.7)$$

$$\int_{\mathcal{V}-\sigma} (\hat{\rho} i_{kl} + \rho \hat{i}_{kl}) \mathfrak{X}_{Kk} \mathfrak{X}_{Ll} \, dv + \int_{\sigma} [\![\hat{G}_{klm}]\!] n_m \mathfrak{X}_{Kk} \mathfrak{X}_{Ll} \, da = 0, \qquad (2.2.8)$$

---

[3]The present derivation differs from the original Eringen–Şuhubi approach. It was first given by Eringen [1992].

where

$$\hat{\rho} \equiv \dot{\rho} + \rho \nabla \cdot \mathbf{v}, \qquad\qquad \hat{R}_k \equiv \rho(v_k - u_k),$$

$$\hat{i}_{kl} \equiv \frac{Di_{kl}}{Dt} - i_{kr}\nu_{lr} - i_{lr}\nu_{kr}, \qquad \hat{G}_{klm} = \rho i_{kl}(v_m - u_m) \qquad (2.2.9)$$

are called *Nonlocal Residuals*.

Postulate of Localization: *In* **local** *continuum mechanics, it is postulated that nonlocal residuals vanish.*

This is equivalent to the postulate that the integral balance laws are valid, not only for the full body, but also for every arbitrary region of the body. Consequently, for the *local law of conservation of mass*, we have

$$\dot{\rho} + \rho \nabla \cdot \mathbf{v} = 0 \qquad \text{in } \mathcal{V} - \sigma, \qquad (2.2.10\text{-a})$$

$$[\![\rho(\mathbf{v} - \mathbf{u})]\!] \cdot \mathbf{n} = 0 \qquad \text{on } \sigma, \qquad (2.2.10\text{-b})$$

and for the *local law of conservation of microinertia*

$$\frac{Di_{kl}}{Dt} - i_{kr}\nu_{lr} - i_{lr}\nu_{kr} = 0 \qquad \text{in } \mathcal{V} - \sigma, \qquad (2.2.11\text{-a})$$

$$[\![\rho i_{kl}(\mathbf{v} - \mathbf{u})]\!] \cdot \mathbf{n} = 0 \qquad \text{on } \sigma. \qquad (2.2.11\text{-b})$$

Here, the jump conditions (2.2.10-b) and (2.2.11-b) are useful for the treatment of the bodies involving discontinuity surfaces, e.g., shock waves. They also give boundary conditions when σ coincides with the surface $\partial \mathcal{V}$ of the body.

The remaining balance laws will be obtained by subjecting the energy balance law to the invariance under the Galilean group of transformations.

Intuitively, it is apparent that the energy balance law should also be valid in a frame-of-reference that is undergoing rigid motions. This is guaranteed by a theorem of Noether for the Hamiltonian, leading to balance laws for each infinitesimal symmetry group (cf. Section 5.10).

Axiom. The energy balance law is form-invariant under the Galilean group of transformations.

Theorem. For each Galilean group of the energy balance law, there corresponds a balance law of microcontinuum mechanics.

Consider a time-dependent rigid motion of the frame-of-reference

$$\tilde{x}_k = Q_{kl}(t)x_l + b_k(t), \qquad (2.2.12)$$

where Q_{kl} is an orthogonal tensor, i.e.,

$$\mathbf{Q}\mathbf{Q}^T = \mathbf{Q}^T\mathbf{Q} = \mathbf{1}, \qquad \det \mathbf{Q} = 1. \qquad (2.2.13)$$

The velocity field in the new frame is given by

$$\tilde{v}_k = Q_{kl}v_l + \dot{Q}_{kl}x_l + \dot{b}_k. \tag{2.2.14}$$

The angular velocity is calculated by

$$\Omega_{kl} = \dot{Q}_{km}Q_{lm}. \tag{2.2.15}$$

The microinertia tensor in the new frame is given by

$$\tilde{i}_{kl} = Q_{km}i_{mn}Q_{ln}. \tag{2.2.16}$$

The material time-rate of this is

$$\frac{D\tilde{i}_{kl}}{Dt} = Q_{km}\frac{Di_{mn}}{Dt}Q_{ln} + 2\dot{Q}_{km}i_{mn}Q_{ln}. \tag{2.2.17}$$

Suppose that at time t, the body is brought back to the original orientation (i.e., $\mathbf{Q} = \mathbf{1}$, having only constant translational and angular velocities, i.e., Ω and $\dot{\mathbf{b}}$ as constants). Thus, at time t, we have

$$\tilde{v}_k = v_k + \Omega_{kl}x_l + \dot{b}_k, \qquad \Omega_{kl} = \dot{Q}_{kl},$$
$$\tilde{\nu}_{kl} = \Omega_{kl} + \nu_{kl},$$
$$\tilde{i}_{kl} = i_{kl}, \qquad \frac{D\tilde{i}_{kl}}{Dt} = \frac{Di_{kl}}{Dt} + 2\Omega_{km}i_{ml}. \tag{2.2.18}$$

Under these transformations, the mass density ρ, the internal energy density ϵ, stress tensors t_{kl}, t_{klm}, and the heat vector q_k are not affected, since the motion is rigid. But the body force f_k, body moments l_{kl}, must be accommodated by corresponding linear and microaccelerations (spin inertia), i.e.,

$$\tilde{\mathbf{f}} - \dot{\tilde{\mathbf{v}}} = \mathbf{f} - \dot{\mathbf{v}}, \qquad \tilde{l}_{kl} - \tilde{\sigma}_{kl} = l_{kl} - \sigma_{kl}. \tag{2.2.19}$$

According to (1.10.14), the new kinetic energy is given by

$$\tilde{K} = \tfrac{1}{2}\tilde{\mathbf{v}} \cdot \tilde{\mathbf{v}} + \tfrac{1}{2}\tilde{i}_{kl}\tilde{\nu}_{mk}\tilde{\nu}_{ml}. \tag{2.2.20}$$

Subtracting the energy balance (2.1.2) in the $\mathbf{x}$-frame from that in the $\tilde{\mathbf{x}}$-frame, we obtain

$$\frac{d}{dt}\int_{\mathcal{V}-\sigma} \rho\left(\tilde{K} - K\right)dv = \int_{\partial\mathcal{V}-\sigma} [t_{kl}\left(\tilde{v}_l - v_l\right) + m_{klm}\left(\tilde{\nu}_{lm} - \nu_{lm}\right)]\,n_k da$$
$$+ \int_{\mathcal{V}-\sigma} \rho\left(\tilde{f}_k\tilde{v}_k - f_k v_k + \tilde{l}_{kl}\tilde{\nu}_{kl} - l_{kl}\nu_{kl}\right)dv. \tag{2.2.21}$$

Calculations give

$$\hat{\hat{i}}_{kl} = \hat{i}_{kl}, \quad \frac{D}{Dt}(\hat{i}_{kl}\nu_{mk}\nu_{ml}) = \hat{i}_{kl}\nu_{mk}\nu_{ml} + 2\sigma_{kl}\nu_{kl},$$

$$\tilde{K} - K = v_k(\Omega_{km}x_m + \dot{b}_k) + \tfrac{1}{2}(\Omega_{kl}x_l + \dot{b}_k)(\Omega_{km}x_m + \dot{b}_k)$$
$$+ i_{kl}\nu_{mk}\Omega_{ml} + \tfrac{1}{2}i_{kl}\Omega_{mk}\Omega_{ml},$$

$$\dot{\tilde{K}} - \dot{K} = (\dot{\tilde{v}}_k\tilde{v}_k - \dot{v}_kv_k) + \hat{i}_{kl}(\nu_{mk}\Omega_{ml} + \tfrac{1}{2}\Omega_{mk}\Omega_{ml})$$
$$+ (\tilde{\sigma}_{kl} - \sigma_{kl})\tilde{\nu}_{kl} - \sigma_{kl}\Omega_{kl},$$

$$\tilde{v}_k - v_k = \Omega_{kl}x_l + \dot{b}_k, \qquad \dot{\tilde{v}}_k - \dot{v}_k = \Omega_{kl}v_l, \qquad \tilde{\nu}_{kl} - \nu_{kl} = \Omega_{kl},$$

$$\tilde{f}_k\tilde{v}_k - f_kv_k = f_k(\tilde{v}_k - v_k) + \tilde{v}_k(\dot{\tilde{v}}_k - \dot{v}_k),$$

$$\tilde{l}_{kl}\tilde{\nu}_{kl} - l_{kl}\nu_{kl} = (\tilde{\sigma}_{kl} - \sigma_{kl})\tilde{\nu}_{kl} + l_{kl}(\tilde{\nu}_{kl} - \nu_{kl}), \tag{2.2.22}$$

where we used (2.2.9) and (2.2.18)–(2.2.20).

Using the transport theorem (2.2.1), we carry out the time-rate of the volume integral on the left-hand side of (2.2.21). By means of the Green–Gauss theorem (2.2.2), we convert the surface integral on the right-hand side of (2.2.21). After using (2.2.10-a)–(2.2.11-b), equation (2.2.21) becomes

$$\int_{\mathcal{V}-\sigma}\left[\left(\Omega_{lm}x_m + \dot{b}_l\right)\hat{f}_l + \Omega_{lm}\hat{l}_{lm}\right]dv$$

$$+ \int_\sigma\left[\left[\left(\Omega_{lm}x_m + \dot{b}_l\right)\hat{F}_{kl} + \Omega_{lm}\hat{L}_{klm}\right]\right]n_k\,da$$

$$- \int_{\mathcal{V}-\sigma}[\hat{\rho}(\tilde{K} - K) + \hat{i}_{kl}(\nu_{mk}\Omega_{ml} + \tfrac{1}{2}\Omega_{mk}\Omega_{ml}]\,dv \tag{2.2.23}$$

$$- \frac{1}{2}\int_\sigma\left[\left[\left(\Omega_{lm}x_m + \dot{b}_l\right)\left(\Omega_{lr}x_r + \dot{b}_l\right)\hat{R}_k + \Omega_{rl}\Omega_{rm}\hat{G}_{lmk}\right]\right]n_k\,da = 0,$$

where

$$\hat{f}_l \equiv t_{kl,k} + \rho(f_l - \dot{v}_l) \qquad \text{in } \mathcal{V} - \sigma,$$

$$\hat{l}_{lm} \equiv m_{klm,k} + t_{ml} - s_{ml} + \rho(l_{lm} - \sigma_{lm}) \qquad \text{in } \mathcal{V} - \sigma,$$

$$\hat{F}_{kl} = t_{kl} - \rho v_l(v_k - u_k) \qquad \text{on } \sigma,$$

$$\hat{L}_{klm} = m_{klm} - \rho i_{rm}\nu_{lr}(v_k - u_k) \qquad \text{on } \sigma, \tag{2.2.24}$$

and $\hat{R}$ and $\hat{G}_{klm}$ are given by (2.2.9).

Here s_{ml} is an arbitrary symmetric tensor

$$s_{ml} = s_{lm}. \tag{2.2.25}$$

It is introduced, since $\Omega_{lm}s_{lm} = 0$ does not contribute to (2.2.23). However, it allows all components of Ω_{lm} to be arbitrary, irrespective of the antisymmetry of Ω_{lm}.

For arbitrary and independent variations of $\dot{b}_l$ and Ω_{lm} the coefficients of these quantities in (2.2.23) must vanish. Hence

$$\int_{\mathcal{V}-\sigma} \hat{f}_l \, dv + \int_{\sigma} \left[\!\left[\hat{F}_{kl} \right]\!\right] n_k \, da = 0, \qquad (2.2.26)$$

$$\int_{\mathcal{V}-\sigma} \hat{l}_{lm} \, dv + \int_{\sigma} \left[\!\left[\hat{L}_{klm} \right]\!\right] n_k \, da = 0. \qquad (2.2.27)$$

The last two integrals in (2.2.23) vanish because of the local mass and microinertia conservation laws (2.2.10-a)–(2.2.11-b).

Equations (2.2.26) and (2.2.27) are, respectively, the *balance of momentum and balance of momentum moments of the nonlocal micromorphic continua.*

The energy balance law (2.1.2) may now be recast. By means of the transport and Green–Gauss theorems, we transform the time-rate of the volume integral on the left-hand side and the surface integral on the right-hand side. This gives

$$\int_{\mathcal{V}-\sigma} (-\hat{\epsilon} + \hat{f}_l v_l + \hat{l}_{lm} \nu_{lm}) \, dv + \int_{\sigma} \left[\!\left[-\hat{E}_k \right]\!\right] n_k \, da = 0, \qquad (2.2.28)$$

where we defined

$$-\hat{\epsilon} \equiv -\rho\dot{\epsilon} + t_{kl}(v_{l,k} - \nu_{lk}) + s_{kl}\nu_{lk} + m_{klm}\nu_{lm,k} + q_{k,k} + \rho h, \quad (2.2.29)$$

$$-\hat{E}_k \equiv t_{kl}v_l + m_{klm}\nu_{lm} + q_k$$
$$- (\rho\epsilon + \tfrac{1}{2}\rho\mathbf{v} \cdot \mathbf{v} + \tfrac{1}{2}\rho i_{rl}\nu_{mr}\nu_{ml})(v_k - u_k). \qquad (2.2.30)$$

Equation (2.2.28) is the *energy balance law of the nonlocal micromorphic continua.*

It is important to note that these three nonlocal balance laws (2.2.26), (2.2.27), and (2.2.28) are obtained on the assumption that the mass and microinertia are conserved locally, i.e., (2.2.10-a, b) and (2.2.11-a, b) are valid. This implies that the material is *inert*, i.e., no mass or inertia is created or destroyed. The extension of these laws to include nonlocality in mass and inertia is highly involved.

We now recall the localization postulate, according to which these nonlocal laws must be valid for any arbitrary volume $\mathcal{V} - \sigma$ and surface σ of a region of the body. This then leads to the *local* balance laws

$$t_{kl,k} + \rho(f_l - \dot{v}_l) = 0 \qquad \text{in } \mathcal{V} - \sigma, \quad (2.2.31\text{-a})$$

$$[\![t_{kl} - \rho v_l(v_k - u_k)]\!]\, n_k = 0 \qquad \text{on } \sigma, \qquad (2.2.31\text{-b})$$

$$m_{klm,k} + t_{ml} - s_{ml} + \rho(l_{lm} - \sigma_{lm}) = 0 \qquad \text{in } \mathcal{V} - \sigma, \quad (2.2.32\text{-a})$$

$$[\![m_{klm} - \rho i_{rm}\nu_{lr}(v_k - u_k)]\!]\, n_k = 0 \qquad \text{on } \sigma, \qquad (2.2.32\text{-b})$$

$$-\rho\dot{\epsilon} + t_{kl}(v_{l,k} - \nu_{lk})$$
$$+s_{kl}\nu_{lk} + m_{klm}\nu_{lm,k} + q_{k,k} + \rho h = 0 \qquad \text{in } \mathcal{V} - \sigma, \quad (2.2.33\text{-a})$$
$$[\![t_{kl}v_l + m_{klm}\nu_{lm} + q_k - (\rho\epsilon + \tfrac{1}{2}\rho\mathbf{v}\cdot\mathbf{v}$$
$$+\tfrac{1}{2}\rho i_{rl}\nu_{mr}\nu_{ml})(v_k - u_k)]\!] n_k = 0 \qquad \text{on } \sigma. \qquad (2.2.33\text{-b})$$

To these are enjoined the local balance laws of mass and microinertia, given by (2.2.10-a, b) and (2.2.11-a, b). This completes the balance laws of micromorphic continua.

Originally, the balance laws (2.2.31-a)–(2.2.33-a) were obtained by Eringen and Şuhubi [1964] and Eringen [1964b, 1968a] by means of a "microscopic space-averaging" process. The law of conservation of microinertia was introduced by Eringen [1964a]. The same equations were also derived by Oevel and Schröter [1981] by means of statistical mechanics. These latter authors also gave expressions for the stress, stress moments, heat, and spin in terms of microscopic variables. If the so-called kinetic fluxes are incorporated into stress and heat, statistical mechanics lead to

$$t_{kl} = t_{kl}^{\text{int.}} + t_{kl}^{\text{kin.}}, \qquad m_{klm} = m_{klm}^{\text{int.}} + m_{klm}^{\text{kin.}}, \qquad q_k = q_k^{\text{int.}} + q_k^{\text{kin.}}, \quad (2.2.34)$$

where

$$m_{klm}^{\text{kin.}} = \frac{1}{\rho}\mathfrak{M}_{kmn}\nu_{ln} \qquad (2.2.35)$$

and the equation of conservation of microinertia is modified to

$$\frac{Di_{kl}}{Dt} - i_{km}\nu_{lm} - i_{lm}\nu_{km} + \frac{1}{\rho}\mathfrak{M}_{mkl,m} = 0. \qquad (2.2.36)$$

With the incorporation of the kinetic fluxes into internal fields, balance laws of micromorphic continua remain unchanged,[4] except for the flux term $\mathfrak{M}_{mkl,m}$ that occurs in (2.2.36). The kinetic terms arise from the internal motions of microelements that contribute to nonsmooth macroscopic fields. As shown by Oevel and Schröter, for smooth macroscopic fields, all kinetic terms drop out from the balance laws. Nevertheless, for certain types of media, the modified form of the microinertia equation (2.2.36) may be valid. In fact, independently, Eringen [1985, 1991] has proposed an equation of the form

$$\frac{Di_{kl}}{Dt} - i_{km}\nu_{lm} - i_{lm}\nu_{km} - f_{kl} = 0 \qquad (2.2.37)$$

in harmony with other balance laws. Separate constitutive equations were obtained by Eringen [1991], applicable to the study of dense suspensions in viscous fluids. A separate discussion for this topic will be found in Volume II. It must be remarked, however, that for smooth macroscopic fields $\mathfrak{M} = \mathbf{0}$ and f_{kl} vanish.

---

[4] A minor difference seems to exist in the contribution of the kinetic $m_{klm}^{\text{kin.}}$ to the energy balance law given by Oevel and Schröter. we enquired about their form from the authors, but have not received an explanation.

B. Balance Laws of Microstretch Continua

Similar to the foregoing analysis, balance laws of microstretch continua may be obtained by imposing the Galilean invariance requirement to the energy equation (2.1.9). Alternatively, by replacing m_{klm}, l_{kl} by (2.1.8), ν_{kl} and σ_{kl} by (1.10.18) and (1.10.19) in the balance laws of micromorphic continua, we obtain those of microstretch continua.

Conservation of mass:

$$\dot{\rho} + \rho \nabla \cdot \mathbf{v} = 0 \qquad \text{in } \mathcal{V} - \sigma, \qquad (2.2.38\text{-a})$$

$$[\![\rho(\mathbf{v} - \mathbf{u})]\!] \cdot \mathbf{n} = 0 \qquad \text{on } \sigma. \qquad (2.2.38\text{-b})$$

Conservation of microstretch inertia:

$$\frac{Dj_0}{Dt} - 2j_0\nu = 0,$$

$$\frac{Dj_{kl}}{Dt} - 2\nu j_{kl} + (\epsilon_{kpr}j_{lp} + \epsilon_{lpr}j_{kp})\nu_r = 0 \qquad \text{in } \mathcal{V} - \sigma, \quad (2.2.39\text{-a})$$

$$[\![\rho j_0(\mathbf{v} - \mathbf{u})]\!] \cdot \mathbf{n} = 0, \qquad [\![\rho j_{kl}(\mathbf{v} - \mathbf{u})]\!] \cdot \mathbf{n} = 0 \qquad \text{on } \sigma. \ (2.2.39\text{-b})$$

Balance of momentum:

$$t_{kl,k} + \rho(f_l - \dot{v}_l) = 0 \qquad \text{in } \mathcal{V} - \sigma, \qquad (2.2.40\text{-a})$$

$$[\![t_{kl} - \rho v_l(v_k - u_k)]\!]\, n_k = 0 \qquad \text{on } \sigma. \qquad (2.2.40\text{-b})$$

Balance of momentum moments:

$$m_{kl,k} + \epsilon_{lmn}t_{mn} + \rho(l_l - \sigma_l) = 0 \qquad \text{in } \mathcal{V} - \sigma,$$

$$m_{k,k} + t - s + \rho(l - \sigma) = 0 \qquad \text{in } \mathcal{V} - \sigma, \qquad (2.2.41\text{-a})$$

where we wrote $t_{kk} = t$ and $s_{kk} = s$,

$$[\![m_{kl} - \rho j_{pl}\nu_p(v_k - u_k)]\!]\, n_k = 0 \qquad \text{on } \sigma,$$

$$[\![m_k - \tfrac{1}{2}\rho j_0\nu(v_k - u_k)]\!]\, n_k = 0 \qquad \text{on } \sigma. \qquad (2.2.41\text{-b})$$

Balance of energy:

$$
\begin{aligned}
& - \rho \dot{\epsilon} + t_{kl}(v_{l,k} + \epsilon_{lkr}\nu_r) + m_{kl}\nu_{l,k} + m_k\nu_{,k} \\
& \quad + (s - t)\nu + q_{k,k} + \rho h = 0 \quad \text{in } \mathcal{V} - \sigma,
\end{aligned}
\tag{2.2.42-a}
$$

$$
\begin{aligned}
& \llbracket t_{kl}v_l + m_{kl}\nu_l + m_k\nu + q_k - \\
& \quad \rho(\epsilon + \tfrac{1}{2}\mathbf{v}\cdot\mathbf{v} + \tfrac{1}{2}j_{mn}\nu_m\nu_n + \tfrac{1}{4}j_0\nu^2)(v_k - u_k)\rrbracket\, n_k = 0 \quad \text{on } \sigma.
\end{aligned}
\tag{2.2.42-b}
$$

C. Balance Laws of Micropolar Continua

Balance laws of micropolar continua are obtained from those of microstretch by setting $\nu = l = \sigma = m_k = 0$ and $t_{kk} = s_{kk}$. This results in:

Conservation of mass:

$$
\dot{\rho} + \rho\nabla\cdot\mathbf{v} = 0 \qquad \text{in } \mathcal{V} - \sigma, \tag{2.2.43-a}
$$

$$
\llbracket \rho(\mathbf{v} - \mathbf{u})\rrbracket\cdot\mathbf{n} = 0 \qquad \text{on } \sigma. \tag{2.2.43-b}
$$

Conservation of microinertia:

$$
\frac{Dj_{kl}}{Dt} + (\epsilon_{kpr}j_{lp} + \epsilon_{lpr}j_{kp})\nu_r = 0 \qquad \text{in } \mathcal{V} - \sigma, \tag{2.2.44-a}
$$

$$
\llbracket \rho j_{kl}(\mathbf{v} - \mathbf{u})\rrbracket\cdot\mathbf{n} = 0 \qquad \text{on } \sigma. \tag{2.2.44-b}
$$

Balance of momentum:

$$
t_{kl,k} + \rho(f_l - \dot{v}_l) = 0 \qquad \text{in } \mathcal{V} - \sigma, \tag{2.2.45-a}
$$

$$
\llbracket t_{kl} - \rho v_l(v_k - u_k)\rrbracket\, n_k = 0 \qquad \text{on } \sigma. \tag{2.2.45-b}
$$

Balance of moments of momentum:

$$
m_{kl,k} + \epsilon_{lmn}t_{mn} + \rho(l_l - \sigma_l) = 0 \qquad \text{in } \mathcal{V} - \sigma, \tag{2.2.46-a}
$$

$$
\llbracket m_{kl} - \rho j_{pl}\nu_p(v_k - u_k)\rrbracket\, n_k = 0 \qquad \text{on } \sigma. \tag{2.2.46-b}
$$

Balance of energy:

$$- \rho \dot{\epsilon} + t_{kl}(v_{l,k} + \epsilon_{lkr} \nu_r)$$
$$+ m_{kl} \nu_{l,k} + q_{k,k} + \rho h = 0 \qquad \text{in } \mathcal{V} - \sigma, \qquad (2.2.47\text{-a})$$
$$[[t_{kl} v_l + m_{kl} \nu_l + q_k$$
$$- \rho(\epsilon + \tfrac{1}{2} \mathbf{v} \cdot \mathbf{v} + \tfrac{1}{2} j_{mn} \nu_m \nu_n)(v_k - u_k)]] n_k = 0 \qquad \text{on } \sigma \quad (2.2.47\text{-b})$$

The balance laws of micromorphic, microstretch and micropolar media are complete. These laws are valid for all types of media (solids, fluids, etc.) irrespective of material constitution. These equations will be *closed* only after the constitutive equations are developed.

Balance laws of both microstretch and micropolar theories were given by Eringen [1966a, b, 1969a, 1971b]. A preliminary version of micropolar balance laws is contained in the papers of Eringen and Şuhubi [1964], and in the early works of the Cosserats [1909]. However, the Cosserats do not define specific spin inertia, nor have they given a conservation law for the microinertia tensor. Their work is concerned with a theory of elasticity. The directors they use are not conducive to the treatment of the material properties, other than isotropic elastic solids.

2.3   Second Law of Thermodynamics

A. Micromorphic Continua

The energy balance law (2.1.2) is formed with a single heat vector $\mathbf{q}$. This means that the effects of the mean spatial moments of $\mathbf{q}$ have been ignored. Consequently, there are no microheat tensors for the micromorphic continuum of grade one. Higher grade micromorphic theories involve such moments (see Eringen [1970b], Grot [1969]). Accordingly, in harmony with a single heat vector $\mathbf{q}$, a single absolute temperature ($\theta > 0$, inf. $\theta = 0$) is posited for micromorphic continuum. It is then clear that the second law of thermodynamics has the classical form

$$\frac{d}{dt} \int_{\mathcal{V} - \sigma} \rho \eta \, dv - \int_{\partial \mathcal{V} - \sigma} \frac{1}{\theta} q_k n_k \, da - \int_{\mathcal{V} - \sigma} \frac{\rho h}{\theta} \geq 0, \qquad (2.3.1)$$

where η is the entropy density per unit mass.

By means of the transport theorem (2.2.1) and the Green–Gauss theorem (2.2.2), (2.3.1) is converted to

$$\int_{\mathcal{V} - \sigma} \hat{\gamma} \, dv + \int_\sigma [[\hat{\Gamma}_k]] n_k \, da \geq 0, \qquad (2.3.2)$$

where

$$\hat{\gamma} \equiv \rho \dot{\eta} - (q_k/\theta)_{,k} - \rho h/\theta,$$

$$\hat{\Gamma}_k \equiv \rho\eta(v_k - u_k) - q_k/\theta. \tag{2.3.3}$$

Expression (2.3.2) is the nonlocal form of the second law of thermodynamics. The *local* forms are obtained by appealing to the postulate of localization

$$\rho\dot{\eta} - (q_k/\theta)_{,k} - \rho h/\theta \geq 0 \qquad \text{in } \mathcal{V} - \sigma, \tag{2.3.4-a}$$
$$[\![\rho\eta(v_k - u_k) - q_k/\theta]\!]\, n_k \geq 0 \qquad \text{on } \sigma. \tag{2.3.4-b}$$

With introduction of the *Helmholtz free energy* ψ by

$$\psi = \epsilon - \theta\eta \tag{2.3.5}$$

and eliminating h between (2.3.4-a) and the energy equation (2.2.33-a), we obtain the *generalized Clausius–Duhem* (C–D) inequality

$$-\rho(\dot{\psi} + \eta\dot{\theta}) + t_{kl}a_{kl} + m_{klm}b_{lmk} + s_{kl}c_{kl} + \frac{1}{\theta}q_k\theta_{,k} \geq 0 \qquad \text{in } \mathcal{V} - \sigma, \tag{2.3.6}$$

where we used (1.8.36).

This inequality plays an important role in the development of constitutive equations.

Definition 1 (Thermodynamic Equilibrium).   The state of the body is said to be in thermodynamic equilibrium, when the C–D inequality vanishes, i.e., (2.3.6) acquires the equal sign.

In accordance with the axiom of causality, the following set of variables constitutes the *dependent variables* set

$$\{\psi, \eta, t_{kl}, m_{klm}, s_{kl}, q_k.\} \tag{2.3.7}$$

We decompose this set into recoverable (static) and dynamic sets:

$$_R\mathfrak{I} \equiv \{\psi, -\rho_R\eta,\ _R\mathbf{t},\ _R\mathbf{m},\ _R\mathbf{s}, \cdot\}, \tag{2.3.8}$$
$$\mathfrak{I} \equiv \{\cdot, -\rho_D\eta,\ _D\mathbf{t},\ _D\mathbf{m},\ _D\mathbf{s}, \mathbf{q}\}, \tag{2.3.9}$$

with

$$\eta = _R\eta + _D\eta, \qquad \mathbf{t} = _R\mathbf{t} + _D\mathbf{t}, \qquad \mathbf{m} = _R\mathbf{m} + _D\mathbf{m}, \qquad \mathbf{s} = _R\mathbf{s} + _D\mathbf{s}. \tag{2.3.10}$$

By virtue of (2.3.5), the introduction of $_D\eta$ allows ψ to be fully static. This simply means that ϵ, which has both static and dynamic parts, is replaced by η.[5]

---

[5]The dynamic entropy density $_D\eta$ was first introduced by Eringen [1966c]. In thermodynamics, it represents a major departure from previous concepts and practices.

It may turn out that some of the members of $_R\mathcal{J}$ or $\mathcal{J}$ may drop out by virtue of the C–D inequality. For example, we already know that the heat vector $\mathbf{q}$ has no static part. Traditionally, the dynamic variables $\mathcal{J}$ are called *thermodynamic fluxes.*

We are free to take

$$-\rho(\dot{\psi} + {}_R\eta\dot{\theta}) + {}_Rt_{kl}a_{kl} + {}_Rm_{klm}b_{lmk} + {}_R s_{kl}c_{kl} = 0, \qquad (2.3.11)$$

which expresses the fact that the recoverable set is in thermal equilibrium. In fact, we shall see that this is a direct consequence of the constitutive structure of $_R\mathcal{J}$, i.e., $_R\mathcal{J}$ is a function of the static independent variables (cf. Section 3.4).

From (2.3.6), it follows that

$$\rho\hat{\eta} \equiv -\rho_D\eta\dot{\theta} + {}_Dt_{kl}a_{kl} + {}_Dm_{klm}b_{lmk} + {}_D s_{kl}c_{lk} + \frac{q_k}{\theta}\theta_{,k} \geq 0. \qquad (2.3.12)$$

This expresses the dissipation of energy. This inequality is typical in irreversible thermodynamics. It has the scalar product form

$$\rho\hat{\eta} \equiv \mathcal{J} \cdot Y \geq 0, \qquad (2.3.13)$$

where Y is called the *thermodynamic force.* It consists of the following set

$$Y \equiv \{\dot{\theta}, a_{lk}, b_{lmk}, c_{lk}, \theta_{,k}/\theta\}. \qquad (2.3.14)$$

B. Microstretch Continua

For microstretch continua, the C–D inequality is obtained by eliminating h among (2.2.42-a), (2.3.4-a), and (2.3.5)

$$-\rho(\dot{\psi} + \eta\dot{\theta}) + t_{kl}a_{kl} + m_{kl}b_{lk} + m_k\nu_{,k} + (s - t)\nu + \frac{q_k}{\theta}\theta_{,k} \geq 0 \quad \text{in } \mathcal{V} - \sigma. \qquad (2.3.15\text{-a})$$

The jump condition (2.3.4-b) remains valid

$$\left[\!\!\left[\rho\eta(\mathbf{v} - \mathbf{u}) - \frac{1}{\theta}\mathbf{q}\right]\!\!\right] \cdot \mathbf{n} \geq 0 \qquad \text{on } \sigma. \qquad (2.3.15\text{-b})$$

In this case, the static variable set satisfies the equation of thermal equilibrium

$$-\rho(\dot{\psi} + {}_R\eta\dot{\theta}) + {}_Rt_{kl}a_{kl} + {}_Rm_{kl}b_{lk} + {}_Rm_k\nu_{,k} + ({}_Rs - {}_Rt)\nu = 0 \qquad (2.3.16)$$

and the dissipation inequality is obtained to read

$$\rho\hat{\eta} \equiv -\rho_D\eta\dot{\theta} + {}_D\mathbf{t}\mathbf{a}^T + {}_D\mathbf{m}\mathbf{b} + {}_Dm_k\nu_{,k} + ({}_Ds - {}_Dt)\nu + \frac{1}{\theta}\mathbf{q}\cdot\nabla\theta \geq 0. \qquad (2.3.17)$$

Again, this has the form (2.3.13) with

$$\begin{aligned}
\mathcal{J} &\equiv \{-\rho_D\eta, \; {}_D\mathbf{t}, \; {}_D\mathbf{m}, \; {}_Dm_k, \; {}_Ds - {}_Dt, \mathbf{q}\}, \\
Y &\equiv \{\dot{\theta}, \mathbf{a}^T, \mathbf{b}, \nu_{,k}, \nu, \nabla\theta/\theta\}.
\end{aligned} \qquad (2.3.18)$$

C. Micropolar Continua

In this case, the C–D inequality is obtained from that of the microstretch continua by setting $s = t, \nu = 0$, and $m_k \equiv 0$,

$$-\rho(\dot{\psi} + \eta\dot{\theta}) + \mathbf{t}\mathbf{a}^T + \mathbf{m}\mathbf{b} + \frac{1}{\theta}\mathbf{q} \cdot \nabla\theta \geq 0 \quad \text{in } \mathcal{V} - \sigma. \tag{2.3.19-a}$$

The jump condition (2.3.15-b) remains intact:

$$\left[\!\!\left[\rho\eta(\mathbf{v} - \mathbf{u}) - \frac{1}{\theta}\mathbf{q}\right]\!\!\right] \cdot \mathbf{n} \geq 0 \qquad \text{on } \sigma. \tag{2.3.19-b}$$

The dissipation inequality reads

$$\rho\hat{\eta} \equiv -\rho_D\eta\dot{\theta} + {}_D\mathbf{t}\mathbf{a}^T + {}_D\mathbf{m}\mathbf{b} + \frac{1}{\theta}\mathbf{q} \cdot \nabla\theta \geq 0 \qquad \text{in } \mathcal{V} - \sigma. \tag{2.3.20}$$

Again, this has the scalar product form (2.3.13) with $\mathcal{J}$ and Y given by

$$\mathcal{J} \equiv \{-\rho_D\eta,\ {}_D\mathbf{t},\ {}_D\mathbf{m}, \mathbf{q}\},$$
$$Y \equiv \{\dot{\theta}, \mathbf{a}^T, \mathbf{b}, \nabla\theta/\theta\}. \tag{2.3.21}$$

The fundamental question here is: Given $\rho\hat{\eta}$, how can we determine $\mathcal{J}$ as a function of Y and perhaps other variables? This question is answered in Section 2.4.

2.4 Dissipation Potential

Both thermodynamic forces Y, given by (2.3.14), and fluxes $\mathcal{J}$ given by (2.3.9), contain 46 scalar components. In the order listed, we may consider Y and $\mathcal{J}$ as vectors in 46-dimensional space, whose scalar product (2.3.13) must be nonnegative. The general solution of an equality of the form (2.3.13) was given by Edelen [1973, 1993].

Theorem (Edelen).   Let $C^{1,0}$ denote the collection of all scalar-valued functions of the arguments $\{Y, \omega\}$ that are of class C^1 in Y and continuous in ω, and let $\hat{C}^{1,0}$ denote all collections $\{\mathcal{J}_\Sigma(Y,\omega)\}$, $K(Y,\omega)$, $1 \leq \Sigma \leq M\}$ such that each of these $M + 1$ functions belongs to $C^{1,0}$. All systems $\{\mathcal{J}, K\} \in C^{1,0}$ that satisfy

$$Y \cdot \mathcal{J}(Y,\omega) \geq K(Y,\omega) \tag{2.4.1}$$

are given by

$$\mathcal{J}(Y,\omega) = \nabla_Y \Phi(Y,\omega) + U(Y,\omega), \tag{2.4.2}$$

$$\Phi(Y,\omega) = \int_0^1 [P(\tau Y, \omega) + K(\tau Y, \omega)]\frac{d\tau}{\tau} + \phi(\omega), \tag{2.4.3}$$

where

$$P(0, \omega) + K(0, \omega) = 0, \tag{2.4.4}$$

$P(Y, \omega)$ is any element of $C^{1,0}$ such that

$$P(Y, \omega) \geq 0, \tag{2.4.5}$$

and *constitutive residuals* $U(Y, \omega)$ is any vector-valued function such that each component belongs to $C^{1,0}$ and

$$Y \cdot U(Y, \omega) = 0. \tag{2.4.6}$$

For proof, see Edelen [1993, p. 86].

In the inequality (2.3.13) $K(Y, \omega) = 0$. However, the inclusion of $K(Y, \omega)$ allows the treatment of *nonlocal* cases. $K(Y, \omega)$ is considered to be a function of zero-mean on the region of the body. This is clear from (2.3.2).

The concept underlying this representation is fairly clear from

$$Y \cdot \mathcal{J}(Y, \omega) = Y \cdot \nabla_Y \Phi = P + K. \tag{2.4.7}$$

The integrability of (2.4.7) is based on a theorem by Poincaré that any closed differential form on a star-shaped domain is an exact differential. Alternatively, if a vector-valued function $\mathcal{J}$ is curlless in the sense that

$$\frac{\partial \mathcal{J}_\Gamma}{\partial Y_\Sigma} = \frac{\partial \mathcal{J}_\Sigma}{\partial Y_\Gamma},$$

then $\mathcal{J}$ admits a potential Φ in a star-shaped domain. This potential can be computed by a line integral from the origin of the domain to the point Y. Hence Φ is given by (2.4.3). From (2.4.6), it is clear that *the vector $U(Y, \omega)$ does not contribute to the dissipation inequality.* If $\mathcal{J}(Y, \omega)$ is of class $C^{2,0}$, then any solution of the inequality (2.4.1) satisfies the symmetry relations

$$\frac{\partial}{\partial Y_\Sigma} \left[\mathcal{J}_\Gamma(Y, \omega) - U_\Gamma(Y, \omega) \right] = \frac{\partial}{\partial Y_\Gamma} \left[\mathcal{J}_\Sigma(Y, \omega) - U_\Sigma(Y, \omega) \right]. \tag{2.4.8}$$

These relations reduce to the Onsager reciprocity relations

$$\frac{\partial \mathcal{J}_\Gamma(Y, \omega)}{\partial Y_\Sigma} = \frac{\partial \mathcal{J}_\Sigma(Y, \omega)}{\partial Y_\Gamma} \tag{2.4.9}$$

if and only if $U(Y, 0) = 0$. In this case, in (2.4.1), $P(Y, \omega) \equiv \rho \hat{\eta} =$ the dissipation energy like (2.3.12).

Corollary. If $K(Y, \omega) = 0$, then the satisfaction of the strong form $Y \cdot \mathcal{J}(Y, \omega) \geq 0$ of the fundamental inequality implies that $\Phi(Y, \omega)$ is given by

$$\Phi(Y, \omega) = \int_0^1 P(\tau Y, \omega) \frac{d\tau}{\tau} + \phi(\omega), \qquad P(Y, \omega) = Y \cdot \mathcal{J}(Y, \omega), \tag{2.4.10}$$

$$\mathcal{J}(Y, \omega) = \nabla_Y \Phi(Y, \omega) + U(Y, \omega), \qquad Y \cdot U(Y, \omega) = 0, \tag{2.4.11}$$

$$\mathcal{J}(0, \omega) = U(0, \omega) = 0, \qquad P(0, \omega) = 0. \tag{2.4.12}$$

$\Phi(0, \omega)$ is the absolute minimum of $\Phi(Y, \omega)$ with respect to Y at each ω, and the symmetry relations (2.4.8) hold if each of the components of $\mathfrak{J}(Y, \omega)$ belongs to $C^{2,0}$.

It is clear that

$$\Phi(0, \omega) = \min_Y \Phi(Y, \omega) \qquad (2.4.13)$$

and that $\Phi(Y, \omega)$ is a *nondecreasing* function on all rays in the vector space of Y that emanate from the origin **O**.

This theorem is fundamental to the development of the dynamic constitutive equations.

Remark. It is clear that, if we set $U = 0$, the orthogonality relation $Y \cdot U(Y, \omega) = 0$ is satisfied automatically. However, with this, a member U of $\mathfrak{J}$, not contributing to the dissipation, is lost.

Edelen gives several such examples. Consider the *Hall* current present in the constitutive equations. For the current

$$\mathbf{J} = \sigma \mathbf{E} + \mu_H \mathbf{E} \times \mathbf{B},$$

which gives

$$\mathbf{J} \cdot \mathbf{E} = \sigma \mathbf{E} \cdot \mathbf{E},$$

since $\mathbf{E} \cdot (\mathbf{E} \times \mathbf{B}) = 0$.

It should be remembered that, in the atomic scale, with the time reversal, **B** changes its sign, so that the constitutive coefficients (such as σ), which depend on the invariants of **E** and **B**, are affected by this sign change.[6] Even in the linear constitutive equations:

$$\mathfrak{J}_\Sigma = K_{\Sigma\Gamma} Y_\Gamma,$$

we see that

$$\Phi = \tfrac{1}{4}(K_{\Sigma\Gamma} + K_{\Gamma\Sigma}) Y_\Sigma Y_\Gamma, \qquad U_\Sigma = \tfrac{1}{2}(K_{\Sigma\Gamma} - K_{\Gamma\Sigma}) Y_\Gamma,$$

so that the Onsager reciprocal relations $K_{\Sigma\Gamma} = K_{\Gamma\Sigma}$ are not necessarily satisfied. However, if we assume that the U's satisfy a weaker relation

$$U(\lambda Y, \omega) = O(\lambda^2), \qquad (2.4.14)$$

then a linear approximation leads to

$$\mathfrak{J}(\lambda Y, \omega) = \lambda \mathfrak{J}_1(Y, \omega) + O(\lambda^2). \qquad (2.4.15)$$

---

[6] The invariance under time reversal is an essential part of the theory of electromagnetic constitutive equations. For the nonlinear theory, this is discussed in Kiral and Eringen [1990].

This then results in the satisfaction of Onsager's relation (2.4.9) for the fluxes $\mathcal{J}(Y, \omega)$.

Onsager reciprocal relations represent a special assumption based on *microscopic time reversal*. For large values of thermodynamic forces, there appears to be no sound physical principle to set $U = 0$. However, for small thermodynamic forces, the linear constitutive equations prevail, since the microscopic time reversal imposes symmetry on the constitutive moduli, i.e., $K_{\Sigma\Gamma} = K_{\Gamma\Sigma}$.

In this book, we shall be concerned mostly with the constitutive equations that are linear in Y_Γ, so that we shall take $U = 0$ in the sense of (2.4.14). Alternatively, use:

Onsager's Postulate: *Onsager reciprocal relations are posited to be valid for the constitutive equations that are linear in the thermodynamic force Y, with proper considerations given to the signs of thermodynamic forces and fluxes arising from microscopic time reversal.*

Nevertheless, it must be noted that, for large thermodynamic forces, constitutive residuals U can make important contributions to the constitutive equations. For example, Hall current and *Second Sound* (temperature waves), that has been observed in He II, cannot be incorporated into constitutive equations without the residuals U.

Memory Functionals

For memory-dependent materials, the dissipation inequality is of the form

$$Y(t) \cdot \mathcal{J}[Y(t-s), \omega] \geq 0, \qquad 0 \leq s < \infty, \qquad (2.4.16)$$

where $Y(t-s)$ is a collection of *histories* of thermodynamic forces and $\mathcal{J}$ are the thermodynamic fluxes. ω stands for a collection of some parameters. Henceforth, we will omit explicitly showing ω. Our interest is to solve (2.4.16) for $\mathcal{J}$, i.e., if z is a member of $\mathcal{J}$, then find F, such that z is a functional of $Y(t-s)$, $0 < s < \infty$, and a function of $Y(t)$, i.e.,

$$z = F[Y(t-s); Y(t)], \qquad 0 < s < \infty. \qquad (2.4.17)$$

Now, if we treat $Y(t-s)$ as a parameter and consider z as a function of Y, then under the condition of the above theorem with $K \equiv 0$, we have the solution (2.4.11), i.e.,

$$\Phi[Y(t-x); Y(t)] = \int_0^1 P[Y(t-s); \tau Y] \frac{d\tau}{\tau}, \qquad (2.4.18\text{-a})$$

$$P[Y(t-x); Y(t)] \equiv Y(t) \cdot \mathcal{J}[Y(t-s); Y(t)], \qquad (2.4.18\text{-b})$$

$$\mathcal{J} = \nabla_{Y(t)} \Phi[Y(t-s); Y(t)] + U[Y(t-s); Y(t)], \qquad (2.4.19\text{-a})$$

$$Y(t) \cdot U[Y(t-s), Y(t)] = 0, \qquad (2.4.19\text{-b})$$

$$\mathfrak{J}|_{Y(t)=0} = U|_{Y(t)=0} = 0, \qquad P[Y(t-s);0] = 0. \tag{2.4.20}$$

$\Phi[Y(t-s);0] = 0$ is a minimum of $\Phi[Y(t-s);Y(t)]$ with respect to $Y(t)$ at each $Y(t-s)$, $s \neq 0$. The solution (2.4.19-a) maybe stated as the material having *instantaneous elasticity*. Under the assumption

$$U[\lambda Y(t-s); \lambda Y] = 0(\lambda^2), \tag{2.4.21}$$

a linear approximation for $\mathfrak{J}$ satisfies Onsager's reciprocal relations (2.4.9) for fixed $Y(t-s)$, $s \neq 0$.

Chapter 2 Problems

2.1. By means of statistical moments of surface tractions and body forces acting on a microelement, express the physical meanings of the stress tensor t_{kl}, stress moments m_{klm}, and body moments l_{kl}.

2.2. Derive balance laws of micromorphic mechanics by means of statistical methods.

2.3. Prove the Green–Gauss and transport theorems (2.2.2) and (2.2.1).

2.4. For a dense suspension, $\hat{i}_{kl}$ in (2.2.9) do not vanish because of the nonsmooth nature of the flow. In this case, equations of balance of momentum moments and energy are modified. Find their new expressions.

2.5. Express C–D inequality (2.3.6) in the material frame-of-reference.

2.6. Verify the solution (2.4.2) of the entropy production inequality (2.4.1).

2.7. Formulate the second law of thermodynamics that includes the temperature θ and its first moment $\langle \theta' \xi_k \rangle$.

3
Constitutive Equations

3.0 Scope

Chapter 3 is concerned with the theory of constitutive equations.[1] According to the axiom of causality, stress, heat, and energy are considered to be *functionals* of strain measures, temperature gradient, temperature and its rate. This constitutes the basis for the development of constitutive equations of both solids and fluids. For fluids, relative strain measures are used. Memory-dependent, nonlocal media must employ *space–time functionals*.

In Section 3.1 the axiom of objectivity is applied to restrict the constitutive response functionals and to establish the general theory of 3M continua. In special cases, constitutive equations of local memory-dependent continua, nonlocal media with no memory, and the local media with memory are obtained.

Section 3.2 presents the constitutive theory for the fluid media. Here we employ the relative strain measures to construct the theory. Constitutive equations of nonlocal and local 3M fluids are formulated.

In Section 3.3 we obtain the nonlinear constitutive equations of 3M thermoelastic solids. Thermodynamic restrictions are imposed. The nonlinear constitutive equations obtained here are used later in Chapters 5, 6, and 7, to construct the linear theory.

---

[1]Most parts of Chapter 3 represent a new approach differing from other published work.

Thermodynamic restrictions on fluids are studied in Section 3.4. Nonlinear constitutive equations are obtained and restricted by the thermodynamic consideration for 3M fluids.

In Section 3.5 we give the nonlinear constitutive equations, for memory-dependent 3M solids, restricted by the second law of thermodynamics, and in Section 3.6, those of memory-dependent 3M fluids. These results are used to obtain constitutive equations of microfluid continua, liquid crystals, and a variety of other fluid media in Volume II.

3.1   Constitutive Equations of Elastic Solids

The fundamental laws of micromorphic continua consist of a system of 20 partial differential equations (2.2.10-a), (2.2.11-a), (2.2.31-a), (2.2.32-a), and (2.2.33-a), and one inequality (2.3.4-a). Given the external loads f_k, l_{kl}, and h, there are 67 unknowns $\rho, v_k, j_{kl}, \nu_{kl}, t_{kl}, m_{klm}, s_{kl}, q_k, \epsilon, \eta$, and θ. Clearly then, the system is highly indeterminate. Forty-seven independent additional equations are needed for the determination of motions and temperatures of a micromorphic body. This is also clear from the fact that the balance equations are valid for all micromorphic bodies irrespective of their physical constitutions. They are valid for fluids, oils, blood, liquid crystals, elastic solids, gases, plastics, polymers, bones, etc.

For bodies of different constitutions, the response of the body to external stimuli is very different. Thus, we must bring the constitutional nature of bodies into our formulation. This can be done either from molecular or continuum viewpoints. Lattice dynamics and methods of statistical mechanics have been successful for the calculations of fields in certain solids and fluids. Especially important, in the molecular approach, is the determination of the macroscopic material moduli as a function of the properties and the spatial arrangements of its constituent atoms and interatomic forces. In this way, we can improve material properties and create new materials that can meet technological demands. However, an atomic approach can only be used for some special materials with simple constitutions, e.g., perfect crystals, dilute gases. For amorphous materials subject to large deformations and fields, the lattice dynamical approach faces major difficulties. A continuum approach, on the other hand, can be used to describe very general and nonlinear deformations and fields.

However, the effects of the atomic constituents on the microscopic parameters are buried in material constants and functions which must be determined by special experiments. In any case, the continuum theory leads us to an efficient and logical way of introducing relevant material moduli and response functions which can be used in atomic approaches, as well as for the investigation of the effects of the atomic parameters.

A general theory of constitutive equations was proposed by Eringen [1966c, 1967a] which has shown that the constitutive independent variables are the nonlocal strain and temperature histories. For micromorphic solids these consist of the set

$$\mathcal{Y}(\mathbf{X}',t') = \{ \mathfrak{C}_{KL}(\mathbf{X}',t'), \Gamma_{KLM}(\mathbf{X}',t'), \mathcal{C}_{KL}(\mathbf{X}',t'), \theta_{,K}(\mathbf{X}',t'),$$
$$\dot{\theta}(\mathbf{X}',t'), \theta(\mathbf{X}',t'); I_{KL}(\mathbf{X}'), \mathbf{X}', t' \}. \tag{3.1.1}$$

The inclusion of $I_{KL}(\mathbf{X}')$ stems from the dependence on $\langle \Xi'_k \Xi'_L \rangle = I_{KL}(\mathbf{X}')$. Here $\mathbf{X}'$ covers all material particles of the body and t' all present and past times.

The *dependent* constitutive variables are those which exclude the source terms f_k, l_{kl}, and h and the variables that are derivable from the motions $(\rho, v_k, \nu_{kl}$, and $j_{kl})$

$$\mathcal{Z} = \{ t_{kl}, m_{klm}, s_{kl}, q_k, \epsilon \}. \tag{3.1.2}$$

The *axiom of determinism* then states that, the dependent set $\mathcal{Z}$, at $(\mathbf{X},t)$, is a *functional* of the independent variable set (3.1.1). Symbolically

$$\mathcal{Z}(\mathbf{X},t) = \mathcal{F}[\mathcal{Y}(\mathbf{X}',t')], \qquad \mathbf{X}' \in B, \qquad -\infty < t' \le t. \tag{3.1.3}$$

This means that each member of (3.1.2) is a functional of *all* members of (3.1.1). This is known as the *Axiom of Equipresence*. In (3.1.3), $\mathcal{F}$ is tensor-valued for t_{kl}, s_{kl}, and m_{klm}; vector-valued for q_k, and scalar-valued for ϵ. Constitutive equation (3.1.3) is subject to the restrictions of the *Axiom of Objectivity* and the second law of thermodynamics. In accordance with Definition 2 of Section 1.9 the Axiom of Objectivity may be stated as:

Axiom of Objectivity (Material-Frame Indifference).
Constitutive equations must be form-invariant with respect to rigid motions and reflections of the spatial frame-of-reference and constant shift of time.

According to (1.9.1), (1.9.6), and (1.9.2), if two motions $\mathbf{x}(\mathbf{X}',t')$, $\tilde{\mathbf{x}}(\mathbf{X}',\tilde{t}')$ of any point $\mathbf{X}'$ of the body at time t', differ by a rigid motion, reflection and constant time shift, they are related to each other by

$$\tilde{x}_k(\mathbf{X}',\tilde{t}') = Q_{kl}(t')x_l(\mathbf{X}',t') + b_k(t'), \tag{3.1.4-a}$$
$$\tilde{\chi}_{kK}(\mathbf{X}',\tilde{t}') = Q_{kl}(t')\chi_{lK}(\mathbf{X}',t'), \tag{3.1.4-b}$$

where $\mathbf{Q}(t')$ and $\tilde{t}$ are subject to

$$\mathbf{Q}(t')\mathbf{Q}^T(t') = \mathbf{Q}^T(t')\mathbf{Q}(t') = \mathbf{1}, \qquad \det \mathbf{Q}(t') = \pm 1,$$
$$\tilde{t}' = t' - a. \tag{3.1.5}$$

Here $\mathbf{Q}^T$ denotes the transpose of $\mathbf{Q}$; $\mathbf{1}$ is the unit tensor, and a is a constant.

A. Micromorphic Elastic Solids

Constitutive equations (3.1.3) for micromorphic elastic solids must obey the transformation laws given by (3.1.4-a, b) for all arbitrary values of $\{Q_{kl}(t')\}, \{b_k(t')\}$, and $\{a\}$. Clearly then, under these transformations, (3.1.2) transforms as

$$\tilde{\mathcal{Z}} = \{\tilde{t}_{kl}, \tilde{m}_{klm}, \tilde{s}_{kl}, \tilde{q}_k, \tilde{\epsilon}\}, \tag{3.1.6}$$

where

$$\{\tilde{\mathbf{t}}, \tilde{\mathbf{s}}\} = \mathbf{Q}\{\mathbf{t}, \mathbf{s}\}\mathbf{Q}^T, \qquad \tilde{\mathbf{q}} = \mathbf{Q}\mathbf{q},$$
$$\tilde{m}_{klm} = Q_{kp}Q_{lq}Q_{mr}m_{pqr}, \qquad \tilde{\epsilon} = \epsilon. \tag{3.1.7}$$

On the other hand, (3.1.1) obeys the invariance under the group $\{\mathbf{Q}(t')\}$ and $\{\mathbf{b}(t')\}$, with the time-shift providing futher restrictions. This restriction is imposed by noting that when $\mathbf{Q}(t') = \mathbf{1}$, $\mathbf{b}(t') = 0$, and upon writing $a = t$ and $\tau' = t - t'$ (3.1.4-a, b) gives

$$\tilde{\mathbf{x}}(\mathbf{X}', \tilde{t}') = \mathbf{x}(\mathbf{X}', t{-}\tau'), \qquad \tilde{\chi}_{kK}(\mathbf{X}', \tilde{t}') = \chi_{kK}(\mathbf{X}', t{-}\tau'), \qquad 0 \le \tau' < \infty.$$

Consequently, from the definitions of the argument functions in (3.1.1), it follows that all we need is to replace t' by $t{-}\tau'$. With this then, constitutive equations (3.1.3) read

$$\mathcal{Z}(\mathbf{X}, t) = \mathcal{F}[\mathcal{Y}(\mathbf{X}', t - \tau')], \qquad \mathbf{X}' \in B, \qquad 0 \le \tau' < \infty. \tag{3.1.8}$$

These are the most general constitutive equations for the *nonlocal, memory-dependent micromorphic elastic solids.*

Several special cases are noted:

(i) **Nonlocal micromorphic solids without memory**. This case follows from (3.1.8) by ignoring the memory effects, i.e., formally setting $\tau' \equiv 0$, i.e.,

$$\mathcal{Z}(\mathbf{X}, t) = \mathcal{F}[\mathcal{Y}(\mathbf{X}')], \qquad \mathbf{X}' \in B. \tag{3.1.9}$$

Here we suppressed the dependence of $\mathcal{Y}$, on the present time t.

(ii) **Local micromorphic solids with memory**. The constitutive equations of local micromorphic continua result when we ignore the space–nonlocality, i.e., consider that $\mathcal{Y}$ depends only on the reference point $\mathbf{X}$. Hence

$$\mathcal{Z}(\mathbf{X}, t) = \mathcal{F}[\mathcal{Y}(t - \tau')], \qquad 0 \le \tau' < \infty. \tag{3.1.10}$$

Here we suppressed the dependence on the reference point $\mathbf{X}$.

(iii) **Local micromorphic solids with no memory**. In this case, only local independent variables are used with no memory effects. Consequently, the response functional $\mathcal{F}$ now becomes a function

$$\mathcal{Z}(\mathbf{X}, t) = f(\mathcal{Y}). \tag{3.1.11}$$

B. Microstretch Elastic Solids

From the decomposition of the members of $\mathcal{Y}(\mathbf{X}', t')$ in (3.1.1) and those of $\mathcal{Z}(\mathbf{X}, t)$ in (3.1.2), it is clear that the constitutive equations of microstretch elastic solids will be of the form

$$\mathcal{Z}(\mathbf{X}, t) = \mathcal{F}[\mathcal{Y}(\mathbf{X}', t - \tau')], \qquad \mathbf{X}' \in B, \qquad 0 \le \tau' < \infty, \qquad (3.1.12)$$

where

$$\mathcal{Z}(\mathbf{X}', t) = \{t_{kl}, m_{kl}, m_k, s, q_k, \epsilon\}, \qquad (3.1.13)$$

$$\mathcal{Y}(\mathbf{X}', t - \tau') = \{\bar{\mathbb{C}}_{KL}(\mathbf{X}', t - \tau'), \Gamma_{KL}(\mathbf{X}', t - \tau'), \Gamma_K(\mathbf{X}', t - \tau'),$$
$$j^2(\mathbf{X}', t - \tau'), \theta_{,K}(\mathbf{X}', t - \tau'), \dot{\theta}(\mathbf{X}', t - \tau'), \theta(\mathbf{X}', t - \tau'),$$
$$J_{KL}(\mathbf{X}'), J_0(\mathbf{X}'), \mathbf{X}', t - \tau'\}, \qquad (3.1.14)$$

where the strain measures are given by (1.5.20).

(i) **Nonlocal microstretch solids without memory**. This is obtained by ignoring the memory

$$\mathcal{Z}(\mathbf{X}, t) = \mathcal{F}[\mathcal{Y}(\mathbf{X}')], \qquad \mathbf{X}' \in B. \qquad (3.1.15)$$

(ii) **Local microstretch solids with memory**. In this case, the non-locality in space is ignored

$$\mathcal{Z}(\mathbf{X}, t) = \mathcal{F}[\mathcal{Y}(t - \tau')], \qquad 0 \le \tau' < \infty. \qquad (3.1.16)$$

(iii) **Local microstretch solids without memory**.

$$\mathcal{Z}(\mathbf{X}, t) = f(\mathcal{Y}). \qquad (3.1.17)$$

C. Micropolar Elastic Solids

For micropolar continua, $j = 1$, $s = t$, $m_k = 0$, and $\Gamma_K = 0$. The constitutive equations are of the form

$$\mathcal{Z}(\mathbf{X}, t) = \mathcal{F}[\mathcal{Y}(\mathbf{X}', t - \tau')], \qquad \mathbf{X}' \in B, \qquad 0 \le \tau' < \infty, \qquad (3.1.18)$$

where

$$\mathcal{Z}(\mathbf{X}', t) = \{t_{kl}, m_{kl}, q_k, \epsilon\}, \qquad (3.1.19)$$

$$\mathcal{Y}(\mathbf{X}', t - \tau') = \{\bar{\mathbb{C}}_{KL}(\mathbf{X}', t - \tau'), \Gamma_{KL}(\mathbf{X}', t - \tau'), \theta_{,K}(\mathbf{X}', t - \tau'),$$
$$\dot{\theta}(\mathbf{X}', t - \tau'), \theta(\mathbf{X}', t - \tau'); J_{KL}(\mathbf{X}'), \mathbf{X}', t - \tau'\}, \quad (3.1.20)$$

where the strain measures are given by (1.5.24).

(i) **Nonlocal micropolar solids without memory.**

$$\mathcal{Z}(\mathbf{X}, t) = \mathcal{F}[\mathcal{Y}(\mathbf{X}')], \qquad \mathbf{X}' \in B. \tag{3.1.21}$$

(ii) **Local micropolar solids with memory.**

$$\mathcal{Z}(\mathbf{X}, t) = \mathcal{F}[\mathcal{Y}(t - \tau')], \qquad 0 \le \tau' < \infty. \tag{3.1.22}$$

(iii) **Local micropolar solids without memory.**

$$\mathcal{Z}(\mathbf{X}, t) = f(\mathcal{Y}). \tag{3.1.23}$$

By means of power series expansions of $\mathcal{Y}(\mathbf{X}', t - \tau')$ about the reference point $\mathbf{X}$ and/or $\tau' = 0$, the constitutive equation can be approximated by functions of gradients and/or time-rates of $\mathcal{Y}$. Such theories are known as *gradient* and/or *rate theories*. Among these, a prominent case involves only the first-order rates replacing $\mathcal{Y}(\mathbf{X}', t - \tau')$ by

$$\mathcal{Y}_R(\mathbf{X}, t) = \{\mathfrak{C}_{KL}, \dot{\mathfrak{C}}_{KL}, \Gamma_{KLM}, \dot{\Gamma}_{KLM}, \mathcal{C}_{KL}, \dot{\mathcal{C}}_{KL}, \theta_{,K}, \theta, \dot{\theta}; I_{KL}, \mathbf{X}, t\}. \tag{3.1.24}$$

The constitutive equations are then of the form

$$\mathcal{Z}(\mathbf{X}, t) = f(\mathcal{Y}_R). \tag{3.1.25}$$

This generalizes Kelvin–Voigt elastic solids to micromorphic continua.

Clearly, a variety of other approximate constitutive equations can be constructed by retaining gradients and/or rates up to certain degrees.

Another important case is that of the micromorphic thermoelastic solids, which involves no rates:

$$\mathcal{Z}(\mathbf{X}, t) = f(\mathfrak{C}_{KL}, \Gamma_{KLM}, \mathcal{C}_{KL}, \theta_{,K}, \theta; I_{KL}, \mathbf{X}). \tag{3.1.26}$$

Equations for microstretch media corresponding to (3.1.25) and (3.1.26) follow from (3.1.16) by retaining the first-order rates of $\mathcal{Y}$, i.e.,

$$\mathcal{Z}(\mathbf{X}, t) = f(\mathcal{Y}, \dot{\mathcal{Y}}) \qquad \text{(microstretch Kelvin–Voigt)}, \tag{3.1.27}$$
$$\mathcal{Z}(\mathbf{X}, t) = f(\mathcal{Y}) \qquad \text{(microstretch, thermoelastic)}, \tag{3.1.28}$$

where $\mathcal{Y}$ and $\dot{\mathcal{Y}}$ are obtained from (3.1.14) with $\mathbf{X}' = \mathbf{X}$ and $\tau' = 0$.

For the micropolar solids, constitutive equations are again in the same form:

$$\mathcal{Z}(\mathbf{X}, t) = f(\mathcal{Y}, \dot{\mathcal{Y}}) \qquad \text{(microstretch Kelvin–Voigt)}, \tag{3.1.29}$$
$$\mathcal{Z}(\mathbf{X}, t) = f(\mathcal{Y}) \qquad \text{(microstretch, thermoelastic)}, \tag{3.1.30}$$

with $\mathcal{Y}$ and $\dot{\mathcal{Y}}$ obtained from (3.1.20) at $\mathbf{X}' = \mathbf{X}$ and $\tau' = 0$ (local media with no memory).

3.2   Constitutive Equations of Fluids

In the development of the constitutive equations of solids, in Section 3.1, the independent variable set $\mathcal{Y}(\mathbf{X}', t')$ was based on the reference configuration $\mathbf{X}$. This is a fundamental characteristic of solids. Indeed, solids possess a reference configuration. Physically, this means that solids remember their reference configuration.

For fluids, however, there is no reference configuration. Indeed fluids take the shape of their container. Alternatively, for fluids, the present configuration is also the reference configuration. Consequently, constitutive equations of fluids must be based on the relative strain measures introduced in Section 1.6. For micromorphic continua, these are given by

$$\mathfrak{C}_{(t)kl}(t'), \qquad \Gamma_{(t)klm}(t'), \qquad \mathcal{C}_{(t)kl}(t'). \tag{3.2.1}$$

Note, however, that when the two spatial configurations at t and t' coincide, we have

$$\mathfrak{C}_{(t)}(t) = \mathbf{1}, \qquad \Gamma_{(t)klm}(t) = 0, \qquad \mathcal{C}_{(t)}(t) = \mathbf{1}. \tag{3.2.2}$$

As can be seen from

$$\det \mathfrak{C}_{(t)kl}(t')\big|_{t'=t} = \left.\frac{\rho(t)}{\rho(t')}\frac{j(t)}{j(t')}\right|_{t'=t} = 1, \tag{3.2.3}$$

the set of independent constitutive variables (3.2.1) excludes the density of fluids. Consequently, we must include the density into the set (3.2.1).

Hence, the set $\mathcal{Y}(\mathbf{X}', t')$, given by (3.1.1) for solids, is to be replaced by the following for fluids:

$$\mathcal{Y}(\mathbf{x}', t') = \{\mathfrak{C}_{(t)kl}(\mathbf{x}', t'), \Gamma_{(t)klm}(\mathbf{x}', t'), \mathcal{C}_{(t)kl}(\mathbf{x}', t'), \theta_{,k}(\mathbf{x}', t'),$$
$$\dot{\theta}(\mathbf{x}', t'), \theta(\mathbf{x}', t'), \rho(\mathbf{x}', t'), i_{kl}(\mathbf{x}', t'); \mathbf{x}', t'\}. \tag{3.2.4}$$

A. Micromorphic Fluids

The constitutive equations of the *nonlocal memory-dependent micromorphic thermofluids* are then expressed formally as

$$\mathcal{Z}(\mathbf{x}, t) = \mathcal{F}[\mathcal{Y}(\mathbf{x}', t - \tau')], \qquad \mathbf{x}' \in B, \qquad 0 \leq \tau' < \infty, \tag{3.2.5}$$

where $\mathcal{Z}$ is the dependent variable set. Here, as usual, $\mathcal{F}$ is a functional of $\mathcal{Y}(\mathbf{x}', t - \tau')$ covering *all* spatial positions of the material points of the body at *all* times t' ($t \leq t' < \infty$). From this general equation, a variety of fluid theories (nonlocal fluids without memory, local fluids with memory, and local fluids without memory, etc.), could be developed along a similar approach to that used in classical field theories. For local micromorphic fluids, however, we do not need the elaborate apparatus of functional analysis. In this case, we recall the material time-rate of the relative strain measures given by (1.8.26) and (1.8.27).

For first-order fluids, the fundamental independent variables are

$$\mathring{y}(\mathbf{x}, t) = \{a_{kl}, b_{klm}, c_{kl}, \theta_{,k}, \dot{\theta}, \theta, \rho, i_{kl}\}. \tag{3.2.6}$$

Note that this set obeys the axiom of objectivity. By the same token, the variables $\mathbf{x}$ and t drop out from the set (3.2.4), when $\mathbf{x}' = \mathbf{x}$, $t' = t$. The constitutive equations of *thermo-micromorphic fluids* are then expressed as

$$\mathcal{Z}(\mathbf{x}, t) = f(\mathring{y}), \tag{3.2.7}$$

where f is now a function, and *each* member of $\mathcal{Z}$ depends on *all* members of $\mathring{y}$. We note that this result may also be obtained from (3.1.24) by letting $X_K \to x_k$. In this case, $\mathfrak{C} \to \mathbf{1}, (\mathfrak{C}, \boldsymbol{\Gamma}) \to \mathbf{0}$, and $\mathcal{Y}_R \to \mathring{y}(x, t)$ and we obtain (3.2.7). This process makes the reference state coincide with the present state. From (3.2.7), we obtain other local fluids.

B. Microstretch Fluids

For microstretch fluids, the deformation-rate tensors a_{kl}, b_{klm}, and ν_{kl} are replaced by (1.8.37), so that

$$\mathring{y}(\mathbf{x}, t) = \{a_{kl}, b_{kl}, \nu_{,k}, \nu, \theta_{,k}, \dot{\theta}, \theta, \rho, j_{kl}, j_0\}. \tag{3.2.8}$$

Constitutive equations read

$$\mathcal{Z}(\mathbf{x}, t) = f(\mathring{y}), \tag{3.2.9}$$

where the constitutive response functions are given by (3.1.13), namely

$$\mathcal{Z}(\mathbf{x}, t) = \{t_{kl}, m_{kl}, m_k, s, q_k, \epsilon\}. \tag{3.2.10}$$

C. Micropolar Fluids

In this case, constitutive equations have the form (3.2.9), with

$$\mathring{y} = \{a_{kl} \equiv v_{l,k} - \epsilon_{klm}\nu_m, b_{kl} = \nu_{k,l}, \theta_{,k}, \dot{\theta}, \theta, \rho, j_{kl}\}, \tag{3.2.11}$$

$$\mathcal{Z}(\mathbf{x}, t) = \{t_{kl}, m_{kl}, q_k, \epsilon\}. \tag{3.2.12}$$

3.3   Thermodynamic Restrictions on Solids

A. Micromorphic Thermoelastic Solids[2]

For local micromorphic elastic solids with no memory, constitutive assumptions (3.1.1) and (3.1.2) show that the independent and dependent variable

---

[2] Eringen and Şuhubi[1964].

sets are

$$\mathcal{Y} = (\mathfrak{C}_{KL}, \Gamma_{KLM}, \mathcal{C}_{KL}, \theta_{,K}, \dot{\theta}, \theta; \mathbf{X}), \qquad (3.3.1)$$

$$\mathcal{Z} = (t_{kl}, m_{klm}, s_{kl}, q_k, \epsilon). \qquad (3.3.2)$$

With the replacement of ϵ by $\epsilon = \psi - \theta\eta$, the dependent variable ψ may be considered a *static* variable, i.e., it does not depend on $\theta_{,K}$ and $\dot{\theta}$, since η is considered to depend on the entire variable set (3.3.1). However, suppose we include $\theta_{,K}$ and $\dot{\theta}$ as well, and employ the C–D inequality (2.3.6) to see what happens, i.e., suppose

$$\psi = \Psi(\mathfrak{C}_{KL}, \Gamma_{KLM}, \mathcal{C}_{KL}, \theta_{,K}, \dot{\theta}, \theta; \mathbf{X}). \qquad (3.3.3)$$

Note that I_{KL} is independent of time; hence, it does not contribute to $\dot{\Psi}$. We calculate

$$\dot{\Psi} = \frac{\partial\Psi}{\partial\mathfrak{C}_{KL}}\dot{\mathfrak{C}}_{KL} + \frac{\partial\Psi}{\partial\Gamma_{KLM}}\dot{\Gamma}_{KLM} + \frac{\partial\Psi}{\partial\mathcal{C}_{KL}}\dot{\mathcal{C}}_{KL} + \frac{\partial\Psi}{\partial\theta_{,K}}\dot{\theta}_{,K} + \frac{\partial\Psi}{\partial\theta}\dot{\theta} + \frac{\partial\Psi}{\partial\dot{\theta}}\ddot{\theta}.$$

Carrying expressions of $\dot{\mathfrak{C}}_{KL}, \dot{\Gamma}_{KLM}$, and $\dot{\mathcal{C}}_{KL}$ from (1.8.18), we have

$$\dot{\Psi} = \frac{\partial\Psi}{\partial\mathfrak{C}_{KL}}a_{kl}x_{k,K}\mathfrak{X}_{Ll} + \frac{\partial\Psi}{\partial\Gamma_{LMK}}b_{klm}\mathfrak{X}_{Kk}\chi_{lL}x_{m,M}$$
$$+ 2\frac{\partial\Psi}{\partial\mathcal{C}_{KL}}c_{kl}\chi_{kK}\chi_{lL} + \frac{\partial\Psi}{\partial\theta_{,K}}\dot{\theta}_{,K} + \frac{\partial\Psi}{\partial\theta}\dot{\theta} + \frac{\partial\Psi}{\partial\dot{\theta}}\ddot{\theta}. \qquad (3.3.4)$$

Substituting this into (2.3.6), we arrange this inequality into

$$-\rho\left(\frac{\partial\Psi}{\partial\theta} + \eta\right)\dot{\theta} + \left(t_{kl} - \rho\frac{\partial\Psi}{\partial\mathfrak{C}_{KL}}x_{k,K}\mathfrak{X}_{Ll}\right)a_{kl}$$
$$+ \left(m_{klm} - \frac{\partial\Psi}{\partial\Gamma_{LMK}}x_{k,K}\mathfrak{X}_{Ll}\chi_{mM}\right)b_{lmk} + \left(s_{kl} - 2\rho\frac{\partial\Psi}{\partial\mathcal{C}_{KL}}\chi_{kK}\chi_{lL}\right)c_{kl}$$
$$-\rho\frac{\partial\Psi}{\partial\theta_{,K}}\dot{\theta}_{,K} - \rho\frac{\partial\Psi}{\partial\dot{\theta}}\ddot{\theta} + \frac{1}{\theta}q_k\theta_{,k} \geq 0. \qquad (3.3.5)$$

This inequality is linear in $\ddot{\theta}, a_{kl}, b_{lmk}, c_{lk}$, and $\dot{\theta}_{,K}$. It must remain in one sign for *all independent variations* of these quantities. But this is impossible unless

$$R\eta = -\frac{\partial\Psi}{\partial\theta}, \qquad t_{kl} = \rho\frac{\partial\Psi}{\partial\mathfrak{C}_{KL}}x_{k,K}\mathfrak{X}_{Ll}, \qquad s_{kl} = 2\rho\frac{\partial\Psi}{\partial\mathcal{C}_{KL}}\chi_{kK}\chi_{lL},$$

$$m_{klm} = \rho\frac{\partial\Psi}{\partial\Gamma_{LMK}}x_{k,K}\mathfrak{X}_{Ll}\chi_{mM}, \qquad \frac{\partial\Psi}{\partial\theta_{,K}} = 0, \qquad \frac{\partial\Psi}{\partial\dot{\theta}} = 0, \quad (3.3.6)$$

and

$$-\rho_D\eta\dot{\theta} + \frac{1}{\theta}q_k\theta_{,k} \geq 0. \qquad (3.3.7)$$

Thus, except for $_D\eta$ and $\mathbf{q}$, all constitutive equations are given in terms of the free energy function Ψ, which also turns out to be independent of the temperature rate and gradient, as expected. Note that (2.3.11) is satisfied identically and (ψ, t_{kl}, m_{klm}, and s_{kl}) turn out to be static variables. The dynamic variables $\mathbf{q}$ and $_D\eta$ are required not to violate the inequality (3.3.7). In this case then

$$Y = (\dot{\theta}, \nabla\theta/\theta), \qquad \mathfrak{J} = (-\rho_D\eta, \mathbf{q}),$$

and we have

$$\rho\hat{\eta} \equiv \mathfrak{J} \cdot \mathbf{Y} = -\rho_D\eta\dot{\theta} + \frac{1}{\theta}\mathbf{q} \cdot \nabla\theta \geq 0. \qquad (3.3.8)$$

The solution of (3.3.8) is given by (2.4.11)

$$-\rho_D\eta = \frac{\partial\Phi}{\partial\dot{\theta}} + U^\eta, \qquad \mathbf{q} = \frac{\partial\Phi}{\partial(\nabla\theta/\theta)} + \mathbf{U}^q, \qquad \dot{\theta}U^\eta + \frac{\nabla\theta}{\theta} \cdot \mathbf{U}^q = 0, \quad (3.3.9)$$

where Φ is the *dissipation function*. As agreed in Section 2.4, following the restrictions of the Onsager postulate, we also set $U^\eta = \mathbf{U}^q = \mathbf{0}$, for the linear theory, so that

$$-\rho_D\eta = \frac{\partial\Phi}{\partial\dot{\theta}}, \qquad q_k = \frac{\partial\Phi}{\partial(\theta_{,k}/\theta)}. \qquad (3.3.10)$$

We note that Φ is a function of the form

$$\Phi = \Phi(\dot{\theta}, \nabla\theta/\theta; \omega), \qquad (3.3.11\text{-a})$$

$$\omega \equiv (\mathfrak{C}_{KL}, \Gamma_{KLM}, \mathcal{C}_{KL}, \theta, \mathbf{X}), \qquad (3.3.11\text{-b})$$

Φ is nonnegative with an absolute minimum at $\nabla\theta/\theta = 0$ and $\dot{\theta} = 0$, i.e.,

$$\Phi(0, \mathbf{0}; \omega) = 0. \qquad (3.3.11\text{-c})$$

Collecting, we have proved:

Theorem. Constitutive equations of micromorphic thermoelastic solids are thermodynamically admissible if and only if they are of the form (3.3.6) and (3.3.9) with ψ given by $\psi = \Psi(\mathfrak{C}_{KL}, \Gamma_{KLM}, \mathcal{C}_{KL}, \theta)$ and Φ by (3.3.11-a). The dissipation potential Φ is nonnegative and it is subject to (3.3.11-c), i.e., it has an absolute minimum at $\nabla\theta/\theta = \mathbf{0}$ and $\dot{\theta} = 0$.

Upon using $\Psi = \epsilon - \theta\eta$ in the energy equation (2.2.33-a), substituting for $\dot{\Psi}$ from (3.3.4), and using (3.3.6), we obtain a reduced form of the energy balance law

$$-\rho\left(\dot{\theta}_D\eta + \theta_D\dot{\eta} + \theta_R\dot{\eta}\right) + \nabla \cdot \mathbf{q} + \rho h = 0. \qquad (3.3.12)$$

B. Microstretch Solids[3]

Constitutive equations of local microstretch elastic solids can be obtained in a similar fashion to those of micromorphic solids. In this case, the independent and dependent variables are

$$\mathcal{Y} = \{\bar{\mathfrak{C}}_{KL}, \Gamma_{KL}, \Gamma_K, j^2, \theta_{,K}, \dot{\theta}, \theta; \mathbf{X}\},$$
$$\mathcal{Z} = \{t_{kl}, m_{kl}, m_k, s, q_k, \eta\}, \tag{3.3.13}$$

where we replaced ϵ by η, in view of $\epsilon = \psi + \theta\eta$. Here, ψ is now a static variable, i.e., it does not depend on $\nabla\theta/\theta$ and $\dot{\theta}$, but η does. As shown in Section A, this will also turn out to be a consequence of the C–D inequality.

A straightforward calculation of $\dot{\Psi}$ by means of (1.8.33) and substitution into (2.3.15-a) results in

$$-\rho\left(\frac{\partial\Psi}{\partial\theta} + \eta\right)\dot{\theta} + \left(t_{kl} - \rho\frac{\partial\Psi}{\partial\bar{\mathfrak{C}}_{KL}}x_{k,K}\bar{\chi}_{lL}\right)a_{kl}$$
$$+ \left(m_{kl} - \rho j^{-2}\frac{\partial\Psi}{\partial\Gamma_{LK}}x_{k,K}\bar{\chi}_{lL}\right)b_{lk} + \left(m_k - \rho\frac{\partial\Psi}{\partial\Gamma_K}x_{k,K}\right)\nu_{,k}$$
$$+ \left(s - t - 3\rho j\frac{\partial\Psi}{\partial j}\right)\nu + \mathbf{q}\cdot\frac{\nabla\theta}{\theta} - \rho\frac{\partial\Psi}{\partial\theta_{,K}}\dot{\theta}_{,K}$$
$$- \rho\frac{\partial\Psi}{\partial\dot{\theta}}\ddot{\theta} \geq 0. \tag{3.3.14}$$

The independent variations of $\mathbf{a}$, $\mathbf{b}$, $\nu_{,k}$, ν, $\dot{\theta}_{,K}$, and $\ddot{\theta}$ and linearity of (3.3.14) in these variables requires that this inequality will not be violated if and only if

$$R\eta = -\frac{\partial\Psi}{\partial\theta}, \qquad t_{kl} = \rho\frac{\partial\Psi}{\partial\bar{\mathfrak{C}}_{KL}}x_{k,K}\bar{\chi}_{lL},$$
$$m_{kl} = \rho j^{-2}\frac{\partial\Psi}{\partial\Gamma_{LK}}x_{k,K}\bar{\chi}_{lL}, \qquad m_k = \rho\frac{\partial\Psi}{\partial\Gamma_K}x_{k,K},$$
$$s - t = 3\rho j\frac{\partial\Psi}{\partial j}, \qquad \frac{\partial\Psi}{\partial\theta_{,K}} = 0, \qquad \frac{\partial\Psi}{\partial\dot{\theta}} = 0, \tag{3.3.15}$$

and

$$-\rho_D\eta\dot{\theta} + \mathbf{q}\cdot\nabla\theta/\theta \geq 0. \tag{3.3.16}$$

Thus, again Ψ comes out to be independent of the dynamic variables $\theta_{,K}$ and $\dot{\theta}$. Hence, the only nonequilibrium part of the C–D inequality is (3.3.16). This, with the use of the Onsager postulate, gives

$$q_k = \frac{\partial\Phi}{\partial(\theta_{,k}/\theta)}, \qquad -\rho_D\eta = \frac{\partial\Phi}{\partial\dot{\theta}}. \tag{3.3.17}$$

---

[3]Eringen [1971b, 1990].

The dissipation potential Φ has the form

$$\Phi = \Phi(\nabla\theta/\theta, \dot\theta; \omega) \geq 0, \qquad (3.3.18\text{-a})$$

$$\omega = (\bar{\mathfrak{C}}_{KL}, \Gamma_{KL}, \Gamma_K, j^2, \theta, \mathbf{X}). \qquad (3.3.18\text{-b})$$

It has an absolute minimum at $\nabla\theta/\theta = \mathbf{0}$ and $\dot\theta = 0$, i.e.,

$$\Phi(\mathbf{0}, \omega) = 0. \qquad (3.3.18\text{-c})$$

This completes the constitutive equations of microstretch elastic solids. By using (3.3.15), the energy balance law (2.2.42-a) of the microstretch elastic solids is reduced to

$$-\rho(\dot\theta_D\eta + \theta_D\dot\eta + \theta_R\dot\eta) + \nabla \cdot \mathbf{q} + \rho h = 0. \qquad (3.3.19)$$

C. Micropolar Solids

For micropolar solids, the procedure is similar to arrive at the thermodynamic restrictions. These results are also obtained from Section B by setting $j = 1$, $\nu = 0$, $s = t$ in (3.3.15). We have

$$R\eta = -\frac{\partial\Psi}{\partial\theta}, \qquad\qquad t_{kl} = \rho\frac{\partial\Psi}{\partial\bar{\mathfrak{C}}_{KL}}x_{k,K}\bar\chi_{lL},$$

$$m_{kl} = \rho\frac{\partial\Psi}{\partial\Gamma_{LK}}x_{k,K}\bar\chi_{lL}, \qquad \psi = \Psi(\bar{\mathfrak{C}}_{KL}, \Gamma_{KL}, \theta, \mathbf{X}), \qquad (3.3.20)$$

for the static part of the constitutive equations and

$$q_k = \frac{\partial\Phi}{\partial(\theta_{,k}/\theta)}, \qquad -\rho_D\eta = \frac{\partial\Phi}{\partial\dot\theta}, \qquad (3.3.21)$$

for the dynamic part. As usual,

$$\Phi = \Phi(\nabla\theta/\theta, \dot\theta; \omega), \qquad (3.3.22\text{-a})$$

$$\omega = (\bar{\mathfrak{C}}_{KL}, \Gamma_{KL}, \theta, \mathbf{X}), \qquad (3.3.22\text{-b})$$

and Φ is an nonnegative function with a minimum at $\nabla\theta/\theta = \mathbf{0}$, $\dot\theta = 0$ for fixed ω, i.e.,

$$\Phi(\mathbf{0}, 0; \omega) = 0. \qquad (3.3.22\text{-c})$$

Again, by substituting $\dot\Psi$ into (2.2.47-a) and using (3.3.20), the energy balance law may be expressed as

$$-\rho(\dot\theta_D\eta + \theta_D\dot\eta + \theta_R\dot\eta) + \nabla \cdot \mathbf{q} + \rho h = 0. \qquad (3.3.23)$$

Incompressible solids

Micromorphic Solids

Since $\rho_0/\rho = \det(x_{k,K})$ and $j = \det \chi_{kK}$, from (1.5.11) we have, for incompressible materials

$$\det \mathfrak{C}_{KL} = \rho_0/\rho j = \text{const.}, \tag{3.3.24}$$

$$\det \mathcal{C}_{KL} = j^2 = \text{const.}, \tag{3.3.25}$$

Of these (3.3.25) expresses *micro-incompressibility*. For macro-incompressible materials, we have ρ =constant. But (3.3.24) shows that, for the incompressible micromorphic materials, we must have both (3.3.24) and (3.3.25) satisfied.

Constitutive equations of incompressible micromorphic solids are obtained by adding two terms with Lagrange multipliers p and ϖ to the free energy ψ, namely

$$\psi_{\text{incompr.}} = -p_0 \det \mathfrak{C}_{KL} - \varpi_0 \det \mathcal{C}_{KL} + \psi. \tag{3.3.26}$$

Upon using (3.3.24), (3.3.25), (1.5.13) and (1.8.18) in

$$\frac{D}{Dt}(\det \mathfrak{C}_{KL}) = \frac{\partial(\det \mathfrak{C}_{KL})}{\partial \mathfrak{C}_{KL}}\dot{\mathfrak{C}}_{KL} = (\rho_0/\rho j)a_{kk},$$

$$\frac{D}{Dt}(\det \mathcal{C}_{KL}) = \frac{\partial(\det \mathcal{C}_{KL})}{\partial \mathcal{C}_{KL}}\dot{\mathcal{C}}_{KL} = 2j^2 c_{kk}. \tag{3.3.27}$$

Constitutive equations are obtained for t_{kl} and s_{kl} replacing those of (3.3.6) by

$$t_{kl} = -p\delta_{kl} + \rho_0\frac{\partial \Psi}{\partial \mathfrak{C}_{KL}}x_{k,K}\bar{\chi}_{l,L}, \quad s_{kl} = -\varpi\delta_{kl} + 2\rho_0\frac{\partial \Psi}{\partial \mathcal{C}_{KL}}\bar{\chi}_{kK}\bar{\chi}_{lL} \tag{3.3.28}$$

where $p = p_0\rho_0/j$, $\varpi = 2\rho j^2\varpi_0$ are arbitrary functions of $\mathbf{x}$ and t.

All other constitutive equations are unchanged.

Similarly we find constitutive equations of incompressible microstretch and micropolar solids.

Microstretch Solids

$$t_{kl} = -p\delta_{kl} + \rho_0\frac{\partial \Psi}{\partial \mathfrak{C}_{KL}}x_{k,K}\bar{\chi}_{l,L}, \quad s - t = -\varpi \tag{3.3.29}$$

Micropolar Solids

$$t_{kl} = -p\delta_{kl} + \rho_0\frac{\partial \Psi}{\partial \mathfrak{C}_{KL}}x_{k,K}\bar{\chi}_{l,L} \tag{3.3.30}$$

3.4 Thermodynamic Restrictions on Fluids

A. Micromorphic Fluids[4]

For local micromorphic thermofluids without memory, constitutive assumptions state that the independent and dependent variables are

$$\dot{y} = \{a_{kl}, b_{klm}, c_{kl}, \theta_{,k}, \theta, \rho^{-1}, i_{kl}\}, \tag{3.4.1}$$

$$\mathcal{Z} = \{t_{kl}, m_{klm}, s_{kl}, q_k, \eta\}. \tag{3.4.2}$$

By virtue of the free energy $\psi = \epsilon - \theta\eta$, ϵ in (3.4.2) is replaced by η, and ψ is a static variable, so that

$$\psi = \Psi(\rho^{-1}, i_{kl}, \theta). \tag{3.4.3}$$

However, η depends on the entire set (3.4.1). The static (recoverable, reversible) members of $\mathcal{Z}$ satisfy the equation of thermodynamic equilibrium (2.3.11), i.e.,

$$-(\rho\dot{\psi} + {}_R\eta\dot{\theta}) + {}_Rt_{kl}a_{kl} + {}_Rm_{klm}b_{lmk} + {}_R\,s_{kl}c_{kl} = 0, \tag{3.4.4}$$

and the dynamic members (3.4.2) must obey the dissipation inequality

$$\rho\hat{\gamma} \equiv -\rho_D\eta\dot{\theta} + {}_Dt_{kl}a_{kl} + {}_Dm_{klm}b_{lmk} + {}_D\,s_{kl}c_{kl} + \tfrac{1}{\theta}q_k\theta_{,k} \geq 0. \tag{3.4.5}$$

From (3.4.3), we calculate

$$\dot{\Psi} = \frac{\partial\Psi}{\partial\rho^{-1}}\rho^{-1}v_{k,k} + \frac{\partial\Psi}{\partial i_{kl}}(i_{kr}\nu_{lr} + i_{lr}\nu_{kr}) + \frac{\partial\Psi}{\partial\theta}\dot{\theta},$$

where we used (2.2.10-a) and (2.2.11-a). Substituting this into (3.4.4), we obtain

$$-\rho\left(\frac{\partial\Psi}{\partial\theta} + {}_R\eta\right)\dot{\theta} + \left({}_Rt_{kl} - \frac{\partial\Psi}{\partial\rho^{-1}}\delta_{kl}\right)v_{l,k}$$

$$+{}_Rm_{klm}b_{lmk} + \left[{}_Rs_{kl} - {}_Rt_{lk} - \rho\left(\frac{\partial\Psi}{\partial i_{rk}} + \frac{\partial\Psi}{\partial i_{kr}}\right)i_{rl}\right]\nu_{kl} = 0.$$

This is satisfied identically by

$$_R\eta = -\frac{\partial\Psi}{\partial\theta}, \qquad {}_Rt_{kl} = -\pi\delta_{kl}, \qquad {}_Rm_{klm} = 0, \qquad {}_Rs_{kl} = -\pi\delta_{kl} - \pi_{kl}, \tag{3.4.6}$$

where π is the *thermodynamic pressure* and π_{kl} is the *micropressure tensor*, defined by

$$\pi \equiv -\frac{\partial\Psi}{\partial\rho^{-1}}, \qquad \pi_{kl} \equiv -\rho\left(\frac{\partial\Psi}{\partial i_{rk}}i_{rl} + \frac{\partial\Psi}{\partial i_{kr}}i_{rl}\right). \tag{3.4.7}$$

---

[4]Eringen [1964a, b], [1972b].

The dissipation inequality (3.4.5) is then integrated, as discussed in Section 2.4, leading to

$$-\rho_D \eta = \frac{\partial \Phi}{\partial \dot{\theta}}, \qquad {}_D t_{kl} = \frac{\partial \Phi}{\partial a_{kl}}, \qquad {}_D m_{klm} = \frac{\partial \Phi}{\partial b_{lmk}},$$

$$_D s_{kl} = \frac{\partial \Phi}{\partial c_{lk}}, \qquad q_k = \frac{\partial \Phi}{\partial (\theta_{,k}/\theta)}, \qquad -\rho_D \eta = \frac{\partial \Phi}{\partial \dot{\theta}}. \qquad (3.4.8)$$

Following the Onsager postulate, we have omitted additional functions U^α which do not contribute to the dissipation.

We also know that the dissipation potential Φ is nonnegative and has an absolute minimum at $\dot{y}_D \equiv (\mathbf{a}, \mathbf{b}, \mathbf{c}, \nabla\theta/\theta, \dot{\theta}) \equiv 0$ irrespective of $\omega \equiv (\theta, \rho^{-1}, i_{kl})$, i.e.,

$$\Phi = \Phi(\dot{y}_D; \omega) \geq 0, \qquad (3.4.9)$$
$$\Phi(\mathbf{0}; \omega) = 0. \qquad (3.4.10)$$

The sets (3.4.6) and (3.4.8) give complete constitutive equations of micromorphic thermofluids with

$$t_{kl} = -\pi \delta_{kl} + {}_D t_{kl}, \qquad m_{klm} = {}_D m_{klm},$$
$$s_{kl} = -\pi \delta_{kl} - \pi_{kl} + {}_D s_{kl}, \qquad \eta = {}_R \eta + {}_D \eta, \qquad q_k. \qquad (3.4.11)$$

Thus, we have proved:

Theorem. Constitutive equations of micromorphic thermofluids are thermodynamically admissible, if and only if:

(a) they are of the form (3.4.6), (3.4.8) with the provision of the Onsager postulate;

(b) ψ and Φ are given by (3.4.3) and (3.4.9);

(c) the dissipation potential Φ is nonnegative and subject to (3.4.10) at its absolute minimum.

B. Microstretch Fluids[5]

For microstretch fluids, the independent and dependent variables are

$$\dot{y} = \{a_{kl}, b_{kl}, \nu_{,k}, \nu, \theta_{,k}, \dot{\theta}, \theta, \rho^{-1}, j_{kl}, j_0\}, \qquad (3.4.12)$$
$$z = \{t_{kl}, m_{kl}, m_k, s - t, q_k, \eta\}. \qquad (3.4.13)$$

The free energy ψ is now considered to be of the following form:

$$\psi = \Psi(\rho^{-1}, j_{kl}, j_0, \theta). \qquad (3.4.14)$$

---

[5]Eringen [1969a].

The axiom of objectivity demands that ψ depends on j_{kl} through its invariants, i.e.,

$$j_1 = j_{kk}, \qquad j_2 = \tfrac{1}{2} j_{kl} j_{lk}, \qquad j_3 = \tfrac{1}{3} j_{kl} j_{lm} j_{mk}. \tag{3.4.15}$$

Calculating $\dot{\Psi}$ and substituting into the thermodynamic equation of equilibrium (2.3.16), we see that this equation is satisfied with the following choices:

$$_R\eta = -\frac{\partial \Psi}{\partial \theta}, \qquad _R t_{kl} = -\pi \delta_{kl}, \qquad _R m_{kl} = 0, \qquad _R m_k = 0,$$

$$_R s - _R t = -\pi_0; \qquad \pi_0 \equiv -2\rho \left(\frac{\partial \Psi}{\partial j_{kl}} j_{kl} + \frac{\partial \Psi}{\partial j_0} j_0 \right),$$

$$\pi \equiv -\frac{\partial \Psi}{\partial \rho^{-1}}. \tag{3.4.16}$$

Here, π_0 represents a thermodynamic *micropressure*.

The dissipation inequality is reduced to

$$\rho \hat{\gamma} \equiv -\rho_D \eta \dot{\theta} + _D t_{kl} a_{kl} + _D m_{kl} b_{lk} + _D m_k \nu_{,k} + (_D s - _D t)\nu + \frac{1}{\theta} q_k \theta_{,k} \geq 0. \tag{3.4.17}$$

This is in the form

$$\rho \hat{\eta} \equiv \mathfrak{J} \cdot Y \geq 0; \tag{3.4.18}$$

$$\mathfrak{J} \equiv (-\rho_D \eta, _D t_{kl}, _D m_{kl}, _D m_k, _D s - _D t, \mathbf{q}),$$
$$Y \equiv (\dot{\theta}, a_{lk}, b_{kl}, \nu_{,k}, \nu, \theta_{,k}/\theta). \tag{3.4.19}$$

The general solution of (3.4.16) is given by (2.4.11), i.e.,

$$-\rho_D \eta = \frac{\partial \Phi}{\partial \dot{\theta}} + U^\eta, \qquad _D \mathbf{t} = \frac{\partial \Phi}{\partial \mathbf{a}} + \mathbf{U^t}, \qquad _D \mathbf{m} = \frac{\partial \Phi}{\partial \mathbf{b}^T} + \mathbf{U^m},$$

$$_D m_k = \frac{\partial \Phi}{\partial \nu_{,k}} + U_k^\nu, \qquad _D s - _D t = \frac{\partial \Phi}{\partial \nu} + U^s, \qquad \mathbf{q} = \frac{\partial \Phi}{\partial (\nabla \theta / \theta)} + \mathbf{U^q}, \tag{3.4.20}$$

where

$$U = \left(U^\eta, \mathbf{U^t}, \mathbf{U^m}, U_k^\nu, U^s, U^q \right) \tag{3.4.21}$$

is subject to

$$Y \cdot U = 0. \tag{3.4.22}$$

Following the Onsager postulate, we also set $U^\alpha = 0$, so that the dynamic constitutive equations reduce to

$$-\rho_D \eta = \frac{\partial \Phi}{\partial \dot{\theta}}, \qquad _D t_{kl} = \frac{\partial \Phi}{\partial a_{kl}}, \qquad _D m_{kl} = \frac{\partial \Phi}{\partial b_{lk}},$$

$$_D m_k = \frac{\partial \Phi}{\partial \nu_{,k}}, \qquad _D s - _D t = \frac{\partial \Phi}{\partial \nu}, \qquad \mathbf{q} = \frac{\partial \Phi}{\partial (\nabla \theta / \theta)}. \tag{3.4.23}$$

The dissipation function Φ is nonnegative,

$$\Phi = \Phi(Y; \theta, \rho^{-1}, j_{kl}, j_0) \geq 0, \tag{3.4.24}$$

and acquires its minimum at $Y = 0$

$$\Phi(\mathbf{0}; \theta, \rho^{-1}, j_{kl}, j_0) = 0. \tag{3.4.25}$$

Thus, we have:

Theorem. Constitutive equations of microstretch thermofluids are thermodynamically admissible if and only if they are of the forms (3.4.14), (3.4.16) and (3.4.20) subject to (3.4.22). With the Onsager postulate, (3.4.20) reduces to (3.4.23). The dissipation function Φ is nonnegative and it acquires its minimum at $Y \equiv (\dot{\theta}, \mathbf{a}, \mathbf{b}, \nu_{,k}, \nu, \theta_{,k}/\theta) = \mathbf{0}$.

C. Micropolar Fluids[6]

For micropolar thermofluids, the independent and dependent variable sets consist of

$$\dot{y} = (\mathbf{a}, \mathbf{b}, \nabla\theta, \dot{\theta}; \theta, \rho, j_{kl}),$$
$$\mathcal{Z} = (\mathbf{t}, \mathbf{m}, \mathbf{q}, \eta). \tag{3.4.26}$$

The same procedures followed in subsections A and B lead to constitutive equations that are thermodynamically admissible. If we note that, for micropolar fluids, $m_k = 0$, $\nu = 0$, we can deduce the constitutive equations from those of microstretch fluids:

Static equations:

$$_R\eta = -\frac{\partial\Psi}{\partial\theta}, \qquad _Rt_{kl} = -\pi\delta_{kl}, \qquad _Rm_{kl} = 0,$$
$$\pi = -\frac{\partial\Psi}{\partial\rho^{-1}}, \qquad \psi = \Psi(\theta, \rho^{-1}, j_{kl}). \tag{3.4.27}$$

Dynamic equations:

$$-\rho_D\eta = \frac{\partial\Phi}{\partial\dot{\theta}}, \qquad _Dt_{kl} = \frac{\partial\Phi}{\partial a_{kl}}, \qquad _Dm_{kl} = \frac{\partial\Phi}{\partial b_{lk}},$$
$$q_k = \frac{\partial\Phi}{\partial(\theta_{,k}/\theta)}, \tag{3.4.28}$$

where we also set $U^\alpha = 0$, following the Onsager postulate,

$$\Phi(\mathbf{a}^T, \mathbf{b}, \nabla\theta/\theta, \dot{\theta}; \theta, \rho^{-1}, j_{kl}) \geq 0, \tag{3.4.29}$$

---

[6]Eringen [1966b], [1972b].

Φ has an absolute minimum at $\mathbf{a}^T = \mathbf{b} = \nabla\theta/\theta = \dot{\theta} = \mathbf{0}$, i.e.,

$$\Phi(\mathbf{0}; \theta, \rho^{-1}, j_{kl}) = 0. \tag{3.4.30}$$

This completes the constitutive equations for all three classes of microfluids.

Incompressible Fluids

Micromorphic Fluids

For the incompressible micromorphic fluids, we have

$$a_{kk} = v_{k,k} - \nu_{kk} = 0 \qquad \nu_{kk} = 0. \tag{3.4.31}$$

Of these $\nu_{kk} = 0$ represent micro-incompressibility. Thus $(3.4.31)_1$ shows that macro-incompressibility cannot be achieved without micro-incompressibility.

In order to obtain constitutive equations of incompressible micromorphic fluids, we set dissipation function as

$$\Phi_{\text{incompr.}} = -pa_{kk} - \varpi\nu_{kk} + \Phi. \tag{3.4.32}$$

This then gives

$$_D t_{kl} = -p\delta_{kl} + \frac{\partial\Phi}{\partial a_{kl}}, \qquad _D s_{kl} = -\varpi\delta_{kl} + \frac{\partial\Phi}{\partial c_{kl}} \tag{3.4.33}$$

and thermodynamic pressure π is indeterminate, hence $_R t_{kl}$ is not needed. All other constitutive equations are unchanged.

Microstretch Fluids

In this case we obtain

$$_D t_{kl} = p\delta_{kl} + \frac{\partial\Phi}{\partial \alpha_{kl}}, \qquad _D s - _D t = -3\varpi\delta_{kl}. \tag{3.4.34}$$

For the micropolar fluid only $(3.4.4)_1$ is valid.

3.5 Thermodynamic Restrictions on Memory-Dependent Solids

A. Micromorphic Solids

In Section 3.1, we have shown that the general constitutive equations of memory-dependent micromorphic solids[7] are of the form

$$\mathcal{Z}(X,t) = \mathcal{F}[\mathcal{Y}(t-\tau')], \qquad 0 \le \tau' < \infty, \qquad (3.5.1)$$

where $\mathcal{Y}$ and $\mathcal{Z}$ are, respectively, the independent and dependent set of variables

$$\mathcal{Y}(t) = \{\mathfrak{C}_{KL}(t), \Gamma_{KLM}(t), \mathcal{C}_{KL}(t), \theta_{,K}, \dot{\theta}, \theta\},$$
$$\mathcal{Z}(\mathbf{X},t) = \{t_{kl}, m_{klm}, s_{kl}, q_k, \eta\}. \qquad (3.5.2)$$

We replace (3.5.1) by

$$\mathcal{Z} = F[Y(t-\tau'); Y(t)], \qquad 0 \le \tau' < \infty \qquad (3.5.3)$$

where

$$Y(t-\tau') = \{\mathfrak{C}_{KL}(t-\tau') + \tau'\dot{\mathfrak{C}}_{KL}(\tau')H(t-\tau'),$$
$$\Gamma_{KLM}(t-\tau') + \tau'\dot{\Gamma}_{KLM}(\tau')H(t-\tau'),$$
$$\mathcal{C}_{KL}(t-\tau') + \tau'\dot{\mathcal{C}}_{KL}(\tau')H(t-\tau'),$$
$$\theta(t-\tau') + \tau'\dot{\theta}(\tau')H(t-\tau'),$$
$$\theta_{,K}(t-\tau') - \tau'\theta_{,K}(\tau')H(t-\tau')\},$$
$$Y(t) = \{\dot{\mathfrak{C}}_{KL}, \dot{\Gamma}_{KLM}, \dot{\mathcal{C}}_{KL}, \dot{\theta}, \theta_{,K}\}, \qquad (3.5.4)$$

casting special attention to the thermodynamic forces at the present time t. In (3.5.3), F is a functional that maps all functions $Y(t-\tau')$ with τ' taking all values in the interval $0 \le \tau' < \infty$ to a member of $\mathcal{Z}$ at $\mathbf{X}$ and t. In (3.5.4), $H(t-\tau')$ is the Heaviside's unit function.

The dissipation inequality then reads

$$Y(t) \cdot \mathcal{J}[Y(t-\tau'); Y(t)] = P[Y(t-\tau'); Y(t)] \ge 0. \qquad (3.5.5)$$

According to (2.4.19-a), the solution for $\mathcal{J}$ is given by

$$\mathcal{J} = \nabla_{Y(t)}\Phi[Y(t-\tau'); Y(t)] + U[Y(t-\tau'), Y(t)], \qquad (3.5.6)$$

with restrictions

$$Y(t) \cdot U[Y(t-s), Y(t)] = 0,$$
$$\mathcal{J}|_{Y(t)=0} = U_{Y(t)=0} = 0, \qquad P[Y(t-s); 0] = 0. \qquad (3.5.7)$$

---

[7]The linear theory was given by Eringen [1964b], [1972a] for micromorphic media. The present section is new. It is valid for the nonlinear theory.

Expression (3.5.6) determines the dynamic fluxes. Specifically,

$$_DT_{KL}(\mathbf{X},t) = {}_Dt_{kl}X_{K,k}\chi_{lL} = \frac{\partial\Phi[Y(t-\tau');\dot{\mathfrak{C}}_{KL}]}{\partial\dot{\mathfrak{C}}_{KL}} + U^{\mathbf{T}}_{KL},$$

$$_Ds_{kl}\mathfrak{X}_{Kk}\mathfrak{X}_{Ll} = \frac{\partial\Phi[\cdot;\dot{\mathfrak{C}}_{KL}]}{\partial\dot{\mathfrak{C}}_{KL}} + U^{S}_{KL},$$

$$_DM_{KLM}(\mathbf{X},t) = {}_Dm_{klm}X_{K,k}\chi_{lL}\mathfrak{X}_{Mm} = \frac{\partial\Phi[\cdot;\dot{\Gamma}_{LMK}]}{\partial\dot{\Gamma}_{LMK}} + U^{\mathbf{M}}_{KLM},$$

$$Q_K = q_k X_{K,k} = \frac{\partial\Phi[\cdot;\theta_{,K}/\theta]}{\partial(\theta_{,K}/\theta)} + U^{Q}_{K},$$

$$-\rho_D\eta = \frac{\partial\Phi[\cdot;\dot\theta]}{\partial\dot\theta} + U^{\eta}, \qquad\qquad (3.5.8)$$

where $U^{T}_{KL},\ldots,U^{\eta}$ are subject to

$$\mathfrak{C}_{KL}U^{T}_{KL} + \mathfrak{C}_{KL}U^{S}_{KL} + \Gamma_{KLM}U^{M}_{LMK} + \frac{\theta_{,K}}{\theta}U^{Q}_{K} + \dot\theta U^{\eta} = 0. \qquad (3.5.9)$$

In constitutive equations (3.5.8), the thermodynamic forces $Y(t-\tau')$, are treated as parameters, so that all gradients are with respect to $Y(t)$ at present time.

If we accept the Onsager postulate, then

$$U^{T}_{KL} = U^{S}_{KL} = U^{M}_{KLM} = U^{Q}_{K} = U^{\eta} = 0, \qquad\qquad (3.5.10)$$

and (3.5.8) give the dynamic constitutive equations of memory-dependent micromorphic thermoelastic solids, satisfying the Onsager reciprocal relations.

The static constitutive equations are given by (3.3.6), namely,

$$_R\eta = -\frac{\partial\Psi}{\partial\theta}, \qquad _Rt_{kl} = \rho\frac{\partial\Psi}{\partial\mathfrak{C}_{KL}}x_{k,K}\mathfrak{X}_{Ll}, \qquad _Rs_{kl} = 2\rho\frac{\partial\Psi}{\partial C_{KL}}\chi_{kK}\chi_{lL},$$

$$_Rm_{klm} = \rho\frac{\partial\Psi}{\partial\Gamma_{LMK}}x_{k,K}\mathfrak{X}_{Ll}\chi_{mM},$$

$$\psi = \Psi(\mathfrak{C}_{KL},\Gamma_{KLM},C_{KL},\theta,\mathbf{X}). \qquad\qquad (3.5.11)$$

So that

$$t_{kl} = {}_Rt_{kl} + {}_Dt_{kl}, \qquad s_{kl} = {}_Rs_{kl} + {}_Ds_{kl}, \qquad m_{klm} = {}_Rm_{klm} + {}_Dm_{klm},$$

$$\eta = {}_R\eta + {}_D\eta, \qquad q_k, \qquad\qquad (3.5.12)$$

where spatial tensors $_Dt_{kl}, _Ds_{kl}, _Dm_{kl}$, and q_k are solved from (3.5.8), with conditions (3.5.10). Hence,

$$_Dt_{kl} = \frac{\partial\Phi[\cdot;\dot{\mathfrak{C}}_{KL}]}{\partial\dot{\mathfrak{C}}_{KL}}x_{k,K}\mathfrak{X}_{Ll}, \qquad _Ds_{kl} = \frac{\partial\Phi[\cdot;\dot{\mathfrak{C}}_{KL}]}{\partial\dot{\mathfrak{C}}_{KL}}\chi_{kK}\chi_{lL},$$

$$_Dm_{klm} = \frac{\partial\Phi[\cdot;\dot{\Gamma}_{LMK}]}{\partial\dot{\Gamma}_{LMK}}x_{k,K}\mathfrak{X}_{Ll}\chi_{mM}, \qquad q_k = \frac{\partial\Phi[\cdot;\theta_{,K}/\theta]}{\partial(\theta_{,K}/\theta)}x_{k,K},$$

$$-\rho_D\eta = \frac{\partial\Phi[\cdot;\dot\theta]}{\partial\dot\theta}. \tag{3.5.13}$$

These results can be transferred to memory-dependent microstretch and micropolar thermoelastic solids quite readily by simply employing the dependent and independent variables appropriate to them.

B. Microstretch Solids

For microstretch solids, the static constitutive equations were given by (3.3.15), namely,

$$_R\eta = -\frac{\partial\Psi}{\partial\theta}, \qquad\qquad _Dt_{kl} = \rho\frac{\partial\Psi}{\partial\bar{\mathfrak{C}}_{KL}}x_{k,K}\bar\chi_{lL},$$

$$m_{kl} = \rho j^{-2}\frac{\partial\Psi}{\partial\Gamma_{LK}}x_{k,K}\bar\chi_{lL}, \qquad _Rm_k = \rho\frac{\partial\Psi}{\partial\Gamma_K}x_{k,K},$$

$$_Rs - _Rt = \rho j\frac{\partial\Psi}{\partial j}. \tag{3.5.14}$$

The dynamic constitutive equations are obtained similar to subsection A:

$$-\rho_D\eta = \frac{\partial\Phi[Y(t-s);\dot\theta]}{\partial\dot\theta}, \qquad\qquad _Dt_{kl} = \frac{\partial\Phi[\cdot;\dot{\bar{\mathfrak{C}}}_{KL}]}{\partial\dot{\bar{\mathfrak{C}}}_{KL}}x_{k,K}\bar\chi_{lL},$$

$$_Dm_{kl} = j^{-2}\frac{\partial\Phi[\cdot;\dot\Gamma_{LK}]}{\partial\dot\Gamma_{LK}}x_{k,K}\bar\chi_{lL}, \qquad _Dm_k = \frac{\partial\Phi[\cdot;\dot\Gamma_K]}{\partial\dot\Gamma_K}x_{k,K},$$

$$_Ds - _Dt = j\frac{\partial\Phi[\cdot;\dot j]}{\partial\dot j}, \qquad\qquad q_k = \frac{\partial\Phi[\cdot;\theta_{,K}/\theta]}{\partial(\theta_{,K}/\theta)}x_{k,K}, \tag{3.5.15}$$

where Φ is a functional of the form

$$\Phi = \Phi[Y(t-\tau');Y(t)], \qquad 0 \le \tau' < \infty, \tag{3.5.16}$$

in which

$$\begin{aligned}
Y(t-\tau') = \{&\bar{\mathfrak{C}}_{KL}(t-\tau') + \tau'\dot{\bar{\mathfrak{C}}}_{KL}(\tau')H(t-\tau'),\\
&\Gamma_{KL}(t-\tau') + \tau'\dot\Gamma_{KL}(\tau')H(t-\tau'),\\
&\Gamma_K(t-\tau') + \tau'\dot\Gamma_K(\tau')H(t-\tau'),\\
&j(t-\tau') + 3j(t')\nu(t')H(t-\tau'),\\
&\theta(t-\tau') + \tau'\dot\theta(\tau')H(t-\tau'),\ \theta_{,K}(t-\tau') - \theta_{,K}(\tau')H(t-\tau')\}\\
Y(t) = \{&\dot{\bar{\mathfrak{C}}}_{KL},\dot\Gamma_{KL},\dot\Gamma_K,\nu,\dot\theta,\theta_{,K}\}.
\end{aligned} \tag{3.5.17}$$

placing special attention to the thermodynamic forces at X, at the present time t. In (3.5.16) Φ is a functional of all function Y in the interval $0 \le \tau' < \infty$ and it is a function of $Y(t)$. Since from (1.8.29) we have $Dj/Dt = 3\nu j$, in (3.5.17) I have replaced Dj/Dt by $3\nu j$.

C. Micropolar Solids[8]

For micropolar continua, from (3.3.20), we have the static parts of constitutive equations

$$_R\eta = -\frac{\partial\Psi}{\partial\theta}, \qquad _Rt_{kl} = \rho\frac{\partial\Psi}{\partial\bar{\mathfrak{C}}_{KL}}x_{k,K}\bar{\chi}_{lL}, \qquad _Rm_{kl} = \rho\frac{\partial\Psi}{\partial\Gamma_{LK}}x_{k,K}\bar{\chi}_{lL}.$$
(3.5.18)

The dynamic parts of constitutive equations are then given by

$$-\rho_D\eta = \frac{\partial\Phi[Y(t-s);\dot{\theta}]}{\partial\dot{\theta}}, \qquad _Dt_{kl} = \frac{\partial\Phi[\cdot;\dot{\bar{\mathfrak{C}}}_{KL}]}{\partial\dot{\bar{\mathfrak{C}}}_{KL}}x_{k,K}\bar{\chi}_{lL},$$

$$_Dm_{kl} = \frac{\partial\Phi[\cdot;\dot{\Gamma}_{LK}]}{\partial\dot{\Gamma}_{LK}}x_{k,K}\bar{\chi}_{lL}, \qquad q_k = \frac{\partial\Phi[\cdot;\theta_{,K}/\theta]}{\partial(\theta_{,K}/\theta)}x_{k,K},$$

$$\Phi = \Phi[Y(t-s);Y(t)] \geq 0,$$
(3.5.19)

where

$$Y(t-\tau') = \{\bar{\mathfrak{C}}_{KL}(t-\tau') + \tau'\dot{\bar{\mathfrak{C}}}_{KL}(\tau')H(t-\tau'),$$
$$\Gamma_{KL}(t-\tau') + \tau'\dot{\Gamma}_{KL}(\tau')H(t-\tau'),$$
$$\theta(t-\tau') + \tau'\dot{\theta}(t-\tau')H(t-\tau'),$$
$$\theta_{,K}(t-\tau') - \theta_{,K}(\tau')H(t-\tau')\},$$
(3.5.20)

$$Y(t) = \{\dot{\bar{\mathfrak{C}}}_{KL}(t), \dot{\Gamma}_{KL}(t), \dot{\theta}(t), \theta_{,K}(t)\}$$
(3.5.21)

In all three microsolids, the dissipation potentials Φ vanish at $Y(t) = 0$, i.e.,

$$\Phi[Y(t-s);0] = 0.$$
(3.5.22)

3.6   Thermodynamic Restrictions on Memory-Dependent Fluids

A. Micromorphic Fluids[9]

In Section 3.2, we have shown that the independent constitutive variables of memory-dependent, nonlocal micromorphic fluids are of the form (3.2.4). For the first-order local, memory-dependent fluids, this translates to the dynamic variable set and the static variable ψ, given by

$$Y(\tau) = \{a_{kl}(\tau), b_{klm}(\tau), c_{kl}(\tau), \theta_{,k}(\tau), \dot{\theta}(\tau)\},$$
$$\psi = \Psi(\rho^{-1}, i_{kl}, \theta).$$
(3.6.1)

[8]The linear theory was given by Eringen [1967b].
[9]The present section is new.

The dynamic constitutive equations of these fluids have the general form

$$\mathcal{Z}(\mathbf{x}, t) = F[Y(t-s), Y(t)], \qquad 0 < s < \infty, \tag{3.6.2}$$

where $\mathcal{Z}$ is the set of dependent variables, given by

$$\mathcal{Z} = \{t_{kl}, m_{klm}, s_{kl}, q_k, \eta\}. \tag{3.6.3}$$

The static parts of the constitutive equations were already obtained by (3.4.6) and (3.4.7)

$$_R\eta = -\frac{\partial \Psi}{\partial \theta}, \qquad _Rt_{kl} = -\pi \delta_{kl}, \qquad _Rm_{klm} = 0,$$

$$_Rs_{kl} + \pi \delta_{kl} = \rho \left(\frac{\partial \Psi}{\partial i_{rk}} i_{rl} + \frac{\partial \Psi}{\partial i_{kr}} i_{rl} \right) = -\pi_{kl}, \qquad \pi = -\frac{\partial \Psi}{\partial \rho^{-1}}. \tag{3.6.4}$$

The dynamic parts of the constitutive equations are obtained in exactly the same way as in the case of memory-dependent solids, except that here $Y(\tau)$ is given by (3.6.1). Hence,

$$-\rho_D\eta = \frac{\partial \Phi[Y(t-s); \dot{\theta}]}{\partial \dot{\theta}}, \qquad _Dt_{kl} = \frac{\partial \Phi[\cdot; a_{kl}]}{\partial a_{kl}},$$

$$_Dm_{klm} = \frac{\partial \Phi[\cdot; b_{lmk}]}{\partial b_{lmk}}, \qquad _Ds_{kl} = 2\frac{\partial \Phi[\cdot; c_{lk}]}{\partial c_{lk}},$$

$$q_k = \frac{\partial \Phi[\cdot; \theta_{,k}/\theta]}{\partial(\theta_{,k}/\theta)}. \tag{3.6.5}$$

The dissipation potential Φ is given by

$$\Phi = \Phi[Y(t-s); Y(t)] \geq 0. \tag{3.6.6-a}$$

It also depends on the static set $\omega = (\theta, \rho^{-1}, i_{kl})$ and has an absolute minimum at $Y(t) = 0$:

$$\Phi[Y(t-s); 0] = 0. \tag{3.6.6-b}$$

B. Microstretch Fluids

For microstretch fluids, static and dynamic independent variables are given in Section 3.4:

$$\psi = \Psi(\rho^{-1}, j_{kl}, j_0, \theta),$$
$$Y(\tau) = \{a_{kl}(\tau), b_{kl}(\tau), \nu_{,k}(\tau), \nu(\tau), \theta_{,k}(\tau), \dot{\theta}(\tau), \theta(\tau)\}. \tag{3.6.7}$$

The static parts of the constitutive equations are already given by (3.4.14):

$$_R\eta = -\frac{\partial \Psi}{\partial \theta}, \qquad _Rt_{kl} = -\pi\delta_{kl}, \qquad _Rm_{kl} = 0, \qquad _Rm_k = 0,$$

$$_Rs - _Rt = -\pi_0, \qquad \pi_0 \equiv -2\rho\left(\frac{\partial \Psi}{\partial j_{kl}}j_{kl} + \frac{\partial \Psi}{\partial j_0}j_0\right),$$

$$\pi \equiv -\frac{\partial \Psi}{\partial \rho^{-1}}. \qquad\qquad (3.6.8)$$

The dynamic parts of the constitutive equations are

$$-\rho_D\eta = \frac{\partial \Phi[Y(t-s);\dot{\theta}]}{\partial \dot{\theta}}, \qquad _Dt_{kl} = \frac{\partial \Phi[\cdot;a_{kl}]}{\partial a_{kl}},$$

$$_Dm_{kl} = \frac{\partial \Phi[\cdot;b_{lk}]}{\partial b_{lk}}, \qquad _Dm_k = \frac{\partial \Phi[\cdot;\nu_{,k}]}{\partial \nu_{,k}},$$

$$_Ds - _Dt = \frac{\partial \Phi[\cdot;\nu]}{\partial \nu}, \qquad q_k = \frac{\partial \Phi[\cdot;\theta_{,k}/\theta]}{\partial(\theta_{,k}/\theta)}, \qquad (3.6.9)$$

where the dissipation potential is of the form

$$\Phi = \Phi[Y(t-s);Y(t)] \geq 0, \qquad 0 < s < \infty, \qquad (3.6.10)$$

which also depends on $\omega = (\rho^{-1}, j_{kl}, j_0, \theta)$, and has an absolute minimum at $Y(t) = 0$:

$$\Phi[Y(t-s);0] = 0. \qquad\qquad (3.6.11)$$

C. Micropolar Fluids

For micropolar fluids, the static and dynamic independent variables are

$$\psi = \Psi(\rho^{-1}, j_{kl}, \theta),$$
$$Y(\tau) = \{a_{kl}(\tau), b_{kl}(\tau), \theta_{,k}(\tau), \dot{\theta}(\tau), \theta(\tau)\}. \qquad (3.6.12)$$

The static parts of the constitutive equations are given by (3.4.27), namely,

$$_R\eta = -\frac{\partial \Psi}{\partial \theta}, \qquad _Rt_{kl} = -\pi\delta_{kl}, \qquad _Rm_{kl} = 0, \qquad \pi = -\frac{\partial \Psi}{\partial \rho^{-1}}. \quad (3.6.13)$$

The dynamic constitutive equations are

$$-\rho_D\eta = \frac{\partial \Phi[Y(t-s);\dot{\theta}]}{\partial \dot{\theta}}, \qquad _Dt_{kl} = \frac{\partial \Phi[\cdot;a_{kl}]}{\partial a_{kl}},$$

$$_Dm_{kl} = \frac{\partial \Phi[\cdot;b_{lk}]}{\partial b_{lk}}, \qquad q_k = \frac{\partial \Phi[\cdot;\theta_{,k}/\theta]}{\partial(\theta_{,k}/\theta)}. \qquad (3.6.14)$$

The dissipation potential Φ is given by

$$\Phi = \Phi[Y(t-s); Y(t)] \geq 0, \qquad 0 < s < \infty. \qquad (3.6.15)$$

It also depends on $\omega \equiv (\theta, \rho^{-1}, j_{kl})$, and has an absolute minimum at $Y(t) = 0$, i.e.,

$$\Phi = \Phi[Y(t-s); 0] = 0. \qquad (3.6.16)$$

Chapter 3 Problems

3.1. Find the linear forms of constitutive equations of micromorphic elastic solids (3.3.6) and (3.3.10).

3.2. Obtain the reduced form of the energy equation (3.3.12).

3.3. Find the linear constitutive equations of micromorphic fluids.

3.4. Construct the linear constitutive equations of memory-dependent micropolar fluids.

3.5. Find the expression of micropressure tensor π_{kl} given by (3.4.7) in terms of the invariants of i_{kl}.

3.6. For micropolar fluids, Φ depends on the invariants of the tensors $\mathbf{a}, \mathbf{b}, \nabla\theta/\theta$, and j_{kl}. List these invariants.

4
Electromagnetic Interactions

4.0 Scope

This chapter is devoted to electromagnetic (E-M) interactions with micromorphic, microstretch, and micropolar (3M) continua.[1] Section 4.1 exhibits the balance laws and jump conditions. This consists of both E-M balance laws and those of microcontinua carrying E-M and mechanical loads.

Nonlinear constitutive equations are obtained in Section 4.2 for 3M thermoelastic solids subject to E-M interactions. In Section 4.3 we develop constitutive equations for 3M thermoviscous fluids subject to E-M effects. Magnetohydrodynamics (MHD) and electric fluids are special cases.

Memory-dependent E-M solids are the subject of Section 4.4. Here, we give nonlinear constitutive equations for memory-dependent 3M-thermo E-M solids. The memory-dependent 3M thermoviscous E-M fluids are considered in Section 4.5.

The constitutive equations obtained in this chapter are used later in Chapter 8 to formulate micropolar piezoelectricity, micropolar magneto-elasticity, and the E-M theory of liquid crystals in Volume II. Thus, with the development of constitutive equations, the 3M theory of E-M continua is closed. Theories are now ready for applications.

---

[1]Constitutive equations developed in this chapter are new and have not appeared before in this unified form.

4.1   Balance Laws

Balance laws of the electromagnetics of microcontinua consist of the mechanical balance laws developed in the previous sections and the E-M balance laws. E-M balance laws consist of Maxwell's equations.[2]

Gauss' law:

$$\nabla \cdot \mathbf{D} = q_e \qquad \text{in } \mathcal{V} - \sigma, \tag{4.1.1-a}$$

$$\mathbf{n} \cdot [\![\mathbf{D}]\!] = w_e \qquad \text{on } \sigma. \tag{4.1.1-b}$$

Faraday's law:

$$\nabla \times \mathbf{E} + \frac{1}{c} \frac{\partial \mathbf{B}}{\partial t} = \mathbf{0} \qquad \text{in } \mathcal{V} - \sigma, \tag{4.1.2-a}$$

$$\mathbf{n} \times \left[\!\left[\mathbf{E} + \frac{1}{c} \mathbf{u} \times \mathbf{B} \right]\!\right] = \mathbf{0} \qquad \text{on } \sigma. \tag{4.1.2-b}$$

Magnetic flux:

$$\nabla \cdot \mathbf{B} = 0 \qquad \text{in } \mathcal{V} - \sigma, \tag{4.1.3-a}$$

$$\mathbf{n} \cdot [\![\mathbf{B}]\!] = 0 \qquad \text{on } \sigma. \tag{4.1.3-b}$$

Ampère's law:

$$\nabla \times \mathbf{H} - \frac{1}{c} \frac{\partial \mathbf{D}}{\partial t} = \frac{1}{c} \mathbf{J} \qquad \text{in } \mathcal{V} - \sigma, \tag{4.1.4-a}$$

$$\mathbf{n} \times \left[\!\left[\mathbf{H} - \frac{1}{c} \mathbf{u} \times \mathbf{D} \right]\!\right] = \mathbf{0} \qquad \text{on } \sigma. \tag{4.1.4-b}$$

---

[2]Eringen and Kafadar [1970] gave a micromorphic E-M theory which introduces balance laws for higher-order electric and magnetic tensors modifying Maxwell's equations. In a theory where the order of magnitude is consistent, the order of mechanical and E-M quantities should be compared, before using these higher-order laws to replace Maxwell's equations with the higher-order micromorphic E-M theory. This latter theory has been applied to superconductivity, Eringen [1971c]. However, presently, this theory remains dormant.

Divergence of (4.1.4-a) upon using (4.1.1-a) leads to the equation servation of charge

$$\frac{\partial q_e}{\partial t} + \nabla \cdot \mathbf{J} = 0. \qquad (4.1.\text{o})$$

The second sets of the above equations are the jump conditions on a discontinuity surface σ. They provide the boundary conditions.

In these equations, c is the velocity of light in a vacuum, $\mathbf{u}$ is the velocity of the moving discontinuity surface σ, sweeping the body in the direction of its unit normal $\mathbf{n}$.

The physical meanings of various symbols are:

$\mathbf{D}$ is the dielectric displacement vector,

$\mathbf{B}$ is the magnetic flux vector, $\qquad \mathbf{E}$ is the electric field vector,

q_e is the volume charge density, $\qquad \mathbf{H}$ is the magnetic field vector,

$\mathbf{J}$ is the current vector. $\qquad w_e$ is the surface charge density,

A. Balance Laws of Micromorphic Continua

With consideration of the first-order E-M fields (i.e., $\mathbf{D}$, $\mathbf{E}$, $\mathbf{B}$, $\mathbf{H}$, $\mathbf{J}$), as in classical electrodynamics, the balance laws are given by:

Conservation of mass:

$$\dot{\rho} + \rho \nabla \cdot \mathbf{v} = 0 \qquad \text{in } \mathcal{V} - \sigma, \qquad (4.1.6\text{-a})$$

$$[\![\rho(\mathbf{v} - \mathbf{u})]\!] \cdot \mathbf{n} = 0 \qquad \text{on } \sigma. \qquad (4.1.6\text{-b})$$

Conservation of microinertia:

$$\frac{Di_{kl}}{Dt} - i_{kr}\nu_{lr} - i_{lr}\nu_{kr} = 0 \qquad \text{in } \mathcal{V} - \sigma, \qquad (4.1.7\text{-a})$$

$$[\![i_{kl}(v_r - u_r)]\!] n_r = 0 \qquad \text{on } \sigma. \qquad (4.1.7\text{-b})$$

Balance of momentum:

$$t_{kl,k} + \rho(f_l - \dot{v}_l) = 0 \qquad \text{in } \mathcal{V} - \sigma, \qquad (4.1.8\text{-a})$$

$$\left[\!\left[t_{kl} + t_{kl}^E + u_k G_l - \rho v_l (v_k - u_k) \right]\!\right] n_k = 0 \qquad \text{on } \sigma. \qquad (4.1.8\text{-b})$$

Balance of momentum moments:

$$m_{klm,k} + t_{ml} - s_{ml} + \rho\left(l_{lm} - \sigma_{lm}\right) = 0 \qquad \text{in } \mathcal{V} - \sigma, \quad \text{(4.1.9-a)}$$

$$[\![m_{klm} - \rho i_{rm}\nu_{lr}(v_k - u_k)]\!]\, n_k = 0 \qquad \text{on } \sigma. \qquad \text{(4.1.9-b)}$$

Balance of energy:

$$-\rho\dot{e} + t_{kl}a_{kl} + m_{klm}b_{lmk} + s_{kl}c_{lk} + q_{k,k} + \rho h + W^E = 0$$
$$\text{in } \mathcal{V} - \sigma, \qquad \text{(4.1.10-a)}$$

$$[\![(t_{kl} + t^E_{kl} + u_k G_l)v_l + m_{klm}\nu_{lm} + q_k - \mathcal{S}_k$$
$$-(\rho\epsilon + \tfrac{1}{2}\rho\mathbf{v}\cdot\mathbf{v} + \tfrac{1}{2}\rho i_{rl}\nu_{mr}\nu_{ml} + \tfrac{1}{2}\mathbf{E}\cdot\mathbf{E} + \tfrac{1}{2}\mathbf{B}\cdot\mathbf{B})(v_k - u_k)]\!]\, n_k = 0$$
$$\text{on } \sigma. \qquad \text{(4.1.10-b)}$$

Second law of thermodynamics:

$$\rho\dot{\eta} - (q_k/\theta)_{,k} - \rho h/\theta \geq 0 \qquad \text{in } \mathcal{V} - \sigma, \qquad \text{(4.1.11-a)}$$

$$\left[\!\left[\rho\eta\left(v_k - u_k\right) - \frac{q_k}{\theta}\right]\!\right] n_k \geq 0 \qquad \text{on } \sigma. \qquad \text{(4.1.11-b)}$$

In the presence of E-M fields, the mechanical loads are modified by the presence of body and surface loads

$$\rho\mathbf{f} = \rho\mathbf{f}^M + \mathbf{F}^E, \qquad \rho\mathbf{l} = \rho\mathbf{l}^M + \mathbf{L}^E, \qquad \text{(4.1.12)}$$

where $\mathbf{f}^M$ and $\mathbf{l}^M$ are the usual mechanical loads. The E-M loads $\mathbf{F}^E$, $\mathbf{L}^E$, and E-M energy W^E are defined by[3]

$$\mathbf{F}^E = q_e\mathbf{E} + \frac{1}{c}\mathbf{J} \times \mathbf{B} + (\nabla\mathbf{E}) \cdot \mathbf{B} + (\nabla\mathbf{B}) \cdot \mathbf{M}$$
$$+ \frac{1}{c}[(\mathbf{P} \times \mathbf{B})v_k]_{,k} + \frac{1}{c}\frac{\partial}{\partial t}(\mathbf{P} \times \mathbf{B}), \qquad \text{(4.1.13-a)}$$

$$L^E_{kl} = P_l\mathcal{E}_k + \mathcal{M}_l B_k,$$

$$W^E = \mathbf{F}^E \cdot \mathbf{v} + \rho\mathcal{E} \cdot \left(\frac{\mathbf{P}}{\rho}\right)^{\!\cdot} - \mathbf{M} \cdot \dot{\mathbf{B}} + \mathcal{J} \cdot \mathcal{E} \qquad \text{(4.1.13-b)}$$

---

[3]$\mathbf{F}^E$ is valid for the micropolar media. L^E_{kl} however takes the form $\mathbf{L}^E = \mathbf{P} \times \mathcal{E} + \mathbf{M} \times \mathbf{B}$. Thus, the form (4.1.13-b) suggested here for L^E_{kl} appears to be compatible with the construct of micromorphic media.

In jump conditions (4.1.8-b) and (4.1.10-b), there appear an E-M str
tensor t_{kl}^E, E-M momentum G_k, and the Poynting vector S_k. These are
defined by

$$t_{kl}^E = P_k \mathcal{E}_l - B_k \mathcal{M}_l + E_k E_l + B_k B_l - \tfrac{1}{2}(E^2 + B^2 - 2\mathbf{M} \cdot \mathbf{B})\delta_{kl}, \quad (4.1.14\text{-a})$$

$$\mathbf{G} = \frac{1}{c}\mathbf{E} \times \mathbf{B}, \qquad\qquad\qquad (4.1.14\text{-b})$$

$$S_k = c(\mathbf{E} \times \mathbf{H})_k + (t_{kl}^E + v_k G_l)v_l - \tfrac{1}{2}(E^2 + B^2 + 2\mathbf{E} \cdot \mathbf{B})v_k. \quad (4.1.14\text{-c})$$

For the derivation of these expressions, we refer the reader to Eringen [1980,
Chap. 10], Eringen and Maugin [1990, Chap. 2], and De Groot and Suttorp
[1972]. In these equations, $\mathbf{P}$ and $\mathbf{M}$ are, respectively, the *polarization* and
magnetization vectors, defined by

$$\mathbf{D} = \mathbf{E} + \mathbf{P}, \qquad \mathbf{B} = \mathbf{H} + \mathbf{M}. \qquad (4.1.15)$$

E-M vectors $\mathbf{D}$, $\mathbf{E}$, $\mathbf{P}$, $\mathbf{B}$, $\mathbf{H}$, $\mathbf{M}$, and $\mathbf{J}$ are all referred to a fixed laboratory frame R_C. The vectors represented by $\mathcal{E}$, $\mathcal{M}$, and $\mathcal{J}$ are referred to a
comoving coordinate frame R_G with material particles of the body. These
are given (nonrelativistically) by

$$\mathcal{E} = \mathbf{E} + \frac{1}{c}\mathbf{v} \times \mathbf{B}, \qquad \mathcal{M} = \mathbf{M} + \frac{1}{c}\mathbf{v} \times \mathbf{P},$$

$$\mathcal{J} = \mathbf{J} - q_e \mathbf{v}. \qquad (4.1.16)$$

The C–D inequality is obtained by eliminating $\rho h/\theta$ between (4.1.10-a)
and (4.1.11-a)

$$- \rho(\dot{\psi} + \eta\dot{\theta}) + t_{kl}a_{kl} + m_{klm}b_{lmk} + s_{kl}c_{lk}$$

$$+ \frac{1}{\theta}q_k\theta_{,k} - \mathbf{P} \cdot \dot{\mathcal{E}} - \mathcal{M} \cdot \dot{\mathbf{B}} + \mathcal{J} \cdot \mathcal{E} \geq 0 \qquad \text{in } \mathcal{V}\text{-}\sigma, \quad (4.1.17)$$

where we introduce Helmholtz's free energy ψ by

$$\psi = \epsilon - \theta\eta - \rho^{-1}\mathcal{E} \cdot \mathbf{P}. \qquad (4.1.18)$$

As discussed before, in view of the presence of *dynamic entropy density* η,
ψ is considered a static dependent variable.

B. Balance Laws of Microstretch Continua

In Section 2.1 we have shown that passage can be made to the balance laws
of microstretch continua from those of micromorphic continua. For this, the
characterization (2.1.8) is crucial. Thus, we have

Conservation of mass:

$$\dot{\rho} + \rho\nabla \cdot \mathbf{v} = 0 \qquad \text{in } \mathcal{V} - \sigma, \qquad (4.1.19\text{-a})$$

$$[\![\rho(\mathbf{v} - \mathbf{u})]\!] \cdot \mathbf{n} = 0 \qquad \text{on } \sigma. \qquad (4.1.19\text{-b})$$

Conservation of microinertia:

$$\frac{Dj_{kl}}{Dt} - 2\nu j_{kl} + (\epsilon_{kpr} j_{lp} + \epsilon_{lpr} j_{kp}) \nu_r = 0 \qquad \text{in } \mathcal{V} - \sigma, \quad (4.1.20\text{-a})$$

$$[\![\rho j_{kl}(\mathbf{v} - \mathbf{u})]\!] \cdot \mathbf{n} = 0 \qquad \text{on } \sigma, \quad (4.1.20\text{-b})$$

$$\frac{Dj_0}{Dt} - 2\nu j_0 = 0, \qquad (4.1.20\text{-c})$$

$$[\![\rho j_0(\mathbf{v} - \mathbf{u})]\!] \cdot \mathbf{n} = 0. \qquad (4.1.20\text{-d})$$

Balance of momentum:

$$t_{kl,k} + \rho(f_l - \dot{v}_l) = 0 \qquad \text{in } \mathcal{V} - \sigma, \quad (4.1.21\text{-a})$$

$$[\![t_{kl} + t_{kl}^E + u_k G_l - \rho v_l(v_k - u_k)]\!] n_k = 0 \qquad \text{on } \sigma. \quad (4.1.21\text{-b})$$

Balance of momentum moments:

$$m_{kl,k} + \epsilon_{lmn} t_{mn} + \rho(l_l - \sigma_l) = 0 \qquad \text{in } \mathcal{V} - \sigma, \quad (4.1.22\text{-a})$$

$$[\![m_{kl} - \rho j_{pl}\nu_p(v_k - u_k)]\!] n_k = 0 \qquad \text{on } \sigma, \quad (4.1.22\text{-b})$$

$$m_{k,k} + t - s + \rho(l - \sigma) = 0 \qquad \text{in } \mathcal{V} - \sigma, \quad (4.1.22\text{-c})$$

$$[\![m_k - \tfrac{1}{2}\rho j_0 \nu(v_k - u_k)]\!] n_k = 0 \qquad \text{on } \sigma. \quad (4.1.22\text{-d})$$

Balance of energy:

$$-\rho\dot{\epsilon} + t_{kl} a_{kl} + m_{kl} b_{lk} + m_k \nu_{,k} + (s - t)\nu + q_{k,k} + \rho h + W^E = 0$$
$$\text{in } \mathcal{V} - \sigma, \quad (4.1.23\text{-a})$$

$$[\![\left(t_{kl} + t_{kl}^E + u_k G_l\right) v_l + m_{kl}\nu_l + m_k\nu + q_k - \mathcal{S}_k - \left(\rho\epsilon + \tfrac{1}{2}\rho\mathbf{v}\cdot\mathbf{v}\right.$$
$$\left. + \tfrac{1}{2}\rho j_{mn}\nu_m\nu_n + \tfrac{1}{4}\rho j_0\nu^2 + \tfrac{1}{2}\mathbf{E}\cdot\mathbf{E} + \tfrac{1}{2}\mathbf{B}\cdot\mathbf{B}\right) (v_k - u_k)]\!] n_k = 0$$
$$\text{on } \sigma. \quad (4.1.23\text{-b})$$

Entropy inequality:

$$\rho\dot{\eta} - (q_k/\theta)_{,k} - \rho h/\theta \geq 0 \qquad \text{in } \mathcal{V} - \sigma, \quad (4.1.24\text{-a})$$

$$[\![\rho\eta(v_k - u_k) - q_k/\theta]\!] n_k \geq 0 \qquad \text{on } \sigma, \quad (4.1.24\text{-b})$$

where applied loads are defined by

$$\rho\mathbf{f} = \rho\mathbf{f}^M + \mathbf{F}^E, \qquad \rho\mathbf{l} = \rho\mathbf{l}^M + \mathbf{L}^E,$$
$$\rho l = \rho l^M + L^E. \tag{4.1.25}$$

Here $\mathbf{F}^E$ is given by (4.1.13-a), W^E by (4.1.13-b) and $\mathbf{L}^E$ and L^E are given by

$$\mathbf{L}^E = \mathbf{P} \times \boldsymbol{\mathcal{E}} + \boldsymbol{\mathcal{M}} \times \mathbf{B}, \qquad L^E = \mathbf{P} \cdot \boldsymbol{\mathcal{E}} + \boldsymbol{\mathcal{M}} \cdot \mathbf{B}. \tag{4.1.26}$$

The E-M stress tensor t_{kl}^E, momentum $\mathbf{G}$, and Poynting vector $\mathcal{S}_k$ are as expressed by (4.1.14-a), (4.1.14-b), and (4.1.14-c). The C-D inequality for the microstretch continua reads

$$-\rho(\dot{\psi} + \eta\dot{\theta}) + t_{kl}a_{kl} + m_{kl}b_{lk} + m_k\nu_{,k} + (s-t)\nu + \frac{1}{\theta}q_k\theta_{,k} - \mathbf{P}\cdot\dot{\boldsymbol{\mathcal{E}}}$$
$$-\boldsymbol{\mathcal{M}}\cdot\dot{\mathbf{B}} + \boldsymbol{\mathcal{J}}\cdot\boldsymbol{\mathcal{E}} \geq 0 \qquad \text{in } \mathcal{V}-\sigma, \tag{4.1.27}$$

where ψ is Helmholtz's free energy, as introduced by (4.1.18).

C. Balance Laws of Micropolar Continua

These laws may be obtained from those of microstretch continua by setting

$$s = t, \quad \nu = l = \sigma = 0, \qquad m_k = 0. \tag{4.1.28}$$

4.2 Constitutive Equations of E-M Solids[4]

A. Micromorphic E-M solids

The independent and dependent constitutive variables of micromorphic elastic solids are, respectively,

$$\mathcal{Y} = \{\mathfrak{C}_{KL}, \Gamma_{KLM}, \mathcal{C}_{KL}, \theta_{,K}, \dot{\theta}, \theta, \boldsymbol{\mathcal{E}}, \mathbf{B}, \mathbf{X}\}, \tag{4.2.1}$$
$$\mathcal{Z} = \{t_{kl}, m_{klm}, s_{kl}, q_k, \eta, \mathbf{P}, \boldsymbol{\mathcal{M}}, \boldsymbol{\mathcal{J}}\}. \tag{4.2.2}$$

Thus, the list of mechanical variables is supplemented by the E-M fields. In (4.2.2), η replaces ϵ, on account of (4.1.18), and ψ is a static dependent variable with the constitutive equation

$$\psi = \Psi\left(\mathfrak{C}_{KL}, \Gamma_{KLM}, \mathcal{C}_{KL}, \theta, \mathcal{E}_K, B_K; \mathbf{X}\right). \tag{4.2.3}$$

Just as in Section 3.3, substituting $\dot{\psi}$ from (4.2.3) into the C–D inequality (4.1.17), we obtain

$$_R\eta = -\frac{\partial\Psi}{\partial\theta}, \qquad _Rt_{kl} = \rho\frac{\partial\Psi}{\partial\mathfrak{C}_{KL}}x_{k,K}\mathfrak{X}_{Ll}, \qquad _Rs_{kl} = 2\rho\frac{\partial\Psi}{\partial\mathcal{C}_{KL}}\chi_{kK}\chi_{lL},$$

---

[4]The present section is new.

$$_R m_{klm} = \rho \frac{\partial \Psi}{\partial \Gamma_{LMK}} x_{k,K} \mathfrak{X}_{Ll} \chi_{mM},$$

$$_R P_k = -\rho \frac{\partial \Psi}{\partial \mathcal{E}_K} x_{k,K}, \qquad _R \mathcal{M}_k = -\rho \frac{\partial \Psi}{\partial B_K} x_{k,K}, \qquad (4.2.4)$$

and the dissipation inequality

$$-\rho_D \eta \dot{\theta} + \mathbf{q} \cdot \frac{\nabla \theta}{\theta} + \boldsymbol{\mathcal{J}} \cdot \boldsymbol{\mathcal{E}} \geq 0. \qquad (4.2.5)$$

In (4.2.3), $\mathcal{E}_K$ and B_K are defined by

$$\mathcal{E}_K = \mathcal{E}_k x_{k,K}, \qquad B_K = B_k x_{k,K}. \qquad (4.2.6)$$

The dynamic part of the constitutive equations is obtained by the solution of (4.2.5), as discussed in Section 2.4,

$$-\rho_D \eta = \frac{\partial \Phi}{\partial \dot{\theta}} + U^{\eta}, \quad _D q_k = \frac{\partial \Phi}{\partial(\theta_{,K}/\theta)} x_{k,K} + U_k^q, \quad _D \mathcal{J}_k = \frac{\partial \Phi}{\partial \mathcal{E}_K} x_{k,K} + U_k^{\mathcal{J}},$$

$$(4.2.7)$$

where Φ is the dissipation potential, given by

$$\Phi = \Phi(Y; \omega); \qquad Y \equiv (\dot{\theta}, \nabla \theta/\theta, \boldsymbol{\mathcal{E}}),$$

$$\omega \equiv (\mathfrak{C}_{KL}, \Gamma_{KLM}, \mathcal{C}_{KL}, \theta, \boldsymbol{\mathcal{E}}, \mathbf{B}; \mathbf{X}), \qquad (4.2.8)$$

and $U = (U^{\eta}, \mathbf{U}^q, \mathbf{U}^{\mathcal{J}})$ is subject to

$$U \cdot Y = 0. \qquad (4.2.9)$$

The complete constitutive equations of micromorphic E-M elastic solids are then given by (4.2.3), (4.2.4), (4.2.7), and (4.2.8) with

$$t_{kl} = _R t_{kl} + _D t_{kl}, \qquad m_{klm} = _R m_{klm}, \qquad \eta = _R \eta + _D \eta,$$

$$P_k = _R P_k, \qquad \mathcal{M}_k = _R \mathcal{M}_k, \qquad q_k = _D q_k, \qquad \mathcal{J}_k = _D \mathcal{J}_k. \quad (4.2.10)$$

Note that $\mathbf{P}$ and $\boldsymbol{\mathcal{M}}$ possess no dynamic parts, and of course $\mathbf{q}$ and $\boldsymbol{\mathcal{J}}$ have no static parts. In memory-dependent materials (e.g., dielectrics with magnetic losses) $\mathbf{P}$ and $\boldsymbol{\mathcal{M}}$ will have dynamic parts as well.

B. Microstretch E-M Solids

By now, the familiar routine, used in Sections 3.3 and 4.2, leads to the constitutive equations of microstretch solids:

$$_R \eta = -\frac{\partial \Psi}{\partial \theta}, \qquad _R t_{kl} = \rho \frac{\partial \Psi}{\partial \overline{\mathfrak{C}}_{KL}} x_{k,K} \overline{\chi}_{lL}, \qquad _R m_{kl} = \rho j^{-2} \frac{\partial \Psi}{\partial \Gamma_{LK}} x_{k,K} \overline{\chi}_{lL},$$

$$_R m_k = \rho \frac{\partial \Psi}{\partial \Gamma_K} x_{k,K}, \qquad _R s - _R t = 3\rho j \frac{\partial \Psi}{\partial j},$$

$$_R P_k = -\rho \frac{\partial \Psi}{\partial \mathcal{E}_K} x_{k,K}, \qquad _R \mathcal{M}_k = -\rho \frac{\partial \Psi}{\partial B_K} x_{k,K},$$

$$\psi = \Psi(\overline{\mathfrak{C}}_{KL}, \Gamma_{KL}, \Gamma_K, j^2, \theta, \mathcal{E}_K, B_K, \mathbf{X}), \qquad (4.2.11)$$

for the static parts, and

$$-\rho_D \eta = \frac{\partial \Phi}{\partial \dot{\theta}} + U^\eta, \qquad q_k = \frac{\partial \Phi}{\partial(\theta_{,K}/\theta)} x_{k,K} + U^q,$$

$$_D \mathcal{J}_k = \frac{\partial \Phi}{\partial \mathcal{E}_K} x_{k,K} + U^\mathcal{J}, \qquad (4.2.12)$$

for the dynamic parts, where

$$\Phi = \Phi(Y; \omega) \geq 0; \qquad Y \equiv (\dot{\theta}, \nabla\theta/\theta, \mathcal{E}),$$
$$\omega = \{\bar{\mathfrak{C}}_{KL}, \Gamma_{KL}, \Gamma_K, j^2, \theta, \mathbf{B}, \mathbf{X}\}. \qquad (4.2.13)$$

The dissipation potential Φ is an increasing function of Y and it has an absolute minimum at $Y = 0$, i.e.,

$$\Phi(0; \omega) = 0. \qquad (4.2.14)$$

The total constitutive equations are the sum of static and dynamic parts as in (4.2.10). The vector $U = (U^\eta, \mathbf{U}^q, \mathbf{U}^\mathcal{J})$ is subject to

$$U \cdot Y = 0. \qquad (4.2.15)$$

C. Micropolar E-M Solids

Constitutive equations, in this case, follow from those of microstretch E-M solids by setting

$$j = 1, \qquad s = t, \qquad m_k = 0,$$

so that

$$t_{kl} = {}_R t_{kl} + {}_D t_{kl}, \qquad m_{kl} = {}_R m_{kl}, \qquad \eta = {}_R \eta + {}_D \eta,$$
$$P_k = {}_R P_k, \qquad \mathcal{M}_k = {}_R \mathcal{M}_k, \qquad q_k = {}_D q_k, \qquad \mathcal{J}_k = {}_D \mathcal{J}_k, \qquad (4.2.16)$$

where static parts read

$$_R \eta = -\frac{\partial \Psi}{\partial \theta}, \qquad _R t_{kl} = \rho \frac{\partial \Psi}{\partial \bar{\mathfrak{C}}_{KL}} x_{k,K} \overline{\mathcal{X}}_{lL}, \qquad _R m_{kl} = \rho \frac{\partial \Psi}{\partial \Gamma_{LK}} x_{k,K} \overline{\mathcal{X}}_{lL},$$

$$_R P_k = -\rho \frac{\partial \Psi}{\partial \mathcal{E}_k} x_{k,K}, \qquad _R \mathcal{M}_k = -\rho \frac{\partial \Psi}{\partial B_k} x_{k,K},$$

$$\Psi = \Psi(\bar{\mathfrak{C}}_{KL}, \Gamma_{KL}, \theta, \mathbf{X}). \qquad (4.2.17)$$

Dynamic parts are given by (4.2.12).

Substituting constitutive equations into the equations of energy, the equations of energy reduce to the following simple form, in all three solids:

$$-\rho(\dot{\theta} {}_D \eta + \theta {}_D \dot{\eta} + \theta {}_R \dot{\eta}) - \nabla \cdot \mathbf{q} - \mathcal{J} \cdot \mathcal{E} - \rho h = 0. \qquad (4.2.18)$$

4.3 Constitutive Equations of E-M Fluids[5]

A. Micromorphic E-M Fluids

For the study of thermodynamic restrictions of micromorphic E-M fluids, the independent and dependent variables listed by (3.4.1) and (3.4.2) are supplemented by E-M quantities, so that

$$\dot{\mathcal{Y}} = (a_{kl}, b_{klm}, c_{kl}, \theta_k, \dot{\theta}, \theta, \rho^{-1}, i_{kl}, \mathcal{E}_k, B_k), \tag{4.3.1}$$

$$\mathcal{Z} = (t_{kl}, m_{klm}, s_{kl}, q_k, \eta, \mathbf{P}, \mathbf{M}, \mathbf{J}). \tag{4.3.2}$$

The free energy is considered to be of the form

$$\psi = \Psi(\theta, \rho^{-1}, i_{kl}, \mathcal{E}_k, B_k). \tag{4.3.3}$$

Following the same procedure used in Section 3.4, the static constitutive equations are obtained to be

$$R\eta = -\frac{\partial \Psi}{\partial \theta}, \qquad Rt_{kl} = -\pi\delta_{kl}, \qquad Rm_{klm} = 0,$$

$$Rs_{kl} + \pi\delta_{kl} = -\pi_{kl}; \qquad \pi \equiv -\frac{\partial \Psi}{\partial \rho^{-1}}, \qquad \pi_{kl} = -\rho\left(\frac{\partial \Psi}{\partial i_{rk}}i_{kr} + \frac{\partial \Psi}{\partial i_{kr}}i_{lr}\right),$$

$$_RP_k = -\rho\frac{\partial \Psi}{\partial \mathcal{E}_k}, \qquad _R\mathcal{M}_k = -\rho\frac{\partial \Psi}{\partial B_k}. \tag{4.3.4}$$

The dynamic parts of constitutive equations are then given by

$$-\rho_D\eta = \frac{\partial \phi}{\partial \dot{\theta}} + U^\eta, \qquad _Dt_{kl} = \frac{\partial \phi}{\partial a_{kl}} + U^t_{kl}, \qquad _Dm_{klm} = \frac{\partial \Phi}{\partial b_{lmk}} + U^m_{lmk},$$

$$_Ds_{kl} = \frac{\partial \Phi}{\partial c_{lk}} + U^s_{lk}, \qquad q_k = \frac{\partial \Phi}{\partial(\theta_{,k}/\theta)} + U^q_k,$$

$$_D\mathcal{J}_k = \frac{\partial \Phi}{\partial \mathcal{E}_k} + U^\mathcal{J}_k, \tag{4.3.5}$$

where Φ is the dissipation potential

$$\Phi(Y; \omega) \geq 0; \qquad Y \equiv (a_{kl}, b_{klm}, c_{kl}, \theta_{,k}/\theta, \dot{\theta}, \mathcal{E}_k),$$

$$\omega \equiv (\rho^{-1}, i_{kl}, \theta, \mathcal{E}_k, B_k). \tag{4.3.6}$$

Φ is an increasing function of Y and has an absolute minimum at $Y = 0$, i.e.,

$$\Phi(0; \omega) = 0, \tag{4.3.7}$$

and $U = (U^\eta, U^t, U^m, U^s, U^q, U^\mathcal{J})$ is subject to

$$U \cdot Y = 0. \tag{4.3.8}$$

---

[5]This section is new.

B. Microstretch E-M Fluids

The method of approach here is exactly the same as in Section 3.4. We simply jot down the results as follows. Static parts of the constitutive equations:

$$_R\eta = -\frac{\partial\Psi}{\partial\theta}, \qquad _Rt_{kl} = -\pi\delta_{kl}, \qquad _Rm_{kl} = 0, \qquad \pi \equiv -\frac{\partial\Psi}{\partial\rho^{-1}},$$

$$_Rm_k = 0, \qquad _Rs - {}_Rt = -\pi_0, \qquad \pi_0 \equiv -2\rho\left(\frac{\partial\Psi}{\partial j_{kl}}j_{kl} + \frac{\partial\Psi}{\partial j_0}j_0\right),$$

$$_RP_k = -\rho\frac{\partial\Psi}{\partial\mathcal{E}_k}, \qquad _R\mathcal{M}_k = -\rho\frac{\partial\Psi}{\partial B_k},$$

$$\psi = \Psi(\theta, \rho^{-1}, j_{kl}, j_0, \mathcal{E}_k, B_k). \tag{4.3.9}$$

Dynamic parts of the constitutive equations:

$$-\rho_D\eta = \frac{\partial\Phi}{\partial\dot\theta} + U^\eta, \qquad _Dt_{kl} = \frac{\partial\Phi}{\partial a_{kl}} + U_{kl}^t, \qquad _Dm_{kl} = \frac{\partial\Phi}{\partial b_{lk}} + U_{kl}^m,$$

$$_Dm_k = \frac{\partial\Phi}{\partial\nu_{,k}} + U^\nu, \qquad _Ds - {}_Dt = \frac{\partial\Phi}{\partial\nu} + U^s,$$

$$q_k = \frac{\partial\Phi}{\partial(\theta_{,k}/\theta)} + U_k^q, \qquad \mathcal{J}_k = \frac{\partial\Phi}{\partial\mathcal{E}_k} + U_k^\mathcal{J}. \tag{4.3.10}$$

The dissipation potential Φ is of the form

$$\Phi = \Phi(Y;\omega) \geq 0, \qquad Y = (\mathbf{a}, \mathbf{b}, \nabla\nu, \nu, \nabla\theta/\theta, \dot\theta, \mathcal{E}),$$

$$\omega = \omega(\rho^{-1}, j_{kl}, j_0, \theta, \mathcal{E}, \mathbf{B}). \tag{4.3.11}$$

Φ is an increasing function of Y with an absolute mimimum at $Y = 0$:

$$\Phi(0;\omega) = 0, \tag{4.3.12}$$

and $U \equiv (U^\eta, U^t, U^m, U^\nu, U^s, U^q, U^\mathcal{J})$ is subject to

$$U \cdot Y = 0. \tag{4.3.13}$$

C. Micropolar E-M Fluids

The constitutive equations of micropolar E-M fluids follow from those of microstretch fluids, by merely setting

$$\nu = 0, \qquad m_k = 0, \qquad \mathbf{U}^\nu = \mathbf{0}, \qquad U^\mathbf{s} = 0, \tag{4.3.14}$$

as a result $s - t \equiv 0$, and the remaining quantities can be read from (4.3.9)–(4.3.13).

4.4   Memory-Dependent E-M Solids[6]

A. Micromorphic E-M Solids

In Section 3.5, we have shown that memory-dependent micromorphic solids are governed by functionals of the type (3.5.4)

$$\mathcal{Z} = F[Y(t - \tau'); Y(t)], \tag{4.4.1}$$

where $\mathcal{Z}$ denotes the dependent variable set. For the E-M solids, it is given by the collection

$$\mathcal{Z} = \{t_{kl}, m_{klm}, s_{kl}, q_k, \eta, P_k, \mathcal{M}_k, \mathcal{J}_k\} \tag{4.4.2}$$

and the independent variables by

$$Y(t - \tau') = \{(3.5.4)_1 \text{ and } \mathcal{E}_K(t - \tau') - \mathcal{E}_K(\tau')H(t - \tau'),$$
$$B_K(t - \tau') - B_K(\tau')H(t - \tau')\},$$
$$Y(t) = \{(3.5.4)_2 \text{ and } \mathcal{E}_K(t), \ B_K(t)\} \tag{4.4.3}$$

$\mathcal{Z}$ may also depend on $\omega = \{\mathfrak{C}_{KL}, \Gamma_{KLM}, \mathcal{C}_{KL}, \theta, \mathbf{X}\}$, but we do not display this in (4.4.1). The free energy ψ depends on $(\mathfrak{C}_{KL}, \Gamma_{KLM}, \mathcal{C}_{KL}, \theta, \varepsilon_K, B_K, \mathbf{X})$. Thus, the static parts of constitutive equations are the same as in (4.2.4). The dynamic parts are obtained as in Section 3.5 supplemented by the constitutive equations for E-M fields. Thus

$$\eta = {}_R\eta + {}_D\eta, \quad t_{kl} = {}_Rt_{kl} + {}_Dt_{kl}, \quad m_{klm} = {}_Rm_{klm} + {}_Dm_{klm},$$
$$s_{kl} = {}_Rs_{kl} + {}_Ds_{kl}, \quad P_k = {}_RP_k + {}_DP_k, \quad \mathcal{M}_k = {}_R\mathcal{M}_k + {}_D\mathcal{M}_k, \tag{4.4.4}$$

with static parts given by

$$_R\eta = -\frac{\partial \Psi}{\partial \theta}, \quad _Rt_{kl} = \rho\frac{\partial \Psi}{\partial \mathfrak{C}_{KL}}x_{k,K}\mathfrak{X}_{Ll}, \quad _Rs_{kl} = 2\rho\frac{\partial \Psi}{\partial \mathcal{C}_{KL}}\chi_{kK}\chi_{lL},$$

$$_Rm_{klm} = \rho\frac{\partial \Psi}{\partial \Gamma_{LMK}}x_{k,K}\mathfrak{X}_{Ll}\chi_{mM}, \quad _RP_k = -\rho\frac{\partial \Psi}{\partial \mathcal{E}_k}x_{k,K},$$

$$_R\mathcal{M}_k = -\rho\frac{\partial \Psi}{\partial B_k}x_{k,K}, \quad \psi = \Psi(\mathfrak{C}_{KL}, \Gamma_{KLM}, \mathcal{C}_{KL}, \mathcal{E}_K, B_K, \theta; \mathbf{X}), \tag{4.4.5}$$

and the dynamic parts by

$$-\rho_D\eta = \frac{\partial \Phi[Y(t - s); \dot{\theta}]}{\partial \dot{\theta}} + U^\eta, \quad q_k = \frac{\partial \Phi[\cdot; \theta_{,k}/\theta]}{\partial (\theta_{,K}/\theta)}x_{k,K} + U_k^q,$$

$$_Dt_{kl} = \frac{\partial\Phi[\cdot;\dot{\mathfrak{C}}_{KL}]}{\partial\dot{\mathfrak{C}}_{KL}}x_{k,K}\mathfrak{X}_{Ll} + U_{kl}^t, \qquad _Ds_{kl} = 2\frac{\partial\Phi[\cdot;\dot{\mathfrak{C}}_{KL}]}{\partial\dot{\mathfrak{C}}_{KL}}\chi_{kK}\chi_{lL} + U_{kl}^s,$$

$$_Dm_{klm} = \frac{\partial\Phi[\cdot;\dot{\Gamma}_{LMK}]}{\partial\dot{\Gamma}_{LMK}}x_{k,K}\mathfrak{X}_{Ll}\chi_{mM} + U_{klm}^m,$$

$$_DP_k = -\frac{\partial\Phi[\cdot;\mathcal{E}_K]}{\partial\mathcal{E}_K}x_{k,K} + U_k^{\mathbf{P}}, \qquad _D\mathcal{M}_k = -\frac{\partial\Phi[\cdot;B_K]}{\partial B_K}x_{k,K} + U_k^{\mathcal{M}},$$

$$_D\mathfrak{J}_k = \frac{\partial\Phi[\cdot;\mathcal{E}_K]}{\partial\mathcal{E}_K}x_{k,K} + U_k^{\mathfrak{J}}. \tag{4.4.6}$$

The dissipation function Φ is of the form

$$\Phi = \Phi[Y(t-\tau');Y(t)] \geq 0, \tag{4.4.7}$$

with a minimum at $Y(t) = 0$, i.e.,

$$\Phi[Y(t-\tau');0] = 0. \tag{4.4.8}$$

The auxiliary dynamic tensors $U = (U^\eta, \mathbf{U}^q, \mathbf{U}^t, \mathbf{U}^s, \mathbf{U}^m, \mathbf{U}^P, \mathbf{U}^{\mathcal{M}}, \mathbf{U}^{\mathfrak{J}})$ are subject to

$$Y \cdot U = 0. \tag{4.4.9}$$

B. Microstretch E-M Solids

The constitutive equations memory-dependent microstretch E-M solids may be expressed formally as:

$$\mathcal{Z} = F[Y(t-\tau');Y(t)], \qquad 0 \leq \tau' < \infty \tag{4.4.10}$$

where

$$\begin{aligned}
Y(t-\tau') &= \{(3.5.17)_1 \text{ and } \mathcal{E}_K(t-\tau') - \mathcal{E}_K(\tau')H(t-\tau'), \\
&\quad B_K(t-\tau') - B_K(\tau')H(t-\tau')\}, \\
Y(t) &= \{(3.5.17)_2 \text{ and } \mathcal{E}_K(t), B_K(t)\} \\
Y(t) &= \{\dot{\mathfrak{C}}_{KL}(t), \dot{\Gamma}_{KL}(t), \dot{\Gamma}_K(t), \nu(t), \dot{\theta}(t)\}.
\end{aligned} \tag{4.4.11}$$

Following the same steps as in Section 4.4, the constitutive equations are obtained to be, the static parts,

$$_R\eta = -\frac{\partial\Psi}{\partial\theta}, \qquad _Rt_{kl} = \frac{\partial\Psi}{\partial\overline{\mathfrak{C}}_{KL}}x_{k,K}\overline{\chi}_{lL}, \qquad _Rm_{kl} = \rho j^{-2}\frac{\partial\Psi}{\partial\Gamma_{LK}}x_{k,K}\overline{\chi}_{lL},$$

$$_Rm_k = \rho\frac{\partial\Psi}{\partial\Gamma_K}x_{k,K}, \qquad _Rs - {_Rt} = 3\rho j\frac{\partial\Psi}{\partial j},$$

$$P_k = -\rho\frac{\partial\Psi}{\partial\mathcal{E}_K}x_{k,K}, \qquad _R\mathcal{M}_k = -\rho\frac{\partial\Psi}{\partial B_K}x_{k,K},$$

$$\Psi = \Psi(\overline{\mathfrak{C}}_{KL}, \Gamma_{KL}, \Gamma_K, j^2, \mathcal{E}_K, B_K; \mathbf{X}), \tag{4.4.12}$$

and the dynamic parts

$$- \rho_D \eta = \frac{\partial \Phi[Y(t-s); \dot{\theta}]}{\partial \dot{\theta}} + U^\eta, \qquad q_k = \frac{\partial \Phi[\cdot; \theta_{,k}/\theta]}{\partial(\theta_{,K}/\theta)} x_{k,K} + U_k^q,$$

$$_D t_{kl} = \frac{\partial \Phi[\cdot; \dot{\overline{\mathfrak{C}}}_{KL}]}{\partial \dot{\overline{\mathfrak{C}}}_{KL}} x_{k,K} \overline{\chi}_{lL} + U_{kl}^t,$$

$$_D m_{kl} = \rho j^{-2} \frac{\partial \Phi[\cdot; \dot{\Gamma}_{LK}]}{\partial \dot{\Gamma}_{LK}} x_{k,K} \overline{\chi}_{lL} + U_{kl}^m, \qquad _D m_k = \frac{\partial \Phi[\cdot; \dot{\Gamma}_K]}{\partial \dot{\Gamma}_K} x_{k,K} + U_k^q,$$

$$_D s - {}_D t = \frac{\partial \Phi[\cdot; \nu]}{\partial \nu} + U^s,$$

$$_D P_k = -\frac{\partial \Phi[\cdot; \mathcal{E}_K]}{\partial \mathcal{E}_K} x_{k,K} + U_k^P, \qquad _D M_k = -\frac{\partial \Phi[\cdot; B_K]}{\partial B_K} x_{k,K} + U_k^M,$$

$$_D \mathfrak{I}_k = \frac{\partial \Phi[\cdot; \mathcal{E}_K]}{\partial \mathcal{E}_K} x_{k,K} + U_k^{\mathfrak{I}}. \qquad (4.4.13)$$

The dissipation potential is given by

$$\Phi = \Phi[Y(t-\tau'); Y(t)] \geq 0, \qquad 0 < \tau' < \infty. \qquad (4.4.14)$$

It possesses an absolute minimum at $Y = 0$, i.e.,

$$\Phi[Y(t-\tau'); 0] = 0. \qquad (4.4.15)$$

The auxiliary vectors $U^\eta, U^q, \dots$ satisfy (4.4.9).

C. Micropolar E-M Solids

The constitutive equations of micropolar E-M solids follow from (4.4.12) and (4.4.13) by setting

$$m_k = 0, \qquad j = 1, \qquad s = t. \qquad (4.4.16)$$

4.5 Memory-Dependent E-M Fluids[7]

Constitutive equations of memory-dependent fluids were derived in Section 3.6, without E-M effects. The method is identical to those of Section 3.6 and fully discussed in Section 2.4. Consequently, here we display only the final results.

---

[7]This section is new.

A. Micromorphic E-M Fluids

Independent and dependent variables are, respectively,

$$Y(\tau) = \left\{ a_{kl}(\tau), b_{klm}(\tau), c_{kl}(\tau), \theta_{,k}(\tau), \dot{\theta}(\tau), \mathcal{E}_k(\tau), B_k(\tau) \right\},$$
$$\mathcal{Z} = \{ t_{kl}, m_{klm}, s_{kl}, q_k, \eta, P_k, \mathcal{M}_k, \mathcal{J}_k \}. \tag{4.5.1}$$

The constitutive equations are expressed formally as

$$\mathcal{Z} = F[Y(t-s); Y(t)]. \tag{4.5.2}$$

The free energy Ψ is considered a static variable

$$\Psi = \Psi(\theta, \rho^{-1}, i_{kl}, \mathcal{E}_k, B_k). \tag{4.5.3}$$

The method developed in Section 2.4 and used in Section 3.6 leads to the following constitutive equations.

Static parts:

$$_R\eta = -\frac{\partial \Psi}{\partial \theta}, \qquad _Rt_{kl} = -\pi\delta_{kl}; \qquad \pi \equiv -\frac{\partial \Psi}{\partial \rho^{-1}}, \qquad _Rm_{kl} = 0,$$

$$_Rs_{kl} + \pi\delta_{kl} = -\pi_{kl}; \qquad \pi_{kl} \equiv -\rho\left(\frac{\partial \Psi}{\partial i_{rk}}i_{kl} + \frac{\partial \Psi}{\partial i_{kr}}i_{lr}\right),$$

$$_RP_k = -\rho\frac{\partial \Psi}{\partial \mathcal{E}_k}, \qquad _R\mathcal{M}_k = -\rho\frac{\partial \Psi}{\partial B_k}. \tag{4.5.4}$$

Dynamic parts:

$$-\rho_D\eta = \frac{\partial \Phi[Y(t-s); \dot{\theta}]}{\partial \dot{\theta}} + U^\eta, \qquad _Dt_{kl} = \frac{\partial \Phi[\cdot; \mathbf{a}]}{\partial a_{kl}} + U^t_{kl},$$

$$_Dm_{klm} = \frac{\partial \Phi[\cdot; b_{lmk}]}{\partial b_{lmk}} + U^m_{lmk}, \qquad \tfrac{1}{2}{}_Ds_{kl} = \frac{\partial \Phi[\cdot;\mathbf{c}]}{\partial c_{lk}} + U^s_{lk},$$

$$_Dq_k = \frac{\partial \Phi[\cdot; \theta_{,k}/\theta]}{\partial(\theta_{,k}/\theta)} + U^q_k, \qquad _DP_k = -\frac{\partial \Phi[\cdot; \mathcal{E}_k]}{\partial \mathcal{E}_k} + U^P_k,$$

$$_D\mathcal{M}_k = -\frac{\partial \Phi[\cdot; B_k]}{\partial B_k} + U^{\mathcal{M}}_k, \qquad _D\mathcal{J}_k = \frac{\partial \Phi[\cdot; \mathcal{E}_k]}{\partial \mathcal{E}_k} + U^{\mathcal{J}}_k. \tag{4.5.5}$$

The dissipation potential Φ has the form

$$\Phi = \Phi[Y(t-s); Y] \geq 0, \qquad \Phi[Y(t-s); 0] = 0. \tag{4.5.6}$$

The auxiliary vectors $U^\eta, U^t, \ldots$ are subject to (4.4.9).

B. Microstretch E-M Fluids

For microstretch E-M fluids, independent and dependent constitutive variables are, respectively,

$$Y(\tau) = \{a_{kl}(\tau), b_{kl}(\tau), \nu_{,k}, \nu(\tau), \theta_{,k}(\tau), \dot{\theta}(\tau), \theta(\tau), \mathcal{E}_k(\tau), B_k(\tau)\},$$
$$\mathcal{Z} = \{t_{kl}, m_{kl}, m_k, s, q_k, \eta, P_k, \mathcal{M}_k, \mathcal{J}_k\}. \tag{4.5.7}$$

The free energy is considered a static variable

$$\Psi = \Psi(\theta, \rho^{-1}, j_{kl}, j_0, \mathcal{E}_k, B_k). \tag{4.5.8}$$

The constitutive equations are of the form

$$\mathcal{Z} = F[Y(t - \tau); Y(t)]. \tag{4.5.9}$$

The method followed in Section 3.6 gives the constitutive equations of microstretch E-M fluids.

Static parts:

$$_R\eta = -\frac{\partial\Psi}{\partial\theta}, \qquad _R t_{kl} = -\pi\delta_{kl}; \qquad \pi \equiv -\frac{\partial\Psi}{\partial\rho^{-1}},$$

$$_R m_{kl} = {}_R m_k = 0, \qquad _R s - {}_R t = -\pi_0; \qquad \pi_0 \equiv -2\rho\left(\frac{\partial\Psi}{\partial j_{kl}}j_{kl} + \frac{\partial\Psi}{\partial j_0^2}j_0^2\right),$$

$$_R P_k = -\rho\frac{\partial\Psi}{\partial\mathcal{E}_k}, \qquad _R\mathcal{M}_k = -\rho\frac{\partial\Psi}{\partial B_k}. \tag{4.5.10}$$

Dynamic parts:

$$-\rho_D\eta = \frac{\partial\Phi[Y(t - s); \dot{\theta}]}{\partial\dot{\theta}} + U^\eta, \qquad _D t_{kl} = \frac{\partial\Phi[\cdot; \mathbf{a}]}{\partial a_{kl}} + U^t_{kl},$$

$$_D m_{kl} = \frac{\partial\Phi[\cdot; \mathbf{b}]}{\partial b_{lk}} + U^m_{kl}, \qquad _D m_k = \frac{\partial\Phi[\cdot; \nu_{,k}]}{\partial\nu_{,k}} + U^\nu_k,$$

$$_D s - {}_D t = \frac{\partial\Phi[\cdot; \nu]}{\partial\nu} + U^s, \qquad q_k = \frac{\partial\Phi[\cdot; \theta_{,k}/\theta]}{\partial(\theta_{,k}/\theta)} + U^q_k,$$

$$_D P_k = -\frac{\partial\Phi[\cdot; \mathcal{E}_k]}{\partial\mathcal{E}_k} + U^P, \qquad _D\mathcal{M}_k = -\frac{\partial\Phi[\cdot; B_k]}{\partial B_k} + U^\mathcal{M}_k,$$

$$_D\mathcal{J}_k = \frac{\partial\Phi[\cdot; \mathcal{E}_k]}{\partial\mathcal{E}_k} + U^\mathcal{J}_k. \tag{4.5.11}$$

The free energy is of the form

$$\Phi = \Phi[Y(t - s); Y(t)] \geq 0, \qquad \Phi[Y(t - s); 0] = 0, \tag{4.5.12}$$

and the auxiliary vector U consisting of $U^\eta, \mathbf{U^t}, \ldots$ satisfies

$$U \cdot Y(t) = 0. \tag{4.5.13}$$

C. Micropolar E-M Fluids

Constitutive equations of these fluids immediately follow from those of microstretch fluids by setting

$$m_k = 0, \qquad s = t, \qquad \nu = 0.$$

Chapter 4 Problems

4.1. Express Maxwell's equations in terms of $\mathbf{B}, \mathcal{E}, \mathbf{D}$, and $\mathcal{H}$.

4.2. Show that equations of energy balance may be expressed as

$$\rho\dot{\epsilon} - t_{kl}a_{kl} - m_{kl}b_{lk} - q_{k,k} - \rho h - \rho\mathcal{E} \cdot (\mathbf{P}/\rho)\dot{} + \mathcal{M} \cdot \dot{\mathbf{B}} - \mathbf{J} \cdot \mathcal{E} = \mathbf{0}.$$

4.3. Show that balance of momentum for micropolar E-M continua can be expressed in the integral form

$$\frac{d}{dt} \int_{\mathcal{V}-\sigma} (\rho v_k + \rho j_{kl}\nu_l + G_k)\, dv = \int_{\partial\mathcal{V}-\sigma} n_l[t_{lk} + t_{lk}^E + v_l G_k]\, da$$
$$+ \int_{\mathcal{V}-\sigma} \rho f_k\, dv.$$

4.4. Show that equations of energy can be expressed in the integral form

$$\frac{d}{dt} \int_{\mathcal{V}-\sigma} \left[\rho\epsilon + \tfrac{1}{2}\rho v^2 + \tfrac{1}{2}\rho j_{kl}\nu_k\nu_l + \tfrac{1}{2}(\mathbf{E}^2 + \mathbf{B}^2)\right] dv$$
$$= \int_{\partial\mathcal{V}-\sigma} v_l[t_{kl}n_k + t_{kl}^E n_k + (\mathbf{v} \cdot \mathbf{n})G_l + \mathbf{n} \cdot (\mathbf{q} - \mathcal{S})]\, da$$
$$+ \int_{\mathcal{V}-\sigma} \rho(\mathbf{f} \cdot \mathbf{v} + \mathbf{l} \cdot \boldsymbol{\nu} + h)\, dv.$$

4.5. Obtain the linear constitutive equations of micropolar E-M elastic solids.

4.6. In a MHD, polarization and magnetizations are negligible. Determine the linear constitutive equations of micropolar E-M fluids.

4.7. (Short-Term Paper) Study the literature, and express the basic equations of plasma consisting of two fluids.

4.8. (Short-Term Paper) Study the literature and express the field equations of a magnetic fluid.

5
Theory of Micropolar Elasticity

5.0   Scope

In the four previous chapters we have given the complete theory of 3M continua, with and without E-M interactions. Balance laws, jump conditions, and nonlinear constitutive equations were obtained, so that the theory is complete and closed. Beginning with Chapter 5 we explore applications of these theories. By means of mathematical solutions and experimental observations, we try to exhibit new physical phenomena predicted by microcontinuum theories. The aim here is not to be exhaustive with the discussion of all problems for this is neither possible nor desirable, as there is a very large volume of literature in the field. It is not desirable since a large number of solutions tend to hide the main purpose, namely, the new physical phenomena that are not in the domain of predictions of classical field theories.

In Section 5.1, *linear* constitutive equations are obtained for the micropolar thermoelastic solids. Material symmetry regulations and stability are discussed in Section 5.2. In Section 5.3 we give the linear constitutive equations of isotropic thermo-micropolar elasticity. A theorem here displays the necessary and sufficient conditions for the material stability. These conditions appear as inequalities on the isotropic material constants.

In Section 5.4 we collect basic equations necessary to formulate boundary-initial value problems. Various boundary conditions (traction, displacement, and mixed) are formulated. Mixed boundary-initial value problems of the linear theory (anisotropic media) are formulated in Section 5.5. Field equations are given for both anisotropic and isotropic micropolar solids. Formulation by means of convolution is discussed. This section also contains the formulation of special cases: plane strain, plane stress, and generalized plane strain.

In Section 5.6 we present the basic equations in orthogonal curvilinear coordinates. In Section 5.7 we prove the uniqueness theorem for the linear theory of micropolar thermoelasticity. Section 5.8 is devoted to the Reciprocal Theorem. In Section 5.9 we establish variational principles, theorems on work and energy, and the principle of minimum potential energy. Hamilton's Principle closes this section. These theorems are useful in theoretical developments, such as uniqueness, stability, and approximate solutions.

Section 5.10 is devoted to the establishment of Noether's theorem and its application to obtain the conservation laws. This section may be of interest only to theoreticians.

Section 5.11 gives the solution of the linear field equations of micropolar elasticity for plane harmonic waves in infinite solids. Here we see, for the first time, the outcome of new physical phenomenon that are not predicted by classical elasticity. The dispersion curves exhibit two scalar waves and two coupled vector waves. Longitudinal and transverse acoustic branches are also predicted by classical elasticity. But the others (a longitudinal optic branch (LO), and a transverse optic branch (TO)) are the two new branches that fall outside the domain of classical elasticity. The fact that these branches are also observed in phonon dispersion experiments assures our faith in micropolar elasticity.

In Section 5.12 we discuss various methods for the determination of material moduli. These include lattice dynamical calculations of a diatomic chain which model the KNO_3 molecule in its argonite structure and grid framework underlying the skeleton of tall buildings. Experimental attempts are described in Section 5.13, for a composite elastic epoxy carrying uniformly distributed rigid aluminum shots and for animal bones.

In Section 5.14 we present several methods of solutions of the field equations in terms of scalar and vector potentials and a generalization of Somigliana representation. Reflection of micropolar waves from the boundary of a half-space is treated in Section 5.15, micropolar surface waves in Section 5.16, and micropolar waves in thick plates in Section 5.17. In Section 5.18 we obtain fundamental solutions: the static and dynamic concentrated load and couple problems. Fundamental solutions are used later on, to express general solutions of boundary-initial value problems by means of the reciprocal theorem. The problems of a sphere and a spherical cavity are treated in Section 5.19.

Discussion of axisymmetric problems for the half-space occupy Section 5.20. In Section 5.21 we present the solution of the penny-shaped crack. Stress concentration around an elliptic hole is presented in Section 5.22 and those of a circular hole in 5.23. In Section 5.26 we discuss the problem of dislocations and disclinations. The solutions presented in these sections

(5.18 to 5.23) exhibit the new physical phenomena inherent in the micropolar elasticity. In this regard, we also witness the nonlinear solitary waves in Section 5.24.

Fundamental solutions of the micropolar elastostatic problem require separate treatment, not easily obtainable from the dynamic case. This is given in Section 5.25 and applied to the case of static dislocations and disclinations in Section 5.26.

Section 5.27 is devoted to the development of the theory of micropolar plates and Section 5.28 to the flexural waves in plates. Here, the predicted flexural wave speed as a function of wavelength agrees with that of the Timoshenko beam theory, all the way from zero to the infinite wavelength. At a zero wavelength it coincides with the Rayleigh surface wave velocity. This is remarkable, in that the classical plate theory leads to infinite phase velocity. The micropolar plate theory also predicts a second class of microrotational waves. These waves are dispersive and possess large phase velocities.

5.1   Linear Constitutive Equations

Constitutive equations of micropolar thermoelastic solids were obtained in Section 3.3. Here, we present the linear theory, Eringen [1966a, 1968a, 1970a]. When deformations, rotations, and thermal changes from the natural state K_C are small, linear approximation is possible. To this end, we introduce displacement vector $\mathbf{u}$, microdisplacement vector $\boldsymbol{\phi}$, and the temperature change T:

$$\boldsymbol{u} = \mathbf{x} - \boldsymbol{X}, \qquad \boldsymbol{\phi} = \boldsymbol{\xi} - \boldsymbol{\Xi},$$
$$T = \theta - T_0, \qquad |T| \ll T_0, \, T_0 > 0, \qquad (5.1.1)$$

where T_0 is a constant ambient temperature, much greater than absolute zero. Essential to the development of the linear theory is the array

$$\boldsymbol{W} = \left(\boldsymbol{u}_{,k}, \boldsymbol{\phi}, \boldsymbol{\phi}_{,k}, T_{,k}, T \right), \qquad (5.1.2)$$

whose norm is defined by

$$\epsilon = ||\mathbf{W}|| = (\mathbf{W} \cdot \mathbf{W})^{1/2}. \qquad (5.1.3)$$

We assume that, in some neighborhood of $\epsilon = 0$, there exist positive constants K and n such that

$$||O(\epsilon^n)|| \leq K\epsilon^n, \qquad K > 0, n > 0. \qquad (5.1.4)$$

With this, for the first-order (linear) approximation, we write

$$x_{k,K} = (\delta_{kl} + u_{k,l}) \, \delta_{lK} + O(\epsilon^2), \qquad v_k \simeq \frac{\partial u_k}{\partial t} + O(\epsilon^2),$$

$$\overline{\chi}_{kK} = (\delta_{kl} + \phi_{kl})\,\delta_{lK} + O(\epsilon^2), \qquad \nu_k = \frac{\partial \phi_k}{\partial t} + O(\epsilon^2),$$

$$T_{,k} = \theta_{,k}, \qquad j_{kl} = J_{kl} + J_{km}\phi_{lm} + J_{lm}\phi_{km} + O(\epsilon^2),$$

$$\frac{\rho}{\rho_0} = 1 - u_{k,k} + O(\epsilon^2), \tag{5.1.5}$$

where

$$\phi_{kl} = -\epsilon_{klm}\phi_m, \qquad \phi_k = -\tfrac{1}{2}\epsilon_{klm}\phi_{lm}. \tag{5.1.6}$$

With these, linear strain tensors (1.5.25) read

$$\epsilon_{kl} = \delta_{kl} - \overline{\mathfrak{c}}_{kl} = u_{l,k} + \epsilon_{lkm}\phi_m, \qquad \gamma_{kl} = \phi_{k,l}. \tag{5.1.7}$$

The material (Lagrangian) strain tensors (1.5.24) are related to these by

$$\overline{\mathfrak{C}}_{KL} - \delta_{KL} \equiv \overline{\mathfrak{E}}_{KL} = \epsilon_{kl}\delta_{kK}\delta_{lL}, \qquad \Gamma_{KL} = \gamma_{kl}\delta_{kK}\delta_{lL}, \tag{5.1.8}$$

where δ_{kK} denote shifters. In a common frame-of-reference for coordinates X_K and x_k, they become Kronecker deltas.

The natural state, K_C, is assumed to be stress and couple-stress free, and $T = T_0$. The linear approximation for the stress, couple-stress, and temperature fields requires a quadratic approximation for the free energy ψ, in its variables, i.e.,

$$\rho_0\psi \equiv \Sigma = \Sigma_0 - \rho_0\eta_0 T - \left(\frac{\rho_0 C_0}{2T_0}\right)T^2 - A_{KL}T\overline{\mathfrak{E}}_{KL} - B_{KL}T\Gamma_{KL}$$

$$+ \tfrac{1}{2}A_{KLMN}\overline{\mathfrak{E}}_{KL}\overline{\mathfrak{E}}_{MN} + \tfrac{1}{2}B_{KLMN}\Gamma_{KL}\Gamma_{MN} + C_{KLMN}\overline{\mathfrak{E}}_{KL}\Gamma_{MN}, \tag{5.1.9}$$

where

$$M \equiv \{\Sigma_0, \eta_0, C_0, A_{KL}, B_{KL}, A_{KLMN} = A_{MNKL},$$

$$B_{KLMN} = B_{MNKL}, C_{KLMN}\}, \tag{5.1.10}$$

are the constitutive moduli (material constants). They depend on ρ_0, T_0, J_{KL}, and $\mathbf{X}$, in general.

The dissipation potential, Φ, given by (3.3.22-a), is approximated by

$$\Phi = \frac{1}{2T_0^2}K_{KL}T_{,K}\,T_{,L} + \frac{M_K}{T_0}T_{,K}\,\dot{T} + \tfrac{1}{2}N\dot{T}^2. \tag{5.1.11}$$

Constitutive equations (3.3.20) and (3.3.21) are then approximated by

$$\Sigma - \Sigma_0 = -\rho\eta_0 T - \frac{\rho C_0}{2T_0}T^2 - A_{kl}T\epsilon_{kl} - B_{kl}T\gamma_{kl} + \tfrac{1}{2}A_{klmn}\epsilon_{kl}\epsilon_{mn}$$

$$+ \tfrac{1}{2}B_{klmn}\gamma_{kl}\gamma_{mn} + C_{klmn}\epsilon_{kl}\gamma_{mn}$$

$$= \tfrac{1}{2}\left[-\rho(\eta + \eta_0)T + t_{kl}\epsilon_{kl} + m_{kl}\gamma_{kl}\right],$$

$$\eta = \eta_0 + \frac{C_0}{T_0}T + \frac{1}{\rho}A_{kl}\epsilon_{kl} + \frac{1}{\rho}B_{kl}\gamma_{lk} = -\frac{1}{\rho}\frac{\partial \Sigma}{\partial T},$$

$$_Rt_{kl} = -A_{kl}T + A_{klmn}\epsilon_{mn} + C_{klmn}\gamma_{mn} = \frac{\partial \Sigma}{\partial \epsilon_{kl}},$$

$$_Rm_{kl} = -B_{lk}T + B_{lkmn}\gamma_{mn} + C_{mnlk}\epsilon_{mn} = \frac{\partial \Sigma}{\partial \gamma_{lk}}, \qquad (5.1.12)$$

and

$$\Phi = \frac{1}{2T_0^2}K_{kl}T_{,k}T_{,l} + \frac{M_k}{T_0}T_{,k}\dot{T} + \tfrac{1}{2}N\dot{T}^2,$$

$$q_k = \frac{1}{T_0}K_{kl}T_{,l} + M_k\dot{T},$$

$$-\rho_D\eta = N\dot{T} + \frac{M_k}{T_0}T_{,k}, \qquad (5.1.13)$$

where

$$\rho \simeq \rho_0(\mathbf{X})|_{\mathbf{X}=\mathbf{x}}, \qquad (A_{KL}, B_{KL}) = (A_{kl}, B_{kl})\delta_{kK}\delta_{lL},$$
$$(A_{klmn}, B_{klmn}, C_{klmn}) = (A_{KLMN}, B_{KLMN}, C_{KLMN})\delta_{kK}\delta_{lL}\delta_{mM}\delta_{nN}.$$
$$(5.1.14)$$

For simplicity, henceforth, we neglect the temperature-rate dependence, so that the dynamic parts of the constitutive equations reduce to

$$\Phi = \frac{1}{2T_0^2}K_{kl}T_{,k}\,T_{,l}\,, \qquad q_k = \frac{1}{T_0}K_{kl}T_{,l}\,, \quad _D\eta = 0. \qquad (5.1.15)$$

At the reference state, $T = 0, \epsilon = \gamma = 0$ and the internal energy density takes the form

$$\Sigma - \Sigma_0 + \rho_0\eta T \equiv U, \qquad (5.1.16)$$

where U, called the *strain energy density*,[1] is given by

$$U = \Sigma - \Sigma_0 + \rho_0\eta T$$
$$= \frac{1}{2}\left(\frac{\rho C_0}{T_0}T^2 + A_{klmn}\epsilon_{kl}\epsilon_{mn} + B_{klmn}\gamma_{kl}\gamma_{mn} + 2C_{klmn}\epsilon_{kl}\gamma_{mn}\right)$$
$$= \tfrac{1}{2}\left[\rho\left(\eta - \eta_0\right)T + t_{kl}\epsilon_{kl} + m_{kl}\gamma_{lk}\right]. \qquad (5.1.17)$$

Constitutive moduli A_{klmn}, B_{klmn}, and the heat conduction coefficient K_{kl} possess the symmetries

$$A_{klmn} = A_{mnkl}, \qquad B_{klmn} = B_{mnkl}, \qquad K_{kl} = K_{lk}. \qquad (5.1.18)$$

---

[1] Usually this terminology is used for the nonthermal case $T = 0$.

5.2   Material Symmetry and Stability

A. Material Symmetry

For the most general anisotropic micropolar solids, the number of independent material moduli is 196, as listed in Table 5.1.

TABLE 5.1. Material moduli for anisotropic solids.

Moduli	C_o	A_{kl}	B_{kl}	A_{klmn}	B_{klmn}	C_{klmn}	K_{kl}
Components	1	9	9	45	45	81	6

For materials that possess symmetry, the number of these components reduces. For example, for the isotropic media, A_{kl} and K_{kl} possess a single component each, B_{kl}, C_{klmn} vanish, and A_{klmn} and B_{klmn} possess three independent components each (see Section 5.3). For crystalline materials, the material symmetry is governed by a group of transformations of the material particles at the reference state. There are 32 crystal groups and five transverse isotropy groups. For centro-symmetric materials, this number reduces to 13. Denoting this group by C_n ($n = 1, 2, \ldots, 13$) and incorporating this group into the *full* isotropy group C_0, Zheng and Spencer [1993] gave tables for the material moduli C_{klmn} for the 14 symmetry groups (C_0, C_1,..., C_{13}) which they call the *mechanics symmetry group*. Coefficients A_{klmn} and B_{klmn} can be derived from the tables of C_{klmn}, with the symmetry considerations (5.1.18). The generators of the 14 mechanics symmetry groups are expressed in terms of rotation tensors $R_j(\theta), j = 0, 1, 2, 3$, that denote rotation around the orthonormal directions e_1, e_2, e_3 associated with the privileged directions of crystals (or transversely isotropic materials) and $e_0 = (e_1 + e_2 + e_3)/\sqrt{3}$ through an angle θ. Let $R_j(\theta)$ be an element of C_n. Then the generators are given in Table 5.2.

TABLE 5.2. The generators of the 14 mechanics symmetry groups.

C_n	Generators of C_n	C_n	Generators of C_n
C_1	I	C_8	$R_3(2\pi/3)$
C_2	$R(\pi)$	C_9	$R_3(2\pi/3)$, $R_1(\pi)$
C_3	$R_3(\pi)$, $R_1(\pi)$	C_{10}	$R_3(\pi/3)$
C_4	$R_3(\pi/2)$	C_{11}	$R_3(\pi/3)$, $R_1(\pi)$
C_5	$R_3(\pi/2)$, $R_1(\pi)$	C_{12}	$R_3(\theta)$ $(0 \le \theta \le 2\pi)$
C_6	$R_3(\pi)$, $R_1(\pi)$, $R_0(2\pi/3)$	C_{13}	$R_3(\theta)$ $(0 \le \theta \le 2\pi)$, $R_1(\pi)$
C_7	$R_3(\pi/2)$, $R_1(\pi/2)$, $R_2(\pi/2)$	C_0	$R_3(\theta)$, $R_1(\phi)$ $(0 \le \theta, \phi \le 2\pi)$

We express C_{klmn} in the form of a 9×9 matrix by writing

$$w_1 = \epsilon_{11}, \quad w_2 = \epsilon_{22}, \quad w_3 = \epsilon_{33}, \quad w_4 = \epsilon_{(23)}, \quad w_5 = \epsilon_{(31)},$$

$$w_6 = \epsilon_{(12)}, \quad w_7 = \epsilon_{[23]}, \quad w_8 = \epsilon_{[31]}, \quad w_9 = \epsilon_{[12]}, \qquad (5.2.1)$$

$$C_{klmn}\epsilon_{mn} = M_{\alpha\beta}w_\beta, \qquad k,l,m,n = 1,2,3,$$
$$\alpha, \beta = 1,2,\ldots,9, \qquad (5.2.2)$$

where $\epsilon_{(kl)}$ and $\epsilon_{[kl]}$ denote, respectively, the symmetric and antisymmetric parts of ϵ_{kl}, i.e.,

$$\epsilon_{(kl)} = \tfrac{1}{2}\left(\epsilon_{kl} + \epsilon_{lk}\right), \qquad \epsilon_{[kl]} = \tfrac{1}{2}\left(\epsilon_{kl} - \epsilon_{lk}\right).$$

With this 9×9 matrix, the mechanics symmetry group $(C_0, \ldots, C_{13})$ is given in Table 5.3.

B. Stability of the Thermodynamic State

Definition. The thermodynamic state of the micropolar body is said to be stable if and only if the internal energy function is nonnegative for all temperatures and strains.

This definition makes sense on the basis of the fact that a solid will be stable, if the strain energy increases (decreases) with increasing (decreasing) strains and temperatures. Thus, the stability condition is

$$U \geq 0 \qquad \text{for all } T, \epsilon_{kl}, \text{ and } \gamma_{kl}, \qquad (5.2.3)$$

where U is given by (5.1.17). This restriction can be expressed as a quadratic form

$$U \equiv \frac{1}{2}\left(\frac{\rho C_0}{T_0}T^2 + \lambda_{\alpha\beta}w_\alpha w_\beta\right) \geq 0, \qquad \alpha, \beta = 1,2\ldots,18, \qquad (5.2.4)$$

where w_α represents the components of a single vector in an 18-dimensional vector space corresponding to nine components each of ϵ_{kl} and γ_{kl}. Here w_1 to w_9 are given by (5.2.1). A similar expression is valid for γ_{kl} starting with w_{10} for γ_{11} and proceeding in the same sequence as in (5.2.1) for the remaining components of γ_{kl}. The 18×18 symmetric matrix consists of linear combinations of A_{klmn}, B_{klmn}, and C_{klmn}. It follows that:

Theorem. Thermo-micropolar elastic solids are stable if and only if $C_0 \geq 0$ and

$$\lambda_{\alpha\beta}w_\alpha w_\beta \geq 0, \qquad \alpha, \beta = 1,2,\ldots,18 \qquad (5.2.5)$$

is positive semidefinite.

The specific forms of restrictions on $\lambda_{\alpha\beta}$ (consequently A_{klmn}, B_{klmn}, and C_{klmn}) follow from a theorem in algebra namely:

Theorem. The quadratic form (5.2.5) will be positive semidefinite if and only if all eigenvalues λ_α, $i = 1,2,\ldots,9$ of $\lambda_{\alpha\beta}$ are nonnegative.

TABLE 5.3. The micropolar elasticity matrices under mechanics symmetry groups C_n^* (after Zheng and Spencer [1993]).

KEY TO NOTATION		
$\cdot$ zero component,	$\ominus$	non-zero component,
$\overset{\cdot}{\underset{\cdot}{\circ}}$ $\frac{1}{2}(M_{11}-M_{12})$,	$\ominus\!\!-\!\!-\!\!\circ$	equal components,
$\ominus\!\!-\!\!-\!\!\circ$ components numerically equal, but opposite in sign.		

$C_0 \; (3-3)$

$C_1 \; (81-51)$

$C_2 \; (41-25)$

$C_3 \; (21-15)$

$C_4 \; (21-13)$

$C_5 \; (11-9)$

*In Table 5.3, the m_1 and m_2 in $Cn(m_1-m_2)$ indicate the numbers of the independent components of $[M_{\alpha\beta}]$ when $M_{\alpha\beta} \neq M_{\alpha\beta}$ and $M_{\beta\alpha} = M_{\alpha\beta}$ respectively.

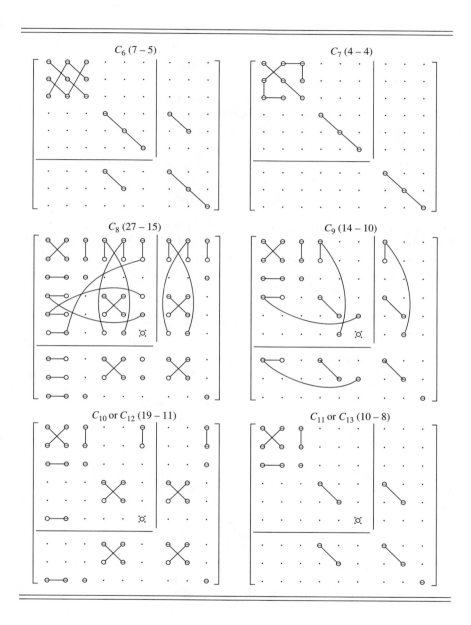

This means that roots of λ_α of the polynomial

$$\det\left(\lambda_{\alpha\beta} - \lambda\delta_{\alpha\beta}\right) = 0 \tag{5.2.6}$$

in λ, must be nonnegative, i.e.,

$$\lambda_\alpha \geq 0, \qquad \alpha = 1, 2, \ldots, 18. \tag{5.2.7}$$

Alternatively, *all invariants of $\lambda_{\alpha\beta}$ must be nonnegative. Equivalently, a quadratic form is called a positive definite form if it is in general positive, and can be zero only if all w_α are zero.*

A set of necessary and sufficient conditions for it to be positive definite is

$$\lambda_{11} > 0, \quad \begin{vmatrix} \lambda_{11} & \lambda_{12} \\ \lambda_{21} & \lambda_{22} \end{vmatrix} > 0, \quad \begin{vmatrix} \lambda_{11} & \lambda_{12} & \lambda_{13} \\ \lambda_{21} & \lambda_{22} & \lambda_{23} \\ \lambda_{31} & \lambda_{32} & \lambda_{33} \end{vmatrix} > 0, \ldots, \quad |\lambda_{\alpha\beta}| > 0.$$
$$\tag{5.2.8}$$

A quadratic form can be essentially positive without being positive definite, e.g., $\det \lambda_{\alpha\beta} = 0$ (cf. Jeffreys and Jeffreys [1950, p. 137]). Thus, any one of these three equivalent theorems may be used to determine the inequalities on the material moduli $\lambda_{\alpha\beta}$. In terms of the original quadratic form (5.1.17), the stability condition (5.2.3) reads

$$U \equiv \frac{1}{2}\left(\frac{\rho C_0}{T_0}T^2 + A_{klmn}\epsilon_{kl}\epsilon_{mn} + B_{klmn}\gamma_{kl}\gamma_{mn} + 2C_{klmn}\epsilon_{kl}\gamma_{mn}\right) \geq 0. \tag{5.2.9}$$

This form is useful in proving uniqueness and existence theorems for the anisotropic micropolar elastic solids, without actually determining the inequalities on each material constant.

5.3   Isotropic Solids

For isotropic materials, the material symmetry group $\{S\}$ is the full group of orthogonal transformations. In this case, all material moduli can be constructed with the exterior products of δ_{kl}. Consequently, we take

$$\begin{aligned}
A_{kl} &= \beta_0\delta_{kl}, & A_{klmn} &= \lambda\delta_{kl}\delta_{mn} + (\mu + \kappa)\delta_{km}\delta_{ln} + \mu\delta_{kn}\delta_{lm}, \\
B_{kl} &= 0, & B_{klmn} &= \alpha\delta_{kl}\delta_{mn} + \beta\delta_{kn}\delta_{lm} + \gamma\delta_{km}\delta_{ln}, \quad (5.3.1) \\
K_{kl}/T_0 &= K\delta_{kl}, & C_{klmn} &= 0.
\end{aligned}$$

In addition, the material is called *spin-isotropic*,[2] if

$$j_{kl} = j\delta_{kl}. \tag{5.3.2}$$

---

[2]In Eringen's terminology, microisotropic, [1966a].

The isotropic material moduli β_0, λ, μ, κ, α, β, γ, and K are functions of $\boldsymbol{x}$ for inhomogeneous materials and they are constants for homogeneous materials. In this case, constitutive equations take the form

$$\Sigma = \Sigma_0 - \rho \left(\eta_0 T + \frac{C_0}{2T_0} T^2 \right) - \beta_0 T \epsilon_{mm}$$
$$+ \tfrac{1}{2} [\lambda \epsilon_{kk} \epsilon_{ll} + (\mu + \kappa) \epsilon_{kl} \epsilon_{kl} + \mu \epsilon_{kl} \epsilon_{lk}]$$
$$+ \tfrac{1}{2} (\alpha \gamma_{kk} \gamma_{ll} + \beta \gamma_{kl} \gamma_{lk} + \gamma \gamma_{kl} \gamma_{kl}),$$

$$\eta = \eta_0 + \left(\frac{C_0}{T_0} \right) T + \left(\frac{\beta_0}{\rho} \right) \epsilon_{mm},$$

$$t_{kl} = -\beta_0 T \delta_{kl} + \lambda \epsilon_{mm} \delta_{kl} + (\mu + \kappa) \epsilon_{kl} + \mu \epsilon_{lk},$$

$$m_{kl} = \alpha \gamma_{mm} \delta_{kl} + \beta \gamma_{kl} + \gamma \gamma_{lk},$$

$$q_k = K T_{,k}. \tag{5.3.3}$$

Thus, for isotropic micropolar elastic solids the number of independent material moduli is 10 as against 196 for the most general anisotropic solids. The internal energy U is given by

$$U = \frac{\rho C_0}{2T_0} T^2 + U_\epsilon + U_\gamma \geq 0, \tag{5.3.4}$$

where the abbreviations U_ϵ and U_γ stand for the quadratic parts of the internal energy functions in ϵ_{kl} and γ_{kl}, i.e.,

$$U_\epsilon = \tfrac{1}{2} [\lambda \epsilon_{kk} \epsilon_{ll} + (\mu + \kappa) \epsilon_{kl} \epsilon_{kl} + \mu \epsilon_{kl} \epsilon_{lk}],$$
$$U_\gamma = \tfrac{1}{2} (\alpha \gamma_{kk} \gamma_{ll} + \beta \gamma_{kl} \gamma_{lk} + \gamma \gamma_{kl} \gamma_{kl}). \tag{5.3.5}$$

For $\epsilon_{kl} = \gamma_{kl} = 0$ from (5.3.4), it follows that

$$C_0 \geq 0. \tag{5.3.6}$$

Since T, ϵ_{kl}, and γ_{kl} are uncoupled from each other, we can study U_ϵ and U_γ separately, e.g.,

$$U_\gamma \geq 0. \tag{5.3.7}$$

We express U_γ in the form

$$U_\gamma = \tfrac{1}{2} \lambda_{\alpha\beta} \gamma_\alpha \gamma_\beta,$$

where

$$\lambda_{11} = \lambda_{22} = \lambda_{33} = \alpha + \beta + \gamma, \qquad \lambda_{12} = \lambda_{23} = \lambda_{31} = \alpha,$$
$$\lambda_{44} = \lambda_{55} = \lambda_{66} = 2(\gamma + \beta),$$
$$\lambda_{77} = \lambda_{88} = \lambda_{99} = 2(\gamma - \beta), \qquad \text{all other } \lambda_{ij} = 0. \tag{5.3.8}$$

From this 9×9 matrix, through the conditions (5.2.7), we extract

$$3\alpha + \beta + \gamma \geq 0, \qquad \gamma + \beta \geq 0, \qquad \gamma - \beta \geq 0. \tag{5.3.9}$$

A similar treatment for the case of $U_\epsilon \geq 0$ gives

$$3\lambda + 2\mu + \kappa \geq 0, \qquad 2\mu + \kappa \geq 0, \qquad \kappa \geq 0. \qquad (5.3.10)$$

From the entropy inequality we have

$$KT_{,k}T_{,k} \geq 0. \qquad (5.3.11)$$

It follows that

$$K \geq 0.$$

Collecting, we have the conditions

$$
\begin{aligned}
3\lambda + 2\mu + \kappa \geq 0, \qquad 2\mu + \kappa \geq 0, \qquad \kappa \geq 0, \\
3\alpha + \beta + \gamma \geq 0, \qquad \gamma + \beta \geq 0, \qquad \gamma - \beta \geq 0, \\
C_0 \geq 0, \qquad K \geq 0,
\end{aligned}
\qquad (5.3.12)
$$

which were first given by Eringen [1968a].

The stipulation of nonnegative kinetic energy density

$$\mathcal{K} = \tfrac{1}{2}\rho\dot{\boldsymbol{u}} \cdot \dot{\boldsymbol{u}} + \tfrac{1}{2}\rho j \dot{\boldsymbol{\phi}} \cdot \dot{\boldsymbol{\phi}} \geq 0 \qquad (5.3.13)$$

implies that

$$\rho \geq 0, \qquad j \geq 0. \qquad (5.3.14)$$

These are all conditions that must be imposed on the material moduli of isotropic solids.

Theorem. The necessary and sufficient conditions that the strain energy U be nonnegative are (5.3.12).

As discussed in Section 5.2, we demand that the same be true for the anisotropic solids.

5.4   Formulation of Problems in Micropolar Elasticity

The formulation of problems of micropolar elasticity requires the knowledge of four different sets of equations:

(i)   the balance laws;

(ii) jump conditions and initial conditions;

(iii) constitutive equations; and

(iv) kinematical relations.

(i) Balance Laws (Valid in $\mathcal{V} - \sigma$)

$$\frac{\partial \rho}{\partial t} + (\rho v_k)_{,k} = 0, \tag{5.4.1}$$

$$\frac{\partial j_{kl}}{\partial t} + j_{kl,m} v_m + (\epsilon_{kpr} j_{lp} + \epsilon_{lpr} j_{kp}) \nu_r = 0, \tag{5.4.2}$$

$$t_{kl,k} + \rho (f_l - \dot{v}_l) = 0, \tag{5.4.3}$$

$$m_{kl,k} + \epsilon_{lmn} t_{mn} + \rho (l_l - \sigma_l) = 0, \tag{5.4.4}$$

$$\rho \theta \dot{\eta} - q_{k,k} - \rho h = 0, \tag{5.4.5}$$

$$q_k \theta_{,k} \geq 0. \tag{5.4.6}$$

(ii) Jump Conditions (and/or Boundary Conditions) and Initial Conditions

$$[\![\rho (\boldsymbol{v} - \boldsymbol{u})]\!] \cdot \boldsymbol{n} = 0, \tag{5.4.7}$$

$$[\![\rho j_{kl}(\boldsymbol{v} - \boldsymbol{u})]\!] \cdot \boldsymbol{n} = 0, \tag{5.4.8}$$

$$[\![t_{kl} - \rho v_l(v_k - u_k)]\!] n_k = 0, \tag{5.4.9}$$

$$[\![m_{kl} - \rho j_{pl} \nu_p(v_k - u_k)]\!] n_k = 0, \tag{5.4.10}$$

$$[\![t_{kl} v_l + m_{kl} \nu_l + q_k - (\rho \epsilon + \tfrac{1}{2} \rho \boldsymbol{v} \cdot \boldsymbol{v} + \tfrac{1}{2} \rho j_{ij} \nu_i \nu_j)(v_k - u_k)]\!] n_k = 0, \tag{5.4.11}$$

$$[\![\rho \eta (\boldsymbol{v} - \boldsymbol{u}) - (\boldsymbol{q}/\theta)]\!] \cdot \boldsymbol{n} \geq 0. \tag{5.4.12}$$

These conditions can be used to obtain a set of boundary conditions on the surface $\partial \mathcal{V}$ of the body. To obtain these boundary conditions, we assume that σ coincides with $\partial \mathcal{V}$. We demonstrate this process for two cases.

(a) Traction Boundary Conditions

We take $\partial \mathcal{V} \equiv \sigma$ and $\boldsymbol{u} \equiv \boldsymbol{v}$. This, from (5.4.9)–(5.4.11), gives

$$\begin{aligned}
t_{kl} n_k &= t_{(\boldsymbol{n})l}, \\
m_{kl} n_k &= m_{(\boldsymbol{n})l}, \\
q_k n_k &= q_{(\boldsymbol{n})},
\end{aligned} \tag{5.4.13}$$

where we expanded boldface brackets in the form

$$[\![A]\!] = A^+ - A^-$$

from positive and negative sides of the surface $\partial \mathcal{V}$ and dropped the superscript $(^-)$ for brevity. Note that (5.4.7) and (5.4.8) produce no new boundary conditions.

(b) Displacement and Mixed Boundary Conditions

Expanding (5.4.7) and (5.4.8), we have

$$\rho^+(\boldsymbol{v}^+ - \boldsymbol{u}) \cdot \boldsymbol{n} = \rho(\boldsymbol{v} - \boldsymbol{u}) \cdot \boldsymbol{n} \equiv P, \qquad (j_{pl}^+ - j_{pl})P = 0. \qquad (5.4.14)$$

Using these in (5.4.9), (5.4.10), and (5.4.11) we obtain

$$[\![t_{kl}]\!]\, n_k - [\![v_l]\!]\, P = 0,$$
$$[\![m_{kl}]\!]\, n_k - j_{pl}\, [\![\nu_p]\!]\, P = 0, \qquad (5.4.15)$$
$$(Pv_l^+ - t_{kl}^- n_k)\, [\![v_l]\!] + (Pj_{kl}\nu_k + m_{kl}^- n_k)\, [\![\nu_l]\!] + [\![q_k]\!]\, n_k$$
$$- [\![\epsilon]\!]\, P - \tfrac{1}{2}P(v^+ + v^-)\, [\![\mathbf{v}]\!] - \tfrac{1}{2}Pj_{ij}^+(\nu_j^+ + \nu_j^-)\, [\![\nu_i]\!] = 0. \qquad (5.4.16)$$

These constitute a set of *mixed boundary conditions*. If we assume that surface tractions are in equilibrium, then the first terms in (5.4.15) vanish, $[\![\mathbf{q}]\!] \cdot \mathbf{n} = 0$ and for $P \neq 0$ we obtain

$$\boldsymbol{v} = \boldsymbol{v}^+, \qquad \boldsymbol{\nu} = \boldsymbol{\nu}^+, \qquad \epsilon = \epsilon^+. \qquad (5.4.17)$$

The displacement conditions result by integrating (5.4.17) with respect to time and using the initial conditions on the displacement fields. we note that the continuity of j_{pl} also follows from (5.4.14) with $P \neq 0$.

There are other sets of admissible boundary conditions. A full discussion requires the proof of the existence and uniqueness theorems. For example, in heat conduction problems, another important boundary condition, replacing (5.4.13)$_3$, is the *radiation condition*

$$\boldsymbol{q} \cdot \boldsymbol{n} + r\, (\theta - \theta_1) = 0, \qquad (5.4.18)$$

where r is a function of the difference $\theta - \theta_1$, between the temperature on the surface of the body and outside it.

(c) Mixed Displacement and Traction Boundary Conditions

A large class of boundary-value problems are covered by considering that the displacement, rotation, and temperature fields are prescribed on some parts of the surface of the body and that the tractions, couples, and heat are prescribed on nonoverlapping parts.

Let $\overline{\mathcal{V}}$ denote a regular region of the Euclidean space occupied by a micropolar body whose boundary is $\partial\mathcal{V}$. The interior of $\overline{\mathcal{V}}$ is denoted by $\mathcal{V}$ and the exterior normal to $\partial\mathcal{V}$ by $\boldsymbol{n}$. Let S_i $(i = 1, 2, \ldots, 6)$ denote subsets of $\partial\mathcal{V}$ such that

$$\overline{S_1} \cup S_2 = \overline{S_3} \cup S_4 = \overline{S_5} \cup S_6 = \partial\mathcal{V},$$
$$S_1 \cap S_2 = S_3 \cap S_4 = S_5 \cap S_6 = 0. \qquad (5.4.19)$$

The boundary condition on these subsurfaces, at the time interval $T^+ = [0, \infty)$ may be expressed as

$$
\begin{array}{ll}
u_k = \hat{u}_k \quad \text{on } \overline{S_1} \times T^+, & t_{kl} n_k = \hat{t}_l \quad \text{on } S_2 \times T^+, \\
\phi_k = \hat{\phi}_k \quad \text{on } \overline{S_3} \times T^+, & m_{kl} n_k = \hat{m}_l \quad \text{on } S_4 \times T^+, \quad (5.4.20) \\
T = \hat{T} \quad \text{on } \overline{S_5} \times T^+, & q_k n_k = \hat{q} \quad \text{on } S_6 \times T^+,
\end{array}
$$

where quantities carrying a carat ($\hat{}$) are prescribed.

(d) Initial Conditions

These conditions usually consist of Cauchy data, expressed by

$$
\begin{array}{ll}
\rho(\boldsymbol{x}, 0) = \rho^0(\boldsymbol{x}), & j(\boldsymbol{x}, 0) = j^0(\boldsymbol{x}), \\
\boldsymbol{u}(\boldsymbol{x}, 0) = \boldsymbol{u}^0(\boldsymbol{x}), & \boldsymbol{v}(\boldsymbol{x}, 0) = \boldsymbol{v}^0(\boldsymbol{x}), \\
\boldsymbol{\phi}(\boldsymbol{x}, 0) = \boldsymbol{\phi}^0(\boldsymbol{x}), & \boldsymbol{\nu}(\boldsymbol{x}, 0) = \boldsymbol{\nu}^0(\boldsymbol{x}) \quad \text{in } \bar{\mathcal{V}}, \quad (5.4.21) \\
T(\boldsymbol{x}, 0) = T^0(\boldsymbol{x}), &
\end{array}
$$

where quantities carrying a superscript (^{0}) are prescribed throughout $\mathcal{V}$. Clearly, other possibilities exist.

(iii) Constitutive Equations

Constitutive equations of micropolar thermoelasticity were obtained in Section 3.3. For the nonlinear anisotropic solids, these are of the form

$$
\rho_0 \psi = \Sigma(\overline{\mathfrak{C}}, \boldsymbol{\Gamma}, \theta, \boldsymbol{X}),
$$

$$
\eta = -\frac{1}{\rho_0} \frac{\partial \Sigma}{\partial \theta},
$$

$$
t_{kl} = \frac{\rho}{\rho_0} \frac{\partial \Sigma}{\partial \overline{\mathfrak{C}}_{KL}} x_{k,K} \bar{X}_{lL},
$$

$$
m_{kl} = \frac{\rho}{\rho_0} \frac{\partial \Sigma}{\partial \Gamma_{LK}} x_{k,K} \bar{X}_{lL},
$$

$$
q_k = Q_K(\overline{\mathfrak{C}}, \boldsymbol{\Gamma}, \nabla \theta, \theta, \boldsymbol{X}) x_{k,K},
$$

$$
\boldsymbol{Q} \equiv \frac{\partial \Phi}{\partial (\nabla \theta / \theta)}, \quad Q_K \theta_{,K} \geq 0. \quad (5.4.22)
$$

For the isotropic solids, Σ is a function of the joint invariants of $\overline{\mathfrak{C}}$ and $\boldsymbol{\Gamma}$.

(iv) Kinematical Relations

$$
\begin{array}{ll}
\boldsymbol{v} = \dot{\boldsymbol{x}}(\boldsymbol{X}, t), & \dot{\boldsymbol{v}} = \dfrac{\partial \boldsymbol{v}}{\partial t} + \boldsymbol{v}_{,k} v_k, \\[2mm]
\nu_k = -\frac{1}{2} \epsilon_{klm} \dot{\bar{X}}_{lK} \bar{X}_{mK}, & \dot{\boldsymbol{\nu}} = \dfrac{\partial \boldsymbol{\nu}}{\partial t} + \boldsymbol{\nu}_{,k} v_k, \\[2mm]
\sigma_k = j_{kl} \dot{\nu}_l + \epsilon_{klm} j_{mn} \nu_l \nu_n, & \overline{\mathfrak{C}}_{KL} = x_{k,K} \chi_{kL}, \\[2mm]
\Gamma_{KL} = \frac{1}{2} \epsilon_{KMN} \bar{X}_{kM,L} \bar{X}_{KN}. &
\end{array}
\quad (5.4.23)
$$

The solution of problems in nonlinear dynamical micropolar thermoelasticity requires the determination of $\rho(\boldsymbol{X},t)$, $\boldsymbol{j}(\boldsymbol{X},t)$, $\boldsymbol{x}(\boldsymbol{X},t)$, $\chi_{kK}(\boldsymbol{X},t)$, and $\theta(\boldsymbol{X},t)$, by solving (5.4.1)–(5.4.5), (5.4.22), (5.4.23) under the boundary conditions (5.4.20) and initial conditions (5.4.21). Clearly, this is a major task and approximations are necessary for nontrivial problems. Moreover, in the absence of existence and uniqueness theorems, it cannot be assured that the problem is *well-posed*.

5.5 Mixed Boundary-Initial Value Problems in Linear Theory

A. Formulation in Terms of $\boldsymbol{u}$, $\boldsymbol{\phi}$, and T Fields

The basic equations of the boundary-initial value problems of the linear theory of anisotropic, micropolar thermoelastodynamics consist of equations of balance, constitutive equations, kinematical relations, boundary, and initial conditions. When combined, these equations lead to field equations for $\boldsymbol{u}$, $\boldsymbol{\phi}$, and T fields (see below).

Equations of balance

$$t_{kl,k} + \rho\left(f_l - \frac{\partial^2 u_l}{\partial t^2}\right) = 0 \quad \text{in } \overline{V} \times T^+, \qquad (5.5.1)$$

$$m_{kl,k} + \epsilon_{lkm}t_{km} + \rho\left(l_l - j_{lk}\frac{\partial^2 \phi_k}{\partial t^2}\right) = 0 \quad \text{in } \overline{V} \times T^+, \qquad (5.5.2)$$

$$\rho T_0 \frac{\partial \eta}{\partial t} - q_{k,k} - \rho h = 0 \quad \text{in } \overline{V} \times T^+. \qquad (5.5.3)$$

Note that, in linear theory, no need arises for the equations of conservation of mass and microinertia. They are considered to be given properties of the body at the natural state.

Constitutive equations

$$t_{kl} = -A_{kl}T + A_{klmn}\epsilon_{mn} + C_{klmn}\gamma_{mn},$$
$$m_{kl} = -B_{lk}T + C_{mnlk}\epsilon_{mn} + B_{lkmn}\gamma_{mn},$$
$$\eta = \eta_0 + \frac{C_0}{T_0}T + \frac{1}{\rho}A_{kl}\epsilon_{kl} + \frac{1}{\rho}B_{kl}\gamma_{kl},$$
$$q_k = \frac{K_{kl}}{T_0}T_{,l}. \qquad (5.5.4)$$

Kinematical relations (strain measures)

$$\epsilon_{kl} = u_{l,k} + \epsilon_{lkm}\phi_m, \qquad \gamma_{kl} = \phi_{k,l}. \qquad (5.5.5)$$

These systems of equations are subject to the following boundary and initial conditions:

Boundary conditions

$$\begin{aligned}
\boldsymbol{u} &= \hat{\boldsymbol{u}} \quad \text{on } \overline{S}_1 \times T^+, & \boldsymbol{t} &\equiv t_{kl}n_k i_l = \hat{\boldsymbol{t}} \quad \text{on } S_2 \times T^+, \\
\boldsymbol{\phi} &= \hat{\boldsymbol{\phi}} \quad \text{on } \overline{S}_3 \times T^+, & \boldsymbol{m} &\equiv m_{kl}n_k i_l = \hat{\boldsymbol{m}} \quad \text{on } S_4 \times T^+, \quad (5.5.6) \\
T &= \hat{T} \quad \text{on } \overline{S}_5 \times T^+, & q &= \mathbf{q} \cdot \mathbf{n} = \hat{q} \quad \text{on } S_6 \times T^+,
\end{aligned}$$

where $\hat{\boldsymbol{u}}, \hat{\boldsymbol{t}}, \hat{\boldsymbol{\phi}}, \hat{\boldsymbol{m}}, \hat{T}$, and $\hat{q}$ are prescribed on surfaces S_1 to S_6, respectively. S_i ($i = 1, 2, \ldots, 6$) are subsurfaces of ∂V of the body, as described by (5.4.19).

Initial conditions

$$\begin{aligned}
\boldsymbol{u}(\boldsymbol{x},0) &= \boldsymbol{u}^0(\boldsymbol{x}), & \dot{u}(\boldsymbol{x},0) &= \boldsymbol{v}^0(\boldsymbol{x}), \\
\boldsymbol{\phi}(\boldsymbol{x},0) &= \boldsymbol{\phi}^0(\boldsymbol{x}), & \dot{\boldsymbol{\phi}}(\boldsymbol{x},0) &= \boldsymbol{\nu}^0(\boldsymbol{x}), \qquad (5.5.7) \\
T(\boldsymbol{x},0) &= T^0(\boldsymbol{x}), & \boldsymbol{x} &\in \overline{V},
\end{aligned}$$

where $\boldsymbol{u}^0, \boldsymbol{v}^0, \boldsymbol{\phi}^0, \boldsymbol{\nu}^0$, and T^0 are prescribed.

The following continuity requirements are assumed:
(a)
$$\{\boldsymbol{u}(\boldsymbol{x},t), \boldsymbol{\phi}(\boldsymbol{x},t)\} \in C^{1,2},$$
$$\{\boldsymbol{\epsilon}(\boldsymbol{x},t), \boldsymbol{\gamma}(\boldsymbol{x},t), \boldsymbol{f}(\boldsymbol{x},t), \boldsymbol{l}(\boldsymbol{x},t), h(\boldsymbol{x},t)\} \in C^{0,0},$$
$$\{\boldsymbol{t}(\boldsymbol{x},t), \boldsymbol{m}(\boldsymbol{x},t), \boldsymbol{q}(\boldsymbol{x},t)\} \in C^{1,0},$$
$$T(\boldsymbol{x},t) \in C^{1,0}, \eta(\boldsymbol{x},t) \in C^{0,1} \quad \text{in } \overline{V} \times T^+, \quad (5.5.8\text{-a})$$
(b)
$$\{\hat{\boldsymbol{u}}, \hat{\boldsymbol{t}}, \hat{\boldsymbol{\phi}}, \hat{\boldsymbol{m}}, \hat{T}, \hat{q}\} \in C^{0,0} \quad \text{on } \partial V \times T^+, \qquad (5.5.8\text{-b})$$
(c)
$$\{\boldsymbol{u}^0, \boldsymbol{v}^0, \boldsymbol{\phi}^0, \boldsymbol{\nu}^0, T^0\} \in C^{0,} \quad \text{on } \overline{V}, \qquad (5.5.8\text{-c})$$
(d)
$$\{A_{kl}, B_{kl}, A_{klmn}, B_{klmn}, C_{klmn}, C_0, K_{kl}, \rho, j_{kl}\} \in C^0, \qquad (5.5.8\text{-d})$$
(e)
$$A_{klmn} = A_{mnkl}, \quad B_{klmn} = B_{mnkl}, \quad K_{kl} = K_{lk}, \quad j_{kl} = j_{lk}, \quad C_0 \geq 0, \qquad (5.5.8\text{-e})$$
(f)
$$A_{klmn}\epsilon_{kl}\epsilon_{mn} + B_{klmn}\gamma_{kl}\gamma_{mn} + 2C_{klmn}\epsilon_{kl}\gamma_{mn} \geq 0, \quad K_{kl}T_{,k}T_{,l} \geq 0. \qquad (5.5.8\text{-f})$$

We use the symbol $C^{i,j}$ to denote the class of functions whose space partial derivatives of order, up to and including i, and whose time derivatives, up to and including j, are continuous.[3]

Definition 1. The collection $S\{u, \phi; \epsilon, \gamma; t, m; q, T\}$ of the ordered set of functions $u, \phi, \epsilon, \gamma, t, m, q$, and T whose continuity requirements are described by (a), is called an admissible state. S is considered to be a linear function space, i.e.,

$$S_1 + S_2 = \{u_1 + u_2, \phi_1 + \phi_2; \epsilon_1 + \epsilon_2; \gamma_1 + \gamma_2; t_1 + t_2, m_1 + m_2,$$
$$q_1 + q_2, T_1 + T_2\},$$
$$\alpha S = \{\alpha u, \alpha \phi, \alpha \epsilon, \alpha \gamma, \alpha t, \alpha m, \alpha q, \alpha T\}. \tag{5.5.9}$$

Admissible state S may be restricted by making additional requirements on the functions. For example, if S meets constitutive equations (5.5.4) and the strain displacement requirements (5.5.5), we say that S is *kinematically admissible*. If a kinematically admissible state meets the boundary and initial conditions and satisfies the equations of motion (5.5.1)–(5.5.3), we call it the *solution of the mixed problem*.

B. Field Equations

Upon substituting (5.5.4) into (5.5.1)–(5.5.3), we obtain a set of coupled, partial differential equations for displacement field u, microrotation field ϕ, and the temperature T

$$[-A_{kl}T + A_{klmn}(u_{n,m} + \epsilon_{nmr}\phi_r) + C_{klmn}\phi_{m,n}]_{,k} + \rho(f_l, -\ddot{u}_l) = 0, \tag{5.5.10}$$

$$[-B_{lk}T + B_{lkmn}\phi_{m,n} + C_{mnlk}(u_{n,m} + \epsilon_{nmr}\phi_r)]_{,k}$$
$$+\epsilon_{lmn}[-A_{mn}T + A_{mnpq}(u_{q,p} + \epsilon_{qpr}\phi_r) + C_{mnpq}\phi_{p,q}]$$
$$+ (\rho l_l - \rho j_{kl}\ddot{\phi}_k) = 0, \tag{5.5.11}$$

$$\rho C_0 \dot{T} + T_0 A_{kl}(\dot{u}_{l,k} + \epsilon_{lkr}\dot{\phi}_r) + T_0 B_{kl}\dot{\phi}_{k,l} - \frac{1}{T_0}(K_{kl}T_{,l})_{,k} - \rho h = 0. \tag{5.5.12}$$

In linear theory, for the time rates, we have

$$\dot{T} = \frac{\partial T}{\partial t}, \quad \dot{u}_{k,l} = \frac{\partial^2 u_k}{\partial x_l \, \partial t}, \quad \text{etc.}$$

The corresponding field equations for the homogeneous and isotropic solids are

$$(\lambda + \mu)u_{l,lk} + (\mu + \kappa)u_{k,ll} + \kappa\epsilon_{klm}\phi_{m,l} - \beta_0 T_{,k} + \rho(f_k - \ddot{u}_k) = 0, \tag{5.5.13}$$

---

[3]In the rest of the book, unless it becomes critical, we shall not always indicate such tiresome, formal continuity requirements. It should be understood that such continuity exists to any order as demanded by the existing expressions, unless otherwise stated.

$$(\alpha + \beta)\phi_{l,lk} + \gamma\phi_{k,ll} + \kappa\epsilon_{klm}u_{m,l} - 2\kappa\phi_k + \rho(l_k - j_{kl}\ddot{\phi}_l) = 0, \quad (5.5.14)$$

$$\rho C_0 \dot{T} + \beta_0 T_0 \dot{u}_{k,k} - K\nabla^2 T - \rho h = 0. \quad (5.5.15)$$

Any boundary-initial value problem in the linear theory of an anisotropic micropolar thermoelastic solid requires the solutions (5.5.10)–(5.5.12) under the boundary and initial conditions. For isotropic solids, it requires the solution of (5.5.13)–(5.5.15).

C. Formulation by Means of Convolution

Definition 2 (Convolution). Let ϕ and ψ be

$$\{\phi(\boldsymbol{x},t), \psi(\boldsymbol{x},t)\} \in C^{0,0} \qquad \text{in } \overline{V} \times T^+,$$

then the function $\theta(\boldsymbol{x},t)$ defined by

$$\theta(\boldsymbol{x},t) = \phi * \psi = \int_0^t \phi(\boldsymbol{x}, t - \tau)\psi(\boldsymbol{x}, \tau)\, d\tau \qquad \text{for all} \quad (\boldsymbol{x},t) \in \overline{V} \times T^+$$
$$(5.5.16)$$

is called the convolution of ϕ and ψ, and is denoted by $\theta = \phi * \psi$.

The elementary properties of the convolution can be deduced from its definition (5.5.16):

(a) *Commutative property: $\phi * \psi = \psi * \phi$;*

(b) *Associative property: $\phi * (\psi * \omega) = (\phi * \psi) * \omega = \phi * \psi * \omega$;*

(c) *Distributive property: $\phi * (\psi + \omega) = \phi * \psi + \phi * \omega$;*

(d) *Titchmarsh's theorem: $\phi * \psi = 0$ on $\overline{V} \times T^+$ implies that either $\phi = 0$ or $\psi = 0$ on $\bar{\nu} \times T^+$.*

We can also show that

$$1 * \phi = \int_0^t \phi\, dt,$$
$$t * \ddot{\phi} = \phi - t\dot{\phi}(\boldsymbol{x}, 0) - \phi(\boldsymbol{x}, 0). \qquad (5.5.17)$$

We now employ the concept of convolution to give an alternative formulation for subsections A and B.

Theorem 1. Let $(\boldsymbol{u}, \boldsymbol{\phi}) \in C^{0,2}, (t_{kl}, m_{kl}) \in C^{1,0}$, and $\eta \in C^{0,1}$. Then $\boldsymbol{u}, \boldsymbol{\phi}, \eta, t_{kl}, m_{kl}$, and q_k satisfy the equations of motion and the initial conditions if and only if

$$t * t_{kl,k} + \rho\tilde{f}_l = \rho u_l, \qquad (5.5.18)$$
$$t * (m_{kl,k} + \epsilon_{lkm}t_{km}) + \rho\tilde{l}_l = \rho j_{kl}\phi_l, \qquad (5.5.19)$$
$$\rho T_o(\eta - \eta_0) - 1 * (q_{k,k} + \rho h) = 0, \qquad (5.5.20)$$

where

$$\tilde{f} = t * f(x, t) + tv^0(x) + u^0(x),$$
$$\tilde{l} = t * l(x, t) + tv^0(x) + \phi^0(x). \qquad (5.5.21)$$

The proof is immediate. The fact that (5.5.18)–(5.5.20) hold, is evident from (5.5.1)–(5.5.3). Conversely, if (5.5.18)–(5.5.20) hold, differentiating twice with respect to time, (5.5.18)–(5.5.20), we obtain (5.5.1)–(5.5.3). The initial conditions are also recovered in this process. This theorem was given by Ignaczak [1963] for the case of classical elastodynamics.

With this theorem, then, the initial conditions on u and ϕ are incorporated into $\tilde{f}$ and $\tilde{l}$. The initial value of η is taken care of with η^0. Thus:

Theorem 2 (Equivalence).   The mixed boundary-initial value problems of the micropolar thermoelastodynamics formulated in subsection A are equivalent to the systems (5.5.18)–(5.5.20), constitutive equations (5.5.4), kinematical relations (5.5.5), and the boundary conditions (5.5.6).

It is simple to see that the alternative formulation given in this subsection is equivalent to the basic equations given in subsection B.

D. Plain Strain

Definition 3. The state of strain at $x_3 = $ constant-planes is called the plain strain, if all field quantities are independent of x_3 and

$$u_3 = \phi_\alpha = 0, \qquad f_3 = l_\alpha = 0,$$
$$t_{(n)3} = m_{(n)\alpha} = q_3 = 0, \qquad \alpha = 1, 2. \qquad (5.5.22)$$

Consequently, u_α, $\phi_3 = \phi$, f_α, $l_3 = l$, $t_{(n)\alpha}$, $m_{(n)3}$, and q_α depend on x_1, x_2, and t only. With these, the basic equations of isotropic micropolar bodies reduce to:

(a) Balance laws:

$$t_{\beta\alpha,\beta} + \rho(f_\alpha - \ddot{u}_\alpha) = 0 \qquad (\alpha, \beta = 1, 2),$$
$$m_{\beta3,\beta} + \epsilon_{\alpha\beta3}t_{\alpha\beta} + \rho(l_3 - j\ddot{\phi}) = 0,$$
$$\rho T_0\dot{\eta} - q_{\alpha,\alpha} - \rho h = 0 \quad \text{in } \overline{V} * T^+. \qquad (5.5.23)$$

(b) Constitutive equations:

$$t_{\beta\alpha} = -\beta_0 T\delta_{\beta\alpha} + \lambda\epsilon_{\rho\rho}\delta_{\beta\alpha} + (\mu + \kappa)\epsilon_{\beta\alpha} + \mu\epsilon_{\alpha\beta},$$
$$m_{\beta3} = m_\beta = \gamma\phi_{,\beta}, \quad q_\alpha = KT_{,\alpha}, \quad \eta = \eta_0 + \frac{C_0}{T_0}T + \frac{\beta_0}{\rho}\epsilon_{\alpha\alpha}. \quad (5.5.24)$$

(c) Strain tensor:

$$\epsilon_{\beta\alpha} = u_{\alpha,\beta} + \epsilon_{\alpha\beta3}\phi. \qquad (5.5.25)$$

(d) Boundary conditions (mixed):

$$u_\alpha = \hat{u}_\alpha(x_1, x_2, t) \quad \text{on } \overline{S}_1 \times T^+, \qquad t_{\beta\alpha}n_\beta = \hat{t}_\alpha(x_1, x_2, t) \quad \text{on } S_2 \times T^+,$$

$$\phi_\alpha = \hat{\phi}(x_1, x_2, t) \quad \text{on } \overline{S}_3 \times T^+, \qquad m_{\alpha 3}n_\alpha = \hat{m}(x_1, x_2, t) \quad \text{on} S_4 \times T^+,$$

$$T = \hat{T}(x_1, x_2, t) \quad \text{on } \overline{S}_5 \times T^+, \qquad q_\alpha n_\alpha = \hat{q}(x_1, x_2, t) \quad \text{on} S_6 \times T^+,$$

$$(5.5.26)$$

where $\hat{u}_\alpha$, $\hat{t}_\alpha$, $\hat{\phi}$, $\hat{m}$, $\hat{T}$, and $\hat{q}$ are prescribed on S_1 to S_6. S_1 to S_6 are pieces of contour, of a cross-section $x_3 = $ const.-plane, as described by (5.4.19).

(e) Initial conditions (in $\mathcal{V}$, $t = 0$):

$$u_\alpha(x_1, x_2, 0) = u_\alpha^0(x_1, x_2), \qquad \dot{u}_\alpha(x_1, x_2, 0) = v_\alpha^0(x_1, x_2),$$

$$\phi(x_1, x_2, 0) = \phi_\alpha^0(x_1, x_2), \qquad \dot{\phi}_\alpha(x_1, x_2, 0) = \nu_\alpha^0(x_1, x_2),$$

$$T(x_1, x_2, 0) = T^0(x_1, x_2), \qquad (5.5.27)$$

where quantities carrying a superscript $(^0)$ are prescribed throughout $\mathcal{V}$ at $t = 0$.

(f) Field equations: Combining (5.5.23) and (5.5.24), we obtain the field equations for the plane strain

$$-\beta T_{,\alpha} + (\lambda + \mu)u_{\beta,\beta\alpha} + (\mu + \kappa)u_{\alpha,\beta\beta} + \kappa\epsilon_{\alpha\beta 3}\phi_{,\beta} + \rho(f_\alpha - \ddot{u}_\alpha)z = 0,$$

$$\gamma\phi_{,\beta\beta} - 2\kappa\phi + \kappa\epsilon_{\alpha\beta 3}u_{\beta,\alpha} + \rho(l - j\ddot{\phi}) = 0,$$

$$\rho C_0 \dot{T} + \beta_0 T_0 \dot{u}_{\alpha,\alpha} - K T_{,\alpha\alpha} - \rho h = 0.$$

$$(5.5.28)$$

E. Plane Stress

Definition 4. The state of stress at $x_3 = $ const.-plane is called plane stress, if all field quantities are independent of x_3, and

$$t_{\alpha 3} = t_{3\alpha} = t_{33} = 0, \quad m_{33} = m_{\alpha\beta} = 0, \qquad \alpha, \beta = 1, 2$$

$$f_3 = l_\alpha = 0, \quad t_{(\mathbf{n})3} = m_{(\mathbf{n})\alpha} = q_3 = 0. \qquad (5.5.29)$$

As a consequence of constitutive equations and balance laws, it follows that u_α, ϕ_3, f_α, l_3, $t_{(\mathbf{n})\alpha}$, $m_{(\mathbf{n})3}$, and q_α depend on x_1, x_2, and t only. Balance equations are of the form (5.5.23). However, in constitutive equations (5.5.24) β_0, λ, and α are to be replaced by $\overline{\beta}_0$, $\overline{\lambda}$, and $\overline{\alpha}$, respectively.

$$\overline{\beta}_0 = \beta_0 \frac{2\mu + \kappa}{\lambda + 2\mu + \kappa}, \quad \overline{\alpha} = \alpha \frac{\beta + \gamma}{\alpha + \beta + \gamma}, \quad \overline{\lambda} = \lambda \frac{2\mu + \kappa}{\lambda + 2\mu + \kappa}. \quad (5.5.30)$$

These results follow by setting $t_{33} = 0$ and $m_{33} = 0$ which give

$$\epsilon_{33} = -\frac{\lambda}{\lambda + 2\mu + \kappa}\epsilon_{\alpha\alpha} + \frac{\beta_0}{\lambda + 2\mu + \kappa}T, \quad \gamma_{33} = -\frac{\alpha}{\alpha + \beta + \gamma}\gamma_{\alpha\alpha}. \quad (5.5.31)$$

When ϵ_{33} and γ_{33} are substituted into the stress constitutive equations, it will show that (5.5.30) is valid. The rest of the constitutive equations (5.5.24) do not change.

F. Generalized Plane Strain

Definition 5. The state of strain is called generalized plane strain if displacement vector $\boldsymbol{u}$, the microrotation vector $\boldsymbol{\phi}$, and temperature T are functions of x_1, x_2, and t, i.e.,

$$u_i = u_i(x_1, x_2, t), \qquad \phi_i = \phi_i(x_1, x_2, t), \qquad T = T(x_1, x_2, t). \qquad (5.5.32)$$

The equations of motion now reduce to

$$t_{\alpha i,\alpha} + \rho(f_i - \ddot{u}_i) = 0, \qquad m_{\alpha i,\alpha} + \epsilon_{ijk}t_{jk} + \rho(l_i - j\ddot{\phi}_i) = 0,$$
$$\rho T_0\dot{\eta} - q_{\alpha,\alpha} - \rho h = 0. \qquad (5.5.33)$$

Constitutive equations follow from (5.3.3) and (5.5.32).

5.6   Curvilinear Coordinates

For the discussion of problems relevant to bodies with curved surfaces, curvilinear coordinates provide convenience. The passage to curvilinear coordinates can be made by some simple interpretations: let g_{kl} be the metric tensor in the curvilinear coordinates x^k. The square of the element of arc length is given by

$$(ds)^2 = g_{kl}(\boldsymbol{x})\, dx^k\, dx^l. \qquad (5.6.1)$$

We observe two simple rules:

(a) the partial differentiation symbol (,) must be replaced by the covariant differentiation sign (;); and

(b) the repeated indices must be placed in a diagonal position.

By using these rules, (5.5.1)–(5.5.3) can be expressed as

$$t^{kl}{}_{;k} + \rho(f^l - \ddot{u}^l) = 0, \qquad (5.6.2)$$
$$m^{kl}{}_{;k} + \epsilon^{lkm}t_{km} + \rho(l^l - j^{lk}\ddot{\phi}_k) = 0, \qquad (5.6.3)$$
$$\rho T_0\dot{\eta} - q^k{}_{;k} - \rho h = 0. \qquad (5.6.4)$$

Naturally, in curvilinear coordinates, covariant, contravariant, mixed forms of the same tensor occur, e.g., t^{kl}, t_{kl}, $t^k{}_l$, and $t_l{}^k$ for the stress tensor. Also the semicolon denotes the covariant partial differentiation. For example,

$$t^{kl}{}_{;m} = t^{kl}{}_{,m} + \{{}^k_{mr}\}t^{rl} + \{{}^l_{mr}\}t^{kr},$$
$$q^k{}_{;l} = q^k{}_{,l} + \{{}^k_{lm}\}q^m,$$
$$u_{k;l} = u_{k,l} - \{{}^m_{kl}\}u_m, \qquad (5.6.5)$$

where $\{_{kl}^m\}$ is the Christoffel symbol of the second kind, defined by

$$\{_{kl}^m\} \equiv \tfrac{1}{2} g^{mn} \left(\frac{\partial g_{kn}}{\partial x^l} + \frac{\partial g_{ln}}{\partial x^k} - \frac{\partial g_{kl}}{\partial x^n} \right),$$
$$g^{mn} = (\text{Cofactor of } g_{mn}) / \det(g_{mn}). \tag{5.6.6}$$

The linear strain measures are expressed as

$$\epsilon_{kl} = u_{l;k} + \epsilon_{lkr}\phi^r, \qquad \gamma_{kl} = \phi_{k;l}. \tag{5.6.7}$$

Often these equations are expressed in terms of the physical components of the vectors and tensors involved. The physical components $t^{(k)}{}_{(l)}$ and $u^{(k)}$ of $t^k{}_l$ and u^k are related to each other by

$$t^k{}_l = t^{(k)}{}_{(l)} \sqrt{g_{\underline{ll}}/g_{\underline{kk}}}, \qquad u^k = u^{(k)}/\sqrt{g_{\underline{kk}}}, \tag{5.6.8}$$

where underscored indices are not summed. For a useful short account on tensor calculus, see Eringen [1962, Appendix, and 1971a].

We give below the explicit expressions of the balance laws (5.6.2)–(5.6.4) and the strain measures (5.6.7).

A. Orthogonal Curvilinear Coordinates

In this case, $g_{kl} = 0$ when $k \neq l$ and

$$(ds)^2 = g_{11}(dx^1)^2 + g_{22}(dx^2)^2 + g_{33}(dx^3)^2,$$
$$g^{\underline{kk}} = 1/g_{\underline{kk}}, \qquad g \equiv \det g_{kl} = g_{11}g_{22}g_{33},$$
$$\{_{lm}^k\} = \frac{1}{2g_{\underline{kk}}} \left(\frac{\partial g_{kk}}{\partial x^m}\delta_{kl} + \frac{\partial g_{mm}}{\partial x^l}\delta_{km} - \frac{\partial g_{ll}}{\partial x^k}\delta_{lm} \right). \tag{5.6.9}$$

Using these, we have

$$\sum_{k=1}^{3} \left\{ \frac{1}{\sqrt{g}} \frac{\partial}{\partial x^k} \left[t^{(k)}{}_{(l)} \frac{\sqrt{g}}{g_{kk}} \right] + \frac{1}{\sqrt{g_{kk}g_{\underline{ll}}}} \frac{\partial \sqrt{g_{\underline{ll}}}}{\partial x^k} t^{(l)}{}_{(k)} - \frac{1}{\sqrt{g_{kk}g_{\underline{ll}}}} \frac{\partial \sqrt{g_{kk}}}{\partial x^l} t^{(k)}{}_{(k)} \right\}$$
$$+ \rho(f^{(l)} - \ddot{u}^{(l)}) = 0, \tag{5.6.10}$$

$$\sum_{k=1}^{3} \left\{ \frac{1}{\sqrt{g}} \frac{\partial}{\partial x^k} \left[m^{(k)}{}_{(l)} \frac{\sqrt{g}}{g_{kk}} \right] + \frac{1}{\sqrt{g_{kk}g_{\underline{ll}}}} \frac{\partial \sqrt{g_{\underline{ll}}}}{\partial x^k} m^{(l)}{}_{(k)} - \frac{1}{\sqrt{g_{kk}g_{\underline{ll}}}} \frac{\partial \sqrt{g_{kk}}}{\partial x^l} m^{(k)}{}_{(k)} \right\}$$
$$+ \sum_{k=1}^{3}\sum_{m=1}^{3} \left[\epsilon_{lkm} \frac{\sqrt{g}}{\sqrt{g_{kk}g_{\underline{ll}}g_{mm}}} t^{(k)}{}_{(m)} \right] + \rho \left[l^{(l)} - \sum_{k=1}^{3} j^{(l)}{}_{(k)}\ddot{\phi}^{(k)} \right] = 0, \tag{5.6.11}$$

$$\rho T_0 \dot{\eta} - \sum_{k=1}^{3} \frac{1}{\sqrt{g}} \frac{\partial}{\partial x^k} \left(\frac{\sqrt{g}q^{(k)}}{\sqrt{g_{kk}}} \right) - \rho h = 0 \tag{5.6.12}$$

Physical components of the linear strain measures are given by

$$
\begin{aligned}
\epsilon^{(k)}{}_{(l)} = {} & \frac{1}{\sqrt{g_{kk}g_{\underline{ll}}}}\left[\frac{\partial}{\partial x^k}\left(v^{(l)}\sqrt{g_{\underline{ll}}}\right) - \frac{\partial\sqrt{g_{\underline{ll}}}}{\partial x^k}v^{(l)} - \frac{\partial\sqrt{g_{kk}}}{\partial x^l}v^{(k)}\right] \\
& + \sum_{r=1}^{3}\left[\frac{1}{\sqrt{g_{kk}g_{rr}}}\frac{\partial\sqrt{g_{\underline{ll}}}}{\partial x^r}v^{(r)}\delta_{kl} + \frac{\sqrt{g}}{\sqrt{g_{kk}g_{\underline{ll}}g_{rr}}}\epsilon_{lkr}\phi^{(r)}\right],
\end{aligned}
$$

$$
\begin{aligned}
\gamma^{(k)}{}_{(l)} = {} & \frac{1}{\sqrt{g_{kk}g_{\underline{ll}}}}\left[\frac{\partial}{\partial x^l}\left(\phi^{(k)}\sqrt{g_{kk}}\right) - \frac{\partial\sqrt{g_{kk}}}{\partial x^l}\phi^{(k)} - \frac{\partial\sqrt{g_{\underline{ll}}}}{\partial x^k}\phi^{(l)}\right] \\
& + \sum_{r=1}^{3}\frac{\sqrt{g_{rr}}}{\sqrt{g_{\underline{ll}}}}\frac{\partial\sqrt{g_{kk}}}{\partial x^r}\phi^{(r)}\delta_{kl}.
\end{aligned}
\tag{5.6.13}
$$

Constitutive equations, as usual, are given by (5.5.4), with replacement of t_{kl}, m_{kl}, q_k, ϵ_{mn}, and γ_{mn} by their physical components. The material moduli remain unchanged, since they are physical quantities.

The physical components of surface tractions and couples are given by

$$
t^{(k)} = t^{(l)}{}_{(k)}n^{(l)}, \qquad m^{(k)} = m^{(l)}{}_{(k)}n^{(l)} \quad \text{on } \partial\mathcal{V},
\tag{5.6.14}
$$

where $n^{(l)}$ is the physical component of the unit normal to the surface $\partial\mathcal{V}$. Equations (5.6.10)–(5.6.13) and the constitutive equations are all that are needed to express the basic equations of the linear micropolar thermoelastic solids in curvilinear coordinates. The passage from the physical components to tensor components is made by (5.6.8).

The field equations of the homogenous isotropic solids, in curvilinear coordinates, can be expressed in vectorial form by also observing the vector identities

$$
u_{k;}{}^{l}{}_{l} = -(\nabla \times \nabla \times \boldsymbol{u})_k + (\nabla\nabla \cdot \boldsymbol{u})_k, \quad u^{l}{}_{;kl} = (\nabla\nabla \cdot \boldsymbol{u})_k,
\tag{5.6.15}
$$

where ∇ is the gradient operator. Using these identities, (5.5.13)–(5.5.15) may be expressed in vectorial forms

$$
(\lambda + 2\mu + \kappa)\nabla\nabla \cdot \boldsymbol{u} - (\mu + \kappa)\nabla \times \nabla \times \boldsymbol{u}
$$
$$
+ \kappa\nabla \times \boldsymbol{\phi} - \beta_0\nabla T + \rho(\boldsymbol{f} - \ddot{\boldsymbol{u}}) = \boldsymbol{0},
\tag{5.6.16}
$$
$$
(\alpha + \beta + \gamma)\nabla\nabla \cdot \boldsymbol{\phi} - \gamma\nabla \times \nabla \times \boldsymbol{\phi}
$$
$$
+ \kappa\nabla \times \boldsymbol{u} - 2\kappa\boldsymbol{\phi} + \rho(\boldsymbol{l} - \boldsymbol{j} \cdot \ddot{\boldsymbol{\phi}}) = \boldsymbol{0},
\tag{5.6.17}
$$
$$
\rho C_0\dot{T} + \beta_0 T_0\nabla \cdot \dot{\boldsymbol{u}} - \nabla \cdot (K\nabla T) - \rho h = 0.
\tag{5.6.18}
$$

The surface tractions and couples have the vectoral forms

$$
\boldsymbol{t}_{(n)} = (-\beta_0 T + \lambda\nabla \cdot \boldsymbol{u})\boldsymbol{n} + (2\mu + \kappa)(\boldsymbol{n} \cdot \nabla)\boldsymbol{u} + \mu\boldsymbol{n} \times \nabla \times \boldsymbol{u} + \kappa\mathbf{n} \times \boldsymbol{\phi},
$$
$$
\boldsymbol{m}_{(n)} = \alpha(\nabla \cdot \boldsymbol{\phi})\boldsymbol{n} + (\beta + \gamma)(\boldsymbol{n} \cdot \nabla)\boldsymbol{\phi} + \gamma\boldsymbol{n} \times \nabla \times \boldsymbol{\phi}.
\tag{5.6.19}
$$

Since expressions of the differential operators (gradient, divergence and curl) in curvilinear coordinates are readily available, these equations give the curvilinear forms of field equations and surface tractions and couples.

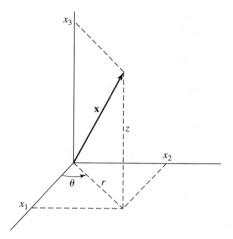

FIGURE 5.6.1.

B. *Cylindrical Coordinates* (r, θ, z) *(Figure 5.6.1):*

$$g_{11} = g_{33} = 1, \qquad g_{22} = r^2, \qquad g = r^2,$$

$$\text{grad } \phi = \frac{\partial \phi}{\partial r} e_r + \frac{1}{r} \frac{\partial \phi}{\partial \theta} e_\theta + \frac{\partial \phi}{\partial z} e_z,$$

$$\text{div } A = \frac{1}{r} \frac{\partial}{\partial r}(r A_r) + \frac{1}{r} \frac{\partial A_\theta}{\partial \theta} + \frac{\partial A_z}{\partial z},$$

$$\text{curl } A = \left(\frac{1}{r} \frac{\partial A_z}{\partial \theta} - \frac{\partial A_\theta}{\partial z} \right) e_r + \left(\frac{\partial A_r}{\partial z} - \frac{\partial A_z}{\partial r} \right) e_\theta$$

$$+ \left[\frac{1}{r} \frac{\partial}{\partial r}(r A_\theta) - \frac{1}{r} \frac{\partial A_r}{\partial \theta} \right] e_z,$$

$$\nabla^2 \phi = \frac{\partial^2 \phi}{\partial r^2} + \frac{1}{r} \frac{\partial \phi}{\partial r} + \frac{1}{r^2} \frac{\partial^2 \phi}{\partial \theta^2} + \frac{\partial^2 \phi}{\partial z^2}. \tag{5.6.20-a}$$

Equations of Motion:

$$\frac{\partial t_{rr}}{\partial r} + \frac{1}{r} \frac{\partial t_{\theta r}}{\partial \theta} + \frac{\partial t_{zr}}{\partial z} + \frac{1}{r}(t_{rr} - t_{\theta\theta}) - \beta_0 \frac{\partial T}{\partial r} + \rho(f_r - \ddot{u}_r) = 0,$$

$$\frac{\partial t_{r\theta}}{\partial r} + \frac{1}{r} \frac{\partial t_{\theta\theta}}{\partial \theta} + \frac{\partial t_{z\theta}}{\partial z} + \frac{1}{r}(t_{r\theta} + t_{\theta r}) - \frac{\beta_0}{r} \frac{\partial T}{\partial \theta} + \rho(f_\theta - \ddot{u}_\theta) = 0,$$

$$\frac{\partial t_{rz}}{\partial r} + \frac{1}{r} \frac{\partial t_{\theta z}}{\partial \theta} + \frac{\partial t_{zz}}{\partial z} + \frac{1}{r} t_{rz} - \beta_0 \frac{\partial T}{\partial z} + \rho(f_z - \ddot{u}_z) = 0,$$

$$\frac{\partial m_{rr}}{\partial r} + \frac{1}{r} \frac{\partial m_{\theta r}}{\partial \theta} + \frac{\partial m_{zr}}{\partial z} + \frac{1}{r}(m_{rr} - m_{\theta\theta}) + t_{\theta z} - t_{z\theta} + \rho(l_r - \ddot{\phi}_r) = 0,$$

$$\frac{\partial m_{r\theta}}{\partial r} + \frac{1}{r} \frac{\partial m_{\theta\theta}}{\partial \theta} + \frac{\partial m_{z\theta}}{\partial z} + \frac{1}{r}(m_{r\theta} + m_{\theta r}) + t_{zr} - t_{rz} + \rho(l_\theta - \ddot{\phi}_\theta) = 0,$$

$$\frac{\partial m_{rz}}{\partial r} + \frac{1}{r}\frac{\partial m_{\theta z}}{\partial \theta} + \frac{\partial m_{zz}}{\partial z} + \frac{1}{r}m_{rz} + t_{r\theta} - t_{\theta r} + \rho(l_z - \ddot{\phi}_z) = 0,$$

$$\rho C_0 \dot{T} + \beta_0 T_0 \left[\frac{1}{r}\frac{\partial}{\partial r}(r\dot{u}_r) + \frac{1}{r}\dot{u}_\theta + \dot{u}_z\right]$$

$$-K\left(\frac{\partial^2 T}{\partial r^2} + \frac{1}{r}\frac{\partial T}{\partial r} + \frac{1}{r^2}\frac{\partial T^2}{\partial \theta^2} + \frac{\partial^2 T}{\partial z^2}\right) - \rho h = 0.$$

$$(5.6.20\text{-b})$$

Physical Components of Strain Tensor:

$$\epsilon_{rr} = \frac{\partial u_r}{\partial r}, \qquad \epsilon_{\theta\theta} = \frac{1}{r}\left(\frac{\partial u_\theta}{\partial \theta} + u_r\right), \qquad \epsilon_{zz} = \frac{\partial u_z}{\partial z},$$

$$\epsilon_{\theta z} = \frac{1}{r}\frac{\partial u_z}{\partial \theta} - \phi_r, \quad \epsilon_{z\theta} = \frac{\partial u_\theta}{\partial z} + \phi_r, \qquad \epsilon_{rz} = \frac{\partial u_z}{\partial r} + \phi_\theta,$$

$$\epsilon_{zr} = \frac{\partial u_r}{\partial z} - \phi_\theta, \quad \epsilon_{r\theta} = \frac{\partial u_\theta}{\partial r} - \phi_z, \qquad \epsilon_{\theta r} = \frac{1}{r}\left(\frac{\partial u_r}{\partial \theta} - u_\theta\right) + \phi_z,$$

$$\gamma_{rr} = \frac{\partial \phi_r}{\partial r}, \qquad \gamma_{\theta\theta} = \frac{1}{r}\left(\frac{\partial \phi_\theta}{\partial \theta} + \phi_r\right), \quad \gamma_{zz} = \frac{\partial \phi_z}{\partial z},$$

$$\gamma_{z\theta} = \frac{1}{r}\frac{\partial \phi_z}{\partial \theta}, \qquad \gamma_{\theta z} = \frac{\partial \phi_\theta}{\partial z}, \qquad \gamma_{zr} = \frac{\partial \phi_z}{\partial r},$$

$$\gamma_{rz} = \frac{\partial \phi_r}{\partial z}, \qquad \gamma_{\theta r} = \frac{\partial \phi_\theta}{\partial r}, \qquad \gamma_{r\theta} = \frac{1}{r}\left(\frac{\partial \phi_r}{\partial \theta} - \phi_\theta\right).$$

$$(5.6.20\text{-c})$$

Constitutive Equations (Linear Isotropic Solids):

$$t_{kl} = \lambda\epsilon_{mm}\delta_{kl} + (\mu + \kappa)\epsilon_{kl} + \mu\epsilon_{lk},$$

$$m_{kl} = \alpha\gamma_{mm}\delta_{kl} + \beta\gamma_{kl} + \gamma\gamma_{lk},$$

$$\boldsymbol{q} = K\left(\frac{\partial T}{\partial r}\boldsymbol{e}_r + \frac{1}{r}\frac{\partial T}{\partial \theta}\boldsymbol{e}_\theta + \frac{\partial T}{\partial z}\boldsymbol{e}_z\right), \qquad (5.6.20\text{-d})$$

where ϵ_{kl} and γ_{kl} are given by (5.6.20-c).

Surface Tractions on ∂V:

$$t_l = t_{kl}n_k, \qquad m_l = m_{kl}n_k, \qquad (5.6.20\text{-e})$$

where $\boldsymbol{n} = (n_r, n_\theta, n_z)$ is the unit exterior normal of ∂V, and t_{kl} and m_{kl} are given by (5.6.20-d).

Field Equations. (see 5.6.16)–(5.6.18).

Inequalities of Material Moduli:

$$3\lambda + 2\mu + \kappa \geq 0, \qquad 2\mu + \kappa \geq 0, \qquad \kappa \geq 0,$$
$$3\alpha + \beta + \gamma \geq 0, \qquad \beta + \gamma \geq 0, \qquad \gamma - \beta \geq 0,$$
$$C_0 \geq 0, \qquad K \geq 0. \qquad (5.6.21)$$

5.7   Uniqueness Theorem

The uniqueness theorem for micropolar elasticity and micropolar thermoe-lasticity were given by Eringen [1966a, 1970a]. For unbounded domains, the uniqueness and stability were discussed by Carbonaro and Russo [1985]. Here, we follow the work of Scalia [1990], who proved the uniqueness theo-rem, without any definiteness assumptions, such as (5.5.8-f), on the material moduli.

First, we establish some results that are useful. For convenience, we define the following functions valid for all $a, b \in I$:

$$K(a,b) = \tfrac{1}{2}\rho \dot{u}_k(a)\dot{u}_k(b) + \tfrac{1}{2}\rho j_{kl}\dot{\phi}_k(a)\dot{\phi}_k(b),$$
$$U(a,b) = \tfrac{1}{2}A_{klmn}\epsilon_{kl}(a)\epsilon_{mn}(b) + \tfrac{1}{2}B_{klmn}\gamma_{kl}(a)\gamma_{mn}(b)$$
$$\qquad + \tfrac{1}{2}C_{klmn}[\epsilon_{kl}(a)\gamma_{mn}(b) + \epsilon_{kl}(b)\gamma_{mn}(a)] + \tfrac{1}{2}\tfrac{\rho C_0}{T_0}T(a)T(b),$$
$$P(a,b) = \int_V \left[\rho\mathbf{f}(a)\cdot\mathbf{u}(b) + \rho\mathbf{l}(a)\cdot\boldsymbol{\phi}(b) + \frac{\rho h(a)}{T_0}T(b) \right] dv$$
$$\qquad + \int_{\partial V} \left[\mathbf{t}(a)\cdot\mathbf{u}(b) + \mathbf{m}(a)\cdot\boldsymbol{\phi}(b) + \frac{1}{T_0}q(a)T(b) \right] da. \qquad (5.7.1)$$

We note that $K(t,t) \equiv K(t)$ is the kinetic density, $U(t,t) \equiv U(t)$ is the strain energy density, and $P(t,t) \equiv P(t)$ is the total power of the applied loads. We set

$$\mathcal{K}(t) = \int_V K(t)\,dv, \qquad \mathcal{U}(t) = \int_V U(t)\,dv. \qquad (5.7.2)$$

Theorem 1.

$$\mathcal{U}(t) - \mathcal{K}(t) = \frac{1}{2}\int_0^t [P(t+s,t-s) - P(t-s,t+s)]\,ds + \int_V [U(0,2t) - K(0,2t)]\,dv.$$
$$(5.7.3)$$

PROOF.   We introduce the notation

$$E(a,b) = t_{kl}(a)\dot{\epsilon}_{kl}(b) + m_{kl}(a)\dot{\gamma}_{lk}(b) + \rho T(a)\dot{\eta}(b). \qquad (5.7.4)$$

From (5.7.1) and (5.1.12), it can be verified that

$$E(t,t) \equiv E(t) = \dot{U}(t). \qquad (5.7.5)$$

Using (5.7.4), we obtain

$$E(t - s, t + s) - E(t + s, t - s) = 2\frac{d}{ds}U(t - s, t + s). \tag{5.7.6}$$

By means of (5.5.1)–(5.5.3) and (5.5.5), we calculate $E(a, b)$ for $a = t - s$ and $b = t + s$

$$
\begin{aligned}
E(t - s, t + s) \\
= \Big[t_{kl}(t - s)\dot{u}_l(t + s) + m_{kl}(t - s)\dot{\phi}_l(t + s) + \frac{1}{T_0}q_k(t + s)T(t - s) \Big]_{,k} \\
+ \rho f_k(t - s)\dot{u}_k(t + s) + \rho l_k(t - s)\dot{\phi}_k(t + s) + \frac{\rho}{T_0}h(t + s)T(t - s) \\
- \frac{1}{T_0^2}K_{kl}T_{,k}(t - s)T_{,l}(t + s) - \rho\ddot{u}_k(t - s)\dot{u}_k(t + s) \\
- \rho j_{kl}\ddot{\phi}_l(t - s)\dot{\phi}_k(t + s).
\end{aligned}
\tag{5.7.7}
$$

Integrating this over the volume and using the divergence theorem, we obtain

$$
\begin{aligned}
\int_V E(t - s, t + s)\, dv = P(t - s, t + s) \\
- \int_V \Big[\rho\ddot{u}_k(t - s)\dot{u}_k(t + s) + \rho j_{kl}\ddot{\phi}_l(t - s)\dot{\phi}_k(t + s) \\
+ \frac{1}{T_0^2}K_{kl}T_{,k}(t - s)T_{,l}(t + s) \Big]\, dv.
\end{aligned}
\tag{5.7.8}
$$

By replacing s by $-s$, we obtain another expression. Upon substituting these expressions into (5.7.6), we have

$$2\int_V \frac{d}{ds}\left[U(t - s, t + s) - K(t - s, t + s) \right] dv = P(t-s,t+s) - P(t+s,t-s). \tag{5.7.9}$$

If we integrate this relation from 0 to τ, we obtain

$$
\begin{aligned}
2\int_V \left[U(t - \tau, t + \tau) - K(t - \tau, t + \tau) - U(t) + K(t) \right] dv \\
= \int_0^\tau \left[P(t - s, t + s) - P(t + s, t - s) \right] ds.
\end{aligned}
\tag{5.7.10}
$$

Setting $t = \tau$ gives the desired result (5.7.3). $\square$

Lemma 1. Functions $\mathcal{U}(t)$ and $\mathcal{K}(t)$ are given by

$$2\mathcal{U}(t) = \mathcal{U}(0) + \mathcal{K}(0) + \int_V [U(0, 2t) - K(0, 2t)]\, dv$$

$$+ \frac{1}{2} \int_0^t \left[P(t+s, t-s) - P(t-s, t+s) + 2P(s,s) \right] ds$$

$$- \frac{1}{T_0^2} \int_0^t K_{kl} T_{,k} T_{,l} \, dv \, ds, \tag{5.7.11}$$

$$2\mathcal{K}(t) = \mathcal{U}(0) + \mathcal{K}(0) - \int_V \left[U(0, 2t) - K(0, 2t) \right] dv$$

$$- \frac{1}{2} \int_0^t \left[P(t+s, t-s) - P(t-s, t+s) - 2P(s,s) \right] ds$$

$$- \frac{1}{T_0^2} \int_0^t \int_V K_{kl} T_{,k} T_{,l} \, dv \, ds, \tag{5.7.12}$$

for all $t \in T^+ = [0, \infty)$.

PROOF.   Recalling (5.7.5), (5.7.8) with $s = 0$, can be written as

$$\dot{\mathcal{U}} + \dot{\mathcal{K}} = P(t) - \frac{1}{T_0} \int_V K_{kl} T_{,k} T_{,l} \, dv. \tag{5.7.13}$$

Integration gives

$$\mathcal{U} + \mathcal{K} = \mathcal{U}(0) + \mathcal{K}(0) + \int_0^t P(s,s) \, ds - \frac{1}{T_0^2} \int_0^t \int_V K_{kl} T_{,k} T_{,l} \, dv \, ds. \tag{5.7.14}$$

Using this equation and (5.7.3), we obtain (5.7.11) and (5.7.12), completing the proof.   □

Theorem 2 (Uniqueness).   It is assumed that:

(i) ρ is strictly positive;

(ii) j_{kl} is positive definite;

(iii) K_{kl} is positive semi-definite; and

(iv) C_0 is strictly positive (or negative).

Then the boundary-initial value problems of linear micropolar elastodynamics has at most one solution.

PROOF.   Suppose that the contrary is valid, and two solutions $\mathbf{u}^{(\alpha)}$, $\dot{\phi}^{(\alpha)}$, $T^{(\alpha)}$, $\alpha = 1, 2$, exist. Let

$$\mathbf{u} = \mathbf{u}^{(1)} - \mathbf{u}^{(2)}, \qquad \phi = \phi^{(1)} - \phi^{(2)}, \qquad T = T^{(1)} - T^{(2)}. \tag{5.7.15}$$

Then, clearly $\mathbf{u}, \phi$, and T satisfy (5.5.1)–(5.5.7) with $\mathbf{f} = \mathbf{l} = \mathbf{0}$, $h = 0$, and $\hat{\mathbf{u}} = \hat{\mathbf{t}} = \hat{\phi} = \hat{\mathbf{m}} = \mathbf{0}$, $\hat{T} = \hat{q} = 0$, $\mathbf{u}^0 = \mathbf{v}^0 = \phi^0 = \nu^0 = \mathbf{0}$, $T^0 = 0$, i.e., homogeneous equations and boundary and initial conditions. With these, (5.7.12) reduces to

$$\int_{\mathcal{V}} \rho(\dot{u}_k \dot{u}_k + \rho j_{kl} \dot{\phi}_k \dot{\phi}_l) \, dv + \frac{1}{T_0^2} \int_0^t \int_{\mathcal{V}} K_{kl} T_{,k} T_{,l} \, dv \, ds = 0. \qquad (5.7.16)$$

By hypotheses (i)–(iv) of the theorem, it follows that

$$\dot{\mathbf{u}} = \mathbf{0}, \qquad \dot{\boldsymbol{\phi}} = \mathbf{0} \quad \text{on } \mathcal{V} \times T^+,$$

$$\int_0^t \int_{\mathcal{V}} K_{kl} T_{,k} T_{,l} \, dv \, ds = 0, \qquad t \in T^+. \qquad (5.7.17)$$

But $\mathbf{u}$ and $\boldsymbol{\phi}$ vanish initially, so that $(5.7.17)_{1,2}$ imply

$$\mathbf{u} = \mathbf{0}, \qquad \boldsymbol{\phi} = \mathbf{0}.$$

In view of this and $(5.7.17)_3$, (5.7.11) reduces to

$$\int_{\mathcal{V}} \rho \frac{C_0}{T_0^2} T^2 \, dv = 0. \qquad (5.7.18)$$

This implies that $T = 0$ on $\mathcal{V} \times T^+$. Hence, the proof of the theorem. $\qquad \square$

The existence theorem for static micropolar elastic solids was given by Ieşan [1971]. Thus, the boundary-value problem, in this case, is *well-posed*, in the sense of Poincaré.

5.8   Reciprocal Theorem

The Reciprocal theorem was discussed by Ieşan [1967], Chandrasekharaiah [1987] and Scalia [1990]. We follow the approach of Scalia.

It is convenient to introduce the following notations:

$$(\omega * f)(\mathbf{x}, t) = \int_0^t \omega(\mathbf{x}, t - \tau) f(\mathbf{x}, \tau) \, d\tau, \qquad (5.8.1)$$

where ω and f are defined over $\mathcal{V} \times T^+$ and they are continuous in time. We also use the notation

$$1(t) = 1, \qquad t \in T^+ = [0, \infty),$$

$$\bar{h}(\mathbf{x}, t) = \int_0^t h(\mathbf{x}, s) \, ds. \qquad (5.8.2)$$

Consider two external systems:

$$L^{(\alpha)} = \{\mathbf{f}^{(\alpha)}, \mathbf{1}^{(\alpha)}, h^{(\alpha)}, \hat{\mathbf{t}}^{(\alpha)}, \hat{\boldsymbol{\phi}}^{(\alpha)}, \hat{\mathbf{m}}^{(\alpha)}, \hat{q}^{(\alpha)}, \mathbf{u}^{0,(\alpha)} \mathbf{v}^{0,(\alpha)} \boldsymbol{\phi}^{0,(\alpha)} \boldsymbol{\nu}^{0,(\alpha)} \eta^{0,(\alpha)}\},$$

$$S^{(\alpha)} = \{\mathbf{u}^{(\alpha)}, \boldsymbol{\phi}^{(\alpha)}, T^{(\alpha)}, \mathbf{t}^{(\alpha)}, \mathbf{m}^{(\alpha)}, \eta^{(\alpha)}, \boldsymbol{\epsilon}^{(\alpha)}, \boldsymbol{\gamma}^{(\alpha)}, \mathbf{q}^{(\alpha)}\}, \qquad \alpha = 1, 2,$$

$$(5.8.3)$$

of which $S^{(\alpha)}$ is the solution corresponding to $L^{(\alpha)}$.

Lemma. Let

$$W_{\alpha\beta}(a,b)$$

$$= \int_{\partial V} \left[t_k^{(\alpha)}(a) u_k^{(\beta)}(b) + m_k^{(\alpha)}(a) \phi_k^{(\beta)}(b) - \frac{1}{T_0} \bar{q}^{(\alpha)}(a) T^{(\beta)}(b) \right] da$$

$$+ \int_V \left[\rho f_k^{(\alpha)}(a) u_k^{(\beta)}(b) + \rho l_k^{(\alpha)}(a) \phi_k^{(\alpha)}(b) + \frac{1}{T_0} \bar{q}_k^{(\alpha)}(a) T_{,k}^{(\beta)}(b) \right.$$

$$\left. - \frac{\rho}{T_0} \bar{h}^{(\alpha)}(a) T^{(\beta)}(b) \right] dv - \int_V [\rho \ddot{u}_k^{(\alpha)}(a) u_k^{(\beta)}(b) + \rho j_{kl} \ddot{\phi}_k^{(\alpha)}(a) \phi_l^{(\beta)}(b)] \, dv,$$

$$\alpha, \beta = 1, 2, \qquad a, b \in T^+. \tag{5.8.4}$$

Then

$$W_{\alpha\beta}(a,b) = W_{\beta\alpha}(b,a). \tag{5.8.5}$$

The quantity $W_{\alpha\beta}(a,b)$, defined by (5.8.4), may be called the *virtual work* of surface, body, thermal, and inertial loads in system $L^{(\alpha)}$, during the displacements, rotations, and temperature changes of $S^{(\beta)}$.

PROOF. Let

$$w_{\alpha\beta}(a,b) = t_{kl}^{(\alpha)}(a) \epsilon_{kl}^{(\beta)}(b) + m_{kl}^{(\alpha)}(a) \gamma_{lk}^{(\beta)}(b) - \frac{1}{T_0} [\bar{q}_{k,k}^{(\alpha)}(a) - \rho \bar{h}^{(\alpha)}(a)] T^{(\beta)}(b).$$

$$\tag{5.8.6}$$

By mere substitution from (5.5.4), we find that

$$w_{\alpha\beta}(a,b) = w_{\beta\alpha}(b,a). \tag{5.8.7}$$

Next, we use (5.5.5) in (5.8.6) to obtain

$$w_{\alpha\beta}(a,b) = \left[t_{kl}^{(\alpha)}(a) u_l^{(\beta)}(b) + m_{kl}^{(\alpha)}(a) \phi_l^{(\beta)}(b) - \frac{1}{T_0} \bar{q}_k^{(\alpha)}(a) T^{(\beta)}(b) \right]_{,k}$$

$$- [\rho \ddot{u}_k^{(\alpha)}(a) - \rho f_k^{(\alpha)}(a)] u_k^{(\beta)}(b) - [\rho j_{kl} \ddot{\phi}_l^{(\alpha)}(a) - \rho l_k^{(\alpha)}(a)] \phi^{(\beta)}(b)$$

$$+ \frac{1}{T_0} \bar{q}_k^{(\alpha)}(a) T_{,k}^{(\beta)}(b) - \frac{\rho}{T_0} \bar{h}^{(\alpha)}(a) T^{(\beta)}(b). \tag{5.8.8}$$

Upon integrating this relation over the volume V and using the Green–Gauss theorem, with the help of (5.8.7), we obtain

$$W_{\alpha\beta}(a,b) = \int_V w_{\alpha\beta}(a,b) \, dv = W_{\beta\alpha}(b,a), \tag{5.8.9}$$

completing the proof. □

Theorem (Reciprocal). If the $S^{(\alpha)}$ is a solution of the linear thermo-micropolar elastic system $L^{(\alpha)}$, $\alpha = 1, 2$, then

$$\int_{\partial V} t * \left[t_k^{(1)} * u_k^{(2)} + m_k^{(1)} * \phi_k^{(2)} - \frac{1}{T_0} 1 * q^{(1)} * T^{(2)} \right] da$$

$$+ \int_V \left[\rho \tilde{f}_k^{(1)} * u_k^{(2)} + \rho \tilde{l}_k^{(1)} \phi_k^{(2)} - \frac{\rho}{T_0} * 1 * h^{(1)} * T^{(2)} \right] dv$$

$$= \int_{\partial V} t * \left[t_k^{(2)} u_k^{(1)} + m_k^{(2)} * \phi_k^{(1)} - \frac{1}{T_0} 1 * q^{(2)} * T^{(1)} \right] da$$

$$+ \int_V \left[\rho \tilde{f}_k^{(2)} * u_k^{(1)} + \rho \tilde{l}_k^{(2)} * \phi_k^{(1)} - \frac{\rho}{T_0} t * 1 * h^{(2)} * T^{(1)} \right] dv, \quad (5.8.10)$$

where

$$\rho \tilde{f}_k^{(\alpha)} = t * \rho f_k^{(\alpha)} + \rho \left(t v_k^{0(\alpha)} + u_k^{0(\alpha)} \right),$$

$$\rho \tilde{l}_k^{(\alpha)} = t * \rho l_k^{(\alpha)} + \rho j_{kl} \left(t v_l^{0(\alpha)} + \phi_l^{0(\alpha)} \right), \quad \alpha = 1, 2. \quad (5.8.11)$$

PROOF.   In (5.8.4), we set $a = \tau$, $b = t - \tau$ and integrate it from 0 to t. This gives

$$\int_0^t W_{12}(\tau, t - \tau) \, d\tau$$

$$= \int_{\partial V} \left[t_k^{(1)} * u_k^{(2)} + m_k^{(1)} * \phi_k^{(2)} - \frac{1}{T_0} 1 * q^{(1)} * T^{(2)} \right] da$$

$$+ \int_V \left[\rho f_k^{(1)} * u_k^{(2)} + \rho l_k^{(1)} * \phi_k^{(2)} + \frac{1}{T_0} 1 * q_k^{(1)} * T_{,k}^{(2)} \right.$$

$$\left. - \frac{\rho}{T_0} 1 * h^{(1)} * T^{(2)} \right] dv - \int_V \left[\rho \ddot{u}_k^{(1)} * u_k^{(2)} + \rho j_{kl} \ddot{\phi}_l^{(1)} \phi_k^{(2)} \right] dv. \quad (5.8.12)$$

But, in view of (5.8.5), we also have

$$\int_0^t W_{12}(\tau, t - \tau) \, d\tau = \int_0^t W_{21}(t - \tau, \tau) \, d\tau. \quad (5.8.13)$$

Taking the convolution of (5.8.12) with t and using

$$t * \ddot{u}_k^{(\alpha)} = u_k^{(\alpha)} - t v_k^{0(\alpha)} - u_k^{0(\alpha)},$$

$$t * \ddot{\phi}_k^{(\alpha)} = \phi_k^{(\alpha)} - t v_k^{0(\alpha)} - \phi_k^{0(\alpha)}, \quad (5.8.14)$$

We arrive at (5.8.10) and the proof of the theorem.   □

Reciprocity relations for the static case were obtained by Ieşan [1981] and applied to earthquake problems.

5.9   Variational Principles

A variational principle for the linear micropolar thermoelastic solid was given by Ieşan [1967]. A Biot-type variational principle was also presented by Chandrasekharaiah [1987].

Theorem 1. Let K be the set of all admissible states and $S \in K$. Define the functional $\Lambda_t(S)$ on K, for each $t \in T^+$, by

$$
\Lambda_t(S) = \frac{1}{2} \int_\mathcal{V} (A_{klmn} t * \epsilon_{kl} * \epsilon_{mn} + B_{klmn} t * \gamma_{kl} * \gamma_{mn}
$$

$$
+ 2C_{klmn} t * \epsilon_{kl} * \gamma_{mn}) \, dv
$$

$$
+ \frac{T_0}{2\rho C_0} \int_\mathcal{V} t * [\rho(\eta - \eta_0) - A_{kl}\epsilon_{kl} - B_{kl}\gamma_{kl}] * [\rho(\eta - \eta_0)
$$

$$
- A_{mn}\epsilon_{mn} - B_{mn}\gamma_{mn}] \, dv + \frac{1}{2} \int_\mathcal{V} (\rho u_k * u_k + \rho j_{kl}\phi_k * \phi_l) \, dv
$$

$$
+ \frac{1}{2T_0} \int_\mathcal{V} \lambda_{kl} t * 1 * q_k * q_l \, dv
$$

$$
- \int_\mathcal{V} t * (t_{kl} * \epsilon_{kl} + m_{kl} * \gamma_{lk}) \, dv - \int_\mathcal{V} t * \rho(\eta - \eta_0) * T \, dv
$$

$$
- \frac{1}{T_0} \int_\mathcal{V} t * 1 * T_{,k} * q_k \, dv - \int_\mathcal{V} \left(t * t_{kl,k} + \rho \tilde{f}_l \right) * u_l \, dv
$$

$$
- \int_\mathcal{V} [t * (m_{kl,k} + \epsilon_{lkr} t_{kr}) + \rho \tilde{l}_l] * \phi_l \, dv
$$

$$
+ \frac{1}{T_0} \int_\mathcal{V} t * 1 * T * \rho h \, dv
$$

$$
+ \int_{S_1} t * t_k * \hat{u}_k \, da + \int_{S_2} t * (t_k - \hat{t}_k) * u_k \, da
$$

$$
+ \int_{S_3} t * m_k * \hat{\phi}_k \, da + \int_{S_4} t * (m_k - \hat{m}_k) * \phi_k \, da
$$

$$
+ \frac{1}{T_0} \int_{S_5} t * 1 * (T - \hat{T}) * q \, da + \frac{1}{T_0} \int_{S_6} t * 1 * T * \hat{q} \, da, \quad (5.9.1)
$$

where $S_1 \cup S_2 = S_3 \cup S_4 = \partial\mathcal{V}$, $S_1 \cap S_2 = S_3 \cap S_4 = 0$, and

$$
t_k = t_{lk} n_l, \qquad m_k = m_{lk} n_l, \qquad q = q_k n_k, \qquad \lambda_{kl} = K_{kl}^{-1},
$$
$$
\rho \tilde{f}_k = t * \rho f_k + \rho(t v_k^0 + u_k^0),
$$
$$
\rho \tilde{l}_k = t * \rho l_k + \rho j_{kl}(t v_l^0 + \phi_l^0). \tag{5.9.2}
$$

Then $\Lambda_t(S)$ is stationary, i.e., for every $S + \epsilon \bar{S} \in K$,

$$
\delta \Lambda_t(S) = \frac{d}{d\epsilon} \Lambda_t(S + \epsilon \bar{S}) \bigg|_{\epsilon=0} = 0 \quad \text{on } K, \quad t \in T^+, \tag{5.9.3}
$$

if and only if S is a solution of the mixed problem.

PROOF.   Let $\bar{S}\{\bar{\mathbf{u}}, \bar{\boldsymbol{\phi}}, \bar{T}, \bar{\boldsymbol{\epsilon}}, \bar{\boldsymbol{\gamma}}, \bar{\eta}, \bar{\mathbf{t}}, \bar{\mathbf{m}}, \bar{\mathbf{q}}\} \in K$ be an arbitrary variation in the state S. Then, the variation of (5.9.1) reads

$$\delta\Lambda_t(S) = \int_v \left(A_{klmn} t * \bar{\epsilon}_{kl} * \epsilon_{mn} + B_{klmn} t * \bar{\gamma}_{kl} * \gamma_{mn} \right.$$
$$\left. + C_{klmn} t * \bar{\epsilon}_{kl} * \gamma_{mn} * \epsilon_{kl} * \bar{\gamma}_{mn} \right) dv$$
$$+ \frac{T_0}{\rho C_0} \int_v t * (\rho\bar{\eta} - A_{kl}\bar{\epsilon}_{kl} - B_{kl}\bar{\gamma}_{kl})$$
$$* \left[\rho(\eta - \eta_0) - A_{mn}\epsilon_{mn} - B_{mn}\gamma_{mn} \right] dv$$
$$+ \int_v \left(\rho u_k * \bar{u}_k + \rho j_{kl}\phi_k * \bar{\phi}_l \right) dv + \frac{1}{T_0} \int_v \lambda_{kl} t * 1 * q_l * \bar{q}_k \, dv$$
$$- \int_v t * \left(\bar{t}_{kl} * \epsilon_{kl} + t_{kl}\bar{\epsilon}_{kl} + \bar{m}_{kl} * \gamma_{lk} + m_{kl} * \bar{\gamma}_{lk} \right) dv$$
$$- \int_v t * \left[\rho\bar{\eta} * T + \rho(\eta - \eta_0) * \bar{T} \right] dv$$
$$- \frac{1}{T_0} \int_v t * 1 * \left(\bar{T}_{,k} * q_k + T_{,k} * \bar{q}_k \right) dv$$
$$- \int_v \left[t * \bar{t}_{kl,k} * u_l + \left(t * t_{kl,k} + \rho\tilde{f}_l \right) * \bar{u}_l \right] dv$$
$$- \int_v \left\{ t * \left(\bar{m}_{kl,k} + \epsilon_{lkr}\bar{t}_{kr} \right) * \phi_l \right.$$
$$\left. + \left[t * \left(m_{kl,k} + \epsilon_{lkr}t_{kr} \right) + \rho\tilde{l}_l \right] * \bar{\phi}_l \right\} dv$$
$$+ \frac{1}{T_0} \int_v t * 1 * \bar{T} * \rho h \, dv + \int_{S_1} t * \bar{t}_k * \hat{u}_k \, da$$
$$+ \int_{S_2} \left[t * \bar{t}_k * u_k + t * (t_k - \hat{t}_k) * \bar{u}_k \right] da$$
$$+ \int_{S_3} t * \bar{m}_k * \hat{\phi}_k \, da$$
$$+ \int_{S_4} \left[t * \bar{m}_k * \phi_k + t * (m_k - \hat{m}_k) * \bar{\phi}_k \right] da$$
$$+ \frac{1}{T_0} \int_{S_5} \left[t * 1 * \bar{T} * q + t * 1 * (T - \hat{T}) * \bar{q} \right] da$$
$$+ \frac{1}{T_0} \int_{S_6} t * 1 * \bar{T} * \hat{q} \, da, \tag{5.9.4}$$

where we used symmetries A_{klmn}, B_{klmn} and the commutative property of convolutions. We note that $\delta(t_{kl,k}) = (\delta t_{kl})_{,k}$. By means of the Green–Gauss theorem, we have

$$\int_v t * u_l * (\bar{t}_{kl})_{,k} \, dv = \int_v t * \left[(u_l * \bar{t}_{kl})_{,k} - u_{l,k} * \bar{t}_{kl} \right] dv$$
$$= \int_{\partial v} t * u_l * \bar{t}_l \, da - \int_v t * u_{l,k} * \bar{t}_{kl} \, dv. \tag{5.9.5}$$

Similar expressions are valid for the volume integrals containing $\bar{m}_{kl,k}$ and $\bar{T}_{,k}$. Using these expressions and collecting the coefficients of $\{\bar{u}, \bar{\phi}, \ldots, \bar{q}\}$, we arrive at

$$
\begin{aligned}
\delta \Lambda_t(S) = \int_v t * \Bigg\{ & A_{klmn}\epsilon_{mn} + C_{klmn}\gamma_{mn} \\
& - \frac{T_0}{\rho C_0} \left[\rho(\eta - \eta_0) - A_{mn}\epsilon_{mn} - B_{mn}\gamma_{mn} \right] A_{kl} - t_{kl} \Bigg\} * \bar{\epsilon}_{kl}\, dv \\
+ \int_v t * \Bigg\{ & B_{lkmn}\gamma_{mn} + C_{mnlk}\epsilon_{mn} \\
& - \frac{T_0}{\rho C_0} \left[\rho(\eta - \eta_0) - A_{mn}\epsilon_{mn} - B_{mn}\gamma_{mn} \right] B_{lk} - m_{kl} \Bigg\} * \bar{\gamma}_{lk}\, dv \\
+ \int_v t * & \frac{T_0}{\rho C_0} \left[\rho(\eta - \eta_0) - \frac{\rho C_0}{T_0}T - A_{kl}\epsilon_{kl} - B_{kl}\gamma_{kl} \right] * \rho\bar{\eta}\, dv \\
+ \frac{1}{T_0} \int_v t & * 1 * (\lambda_{kl}q_l - T_{,k}) * \bar{q}_k\, dv \\
+ \int_v t * & (u_{l,k} + \epsilon_{lkr}\phi_r - \epsilon_{kl}) * \bar{t}_{kl}\, dv \\
+ \int_v t * & (\phi_{l,k} - \gamma_{lk}) * \bar{m}_{kl}\, dv - \int_v \left(t * t_{kl,k} + \rho\tilde{f}_l - \rho u_l \right) * \bar{u}_l\, dv \\
- \int_v \Big[t & * (m_{kl,k} + \epsilon_{lkr}t_{kr}) + \rho\tilde{l}_l - \rho j_{kl}\phi_k \Big] * \bar{\phi}_l\, dv \\
+ \frac{1}{T_0} \int_v t & * [1 * (q_{k,k} + \rho h) - \rho T_0(\eta - \eta_0)] * \bar{T}\, dv \\
+ \int_{S_1} t * & (\hat{u}_k - u_k) * \bar{t}_k\, da + \int_{S_2} t * (t_k - \hat{t}_k) * \bar{u}_k\, da \\
+ \int_{S_3} t * & (\hat{\phi}_k - \phi_k) * \bar{m}_k\, da \\
+ \int_{S_4} t * & (m_k - \hat{m}_k) * \bar{\phi}_k\, da + \frac{1}{T_0} \int_{S_5} t * 1 * (T - \hat{T}) * \bar{q}\, da \\
+ \frac{1}{T_0} \int_{S_6} t & * 1 * (\hat{q} - q) * \bar{T}\, da.
\end{aligned}
\tag{5.9.6}
$$

It is now clear that if S is a solution of the mixed problem, then in view of (5.5.1)–(5.5.7), $\delta\Lambda_t(S) = 0$. This proves the sufficiency condition. The proof of necessity requires a subtler approach, namely: we must stipulate that the members of $\bar{S}$ can be selected completely arbitrarily and independently from one another. Then, again vanishing $\delta\Lambda_t(S)$ requires that coefficients of the members of $\bar{S}$ must vanish separately. This completes the proof.   □

This theorem constitutes the most general variational principle for the mixed problem in that the admissible states meet none of the relations satisfied by the field variables, except some smoothness and symmetry conditions. It is possible to construct other variational principles of lesser complexity by restricting the set of admissible states to a smaller set.

Definition 1. An admissible state S is called kinematically admissible, if it meets strain-displacement relations (5.5.5), constitutive equations (5.5.4), and boundary conditions (5.5.6)$_{1,3,5}$ on $\mathbf{u}$, $\boldsymbol{\phi}$, and T.

Theorem 2. Let U be the set of all kinematically admissible fields. Let $\{\mathbf{U}, \boldsymbol{\phi}, T\} \in U$, and for each $t \in T^+$ define the functional $\Phi\{U\}$ by

$$\Phi(U) = \frac{1}{2} \int_V \{t * [t_{kl} * \epsilon_{kl} + m_{kl} * \gamma_{lk} + \rho(\eta - \eta_0) * T]$$
$$+ \rho u_k * u_k + \rho j_{kl} \phi_l * \phi_k - 2\rho \tilde{f}_k * u_k - 2\rho \tilde{l}_k * \phi_k\} \, dv$$
$$+ \int_V t * \left[-\rho(\eta - \eta_0) * T + \frac{1}{T_0} * \rho h * T + \frac{1}{2T_0} * q_k * T_{,k} \right] \, dv$$
$$- \int_{S_2} t * \hat{t}_k * u_k \, da - \int_{S_4} t * \hat{m}_k * \phi_k \, da.$$
$$- \frac{1}{T_0} \int_{S_6} t * 1 * T * \hat{q} \, da. \tag{5.9.7}$$

Then,
$$\delta\Phi(U) = 0 \quad \text{over } U, \qquad t \in T^+, \tag{5.9.8}$$

if and only if U is a field corresponding to the solution of the mixed problem.

PROOF.   Because of the fact that constitutive equations (5.5.4) are satisfied, the variation of Φ gives

$$\delta\Phi(U) = \int_V [t * (t_{kl} * \bar{\epsilon}_{kl} + m_{kl} * \bar{\gamma}_{lk} + \rho\bar{\eta} * T)$$
$$+ \rho u_k * \bar{u}_k + \rho j_{kl} \phi_l * \bar{\phi}_k - \rho \tilde{f}_k * \bar{u}_k - \rho \tilde{l}_k * \bar{\phi}_k \Big] \, dv$$
$$+ \int_V t * \left[-\rho(\eta - \eta_0) * \bar{T} - \rho\bar{\eta} * T + \frac{1}{T_0}\rho h * \bar{T} + \frac{1}{T_0} q_k * \bar{T}_{,k} \right] \, dv$$
$$- \int_{S_2} t * \hat{t}_k * \bar{u}_k \, da - \int_{S_4} t * \hat{m}_k * \bar{\phi}_k \, da - \frac{1}{T_0} \int_{S_6} t * 1 * \bar{T} * \hat{q} \, da.$$
$$\tag{5.9.9}$$

For the variations $\bar{\epsilon}_{kl}$ and $\bar{\gamma}_{lk}$ we have

$$\bar{\epsilon}_{kl} = \bar{u}_{l,k} + \epsilon_{lkm}\bar{\phi}_m, \qquad \bar{\gamma}_{lk} = \bar{\phi}_{l,k}.$$

Using these, we replace

$$t_{kl} * \bar{\epsilon}_{kl} = (t_{kl}\bar{u}_l)_{,k} - t_{kl,k} * \bar{u}_l,$$
$$m_{kl} * \bar{\gamma}_{lk} = (m_{kl} * \bar{\phi}_l)_{,k} - m_{kl,k} * \bar{\phi}_l,$$
$$q_k * \bar{T}_{,k} = (q_k * \bar{T})_{,k} - q_{k,k}\bar{T}.$$

Using these and converting the first term in each to a surface integral, (5.9.9) reduces to

$$\delta\Phi(U) = \int_{\mathcal{V}}\left\{-(t * t_{kl,k} + \rho\tilde{f}_l - \rho u_l) * \bar{u}_l - [t * (m_{kl,k} + \epsilon_{lkm}t_{km})\right.$$
$$+ \rho\tilde{l}_l - \rho j_{kl}\phi_k] * \tilde{\phi}_l - \frac{1}{T_0}t * [\rho T_0(\eta - \eta_0) - q_{k,k} - 1 * \rho h] * \bar{T}\Big\} dv$$
$$+ \int_{S_2} t * (t_k - \hat{t}_k) * \bar{u}_k \, da + \int_{S_4} t * (m_k - \hat{m}_k) * \phi_k \, da$$
$$+ \frac{1}{T_0}\int_{S_6} t * 1 * (q - \hat{q}) * \bar{T} \, da. \qquad (5.9.10)$$

□

From this, it is clear that if the kinematically admissible field U is a solution of the mixed problem, then (5.9.8) is fulfilled. Conversely, for arbitrary and independent variations of $\bar{\mathbf{u}}, \bar{\boldsymbol{\phi}}$, and $\bar{T}$, equations of motion, boundary and initial conditions are satisfied, leading to (5.9.8). Thus, (5.9.8) constitutes both the necessary and sufficient conditions for the solution of the mixed problem.

For nonthermal and static cases, this theorem implies the principle of work and energy.

Theorem 3 (Work and Energy). Let U be a set of all kinematically admissible, static, nonthermal fields. Let $(\mathbf{u}, \boldsymbol{\phi}) \in U$. Then,

$$\int_{\mathcal{V}}(t_{kl}\epsilon_{kl} + m_{kl}\gamma_{lk}) \, dv = \int_{\mathcal{V}}(\rho\mathbf{f}\cdot\mathbf{u} + \rho\mathbf{l}\cdot\boldsymbol{\phi}) \, dv + \int_{\partial\mathcal{V}}(\hat{t}_k u_k + \hat{m}_k\phi_k) \, da. \quad (5.9.11)$$

Note that the left-hand side of (5.9.11) is twice the total strain energy and the right-hand side is the work of the applied loads.

PROOF. Multiply the equation of motion (5.4.3) by u_l, and (5.4.4) by ϕ_l and integrate over the volume

$$\int_{\mathcal{V}}[(t_{kl,k} + \rho f_l) u_l + (m_{kl,k} + \epsilon_{lkm}t_{km} + \rho l_l) \phi_l] \, dv = 0.$$

Using the identities

$$t_{kl,k}u_l = (t_{kl}u_l)_{,k} - t_{kl}u_{l,k},$$
$$m_{kl,k}\phi_l = (m_{kl}\phi_l)_{,k} - m_{kl}\phi_{l,k}, \qquad (5.9.12)$$

and the divergence theorem, we have

$$\int_{\mathcal{V}}[t_{kl}(u_{l,k} + \epsilon_{lkm}\phi_m) + m_{kl}\phi_{l,k}] \, dv = \int_{\mathcal{V}}\rho(\mathbf{f}\cdot\mathbf{u} + \mathbf{l}\cdot\boldsymbol{\phi}) \, dv$$
$$+ \int_{\partial\mathcal{V}}(t_{kl}n_k u_l + m_{kl}n_k\phi_l) \, da,$$

which is identical to (5.9.11). Hence, the proof of the theorem. □

Theorem 4 (Principle of Minimum Potential Energy).   Let U be a set of all kinematically admissible, static, nonthermal fields. Let $(\mathbf{u}, \boldsymbol{\phi}) \subset U$ be the solution of the field equation, and

$$
F(\mathbf{u}, \boldsymbol{\phi}) \equiv \int_{\mathcal{V}} \left[t_{kl}\epsilon_{kl} + m_{kl}\gamma_{lk} - \rho(\mathbf{f} \cdot \mathbf{u} + \mathbf{l} \cdot \boldsymbol{\phi}) \right] dv
$$

$$
- \int_{\partial \mathcal{V}} \left(\hat{t}_k u_k + \hat{m}_k \phi_k \right) da. \tag{5.9.13}
$$

Then, the functional F assumes the smallest value, among all couples $(\mathbf{u}, \boldsymbol{\phi})$, belonging to U, i.e.,

$$
F(\mathbf{u}', \boldsymbol{\phi}') \leq F(\mathbf{u}, \boldsymbol{\phi}) \qquad \textit{for all } (\mathbf{u}, \boldsymbol{\phi}) \subset U. \tag{5.9.14}
$$

Moreover, the equality holds, if and only if, $\mathbf{u} = \mathbf{u}'$, $\boldsymbol{\phi} = \boldsymbol{\phi}'$.

PROOF.   Let $\bar{\mathbf{u}} = \mathbf{u} - \mathbf{u}'$, $\bar{\boldsymbol{\phi}} = \boldsymbol{\phi} - \boldsymbol{\phi}'$, $\bar{t}_{kl} = t_{kl} - t'_{kl}$, $\bar{m}_{kl} = m_{kl} - m'_{kl}$, and form

$$
F(\mathbf{u}, \boldsymbol{\phi}) - F(\mathbf{u}', \boldsymbol{\phi}') = - \int_{\mathcal{V}} [t'_{kl}\epsilon'_{kl} + m'_{kl}\gamma'_{lk} - \rho(\mathbf{f} \cdot \mathbf{u}' + \mathbf{l} \cdot \boldsymbol{\phi}')] \, dv
$$

$$
+ \int_{\partial \mathcal{V}} (\hat{t}_k u'_k + \hat{m}_k \phi'_k) \, da, \tag{5.9.15}
$$

since $F(\mathbf{u}, \boldsymbol{\phi}) = 0$, by Theorem 3. On the right-hand side, we substitute $\mathbf{f}$ and $\mathbf{l}$ from the equations of motion. Using (5.9.12), the divergence theorem and the boundary conditions on tractions, we obtain

$$
F(\mathbf{u}, \boldsymbol{\phi}) - F(\mathbf{u}', \boldsymbol{\phi}') = \int_{\mathcal{V}} (\bar{t}_{kl}\epsilon'_{kl} + \bar{m}_{kl}\gamma'_{lk}) \, dv. \tag{5.9.16}
$$

But, the integrand on the right-hand side is positive semidefinite and it vanishes when $\mathbf{u}' = \mathbf{u}$, $\boldsymbol{\phi}' = \boldsymbol{\phi}$. This means that (5.9.14) is valid, completing the proof. $\square$

Theorem 1 was extended to micropolar elastic solids by Russo [1985], occupying unbounded regions. However, Russo ignored the constitutive moduli C_{klmn}.

The importance of variational principles and minimum potential energy stems from the fact that approximate solutions can be constructed by means of these theorems. To this end, we can select some functions for the fields (e.g., $\mathbf{u}$ and $\boldsymbol{\phi}$) that satisfy some of the boundary conditions. The unknown parameters of these functions can then be determined by the variations of functionals or minimization of the strain energy. These systems of equations (usually linear) when solved, give approximate solutions for the problem. The situation here is completely analogous to those encountered in classical elasticity.

Hamilton's Principle

Extension of the variational principle to the space–time domain results in Hamilton's Principle. Specifically, let $\mathcal{K}(t), \mathcal{U}(t)$, and $\delta P(t)$ denote, respectively, the total kinetic energy, the internal energy, and the work of the external loads, during the variations $\delta\mathbf{u}$ and $\delta\boldsymbol{\phi}$ of the displacement and rotation fields, i.e.,

$$\mathcal{K}(t) = \int_V K(t)\, dv, \qquad \mathcal{U}(t) = \int_V U(t)\, dv,$$

$$\mathcal{L}[u] = \int_{t_1}^{t_2} \int_V L[\mathbf{x}, \mathbf{u}]\, dv\, dt = \int_{t_1}^{t_2} [\mathcal{U}(t) - \mathcal{K}(t)]\, dt,$$

$$\delta P(t) = \int_V (\rho\mathbf{f} \cdot \delta\mathbf{u} + \rho\mathbf{l} \cdot \delta\boldsymbol{\phi})\, dv + \int_V (\mathbf{t}_{(\mathbf{n})} \cdot \delta\mathbf{u} + \mathbf{m}_{(\mathbf{n})} \cdot \delta\boldsymbol{\phi})\, da, \quad (5.9.17)$$

where $K(t)$ and $U(t)$ are defined by

$$K(t) = \tfrac{1}{2}\rho\dot{\mathbf{u}} \cdot \dot{\mathbf{u}} + \tfrac{1}{2}\rho j_{kl}\dot{\phi}_k\dot{\phi}_l,$$
$$U(t) = \tfrac{1}{2}t_{kl}\epsilon_{kl} + \tfrac{1}{2}m_{kl}\gamma_{lk}. \qquad (5.9.18)$$

We ignore the thermal effects.

Hamilton's principle states that

$$\delta\mathcal{L}[u] = \int_{t_1}^{t_2} \delta P(t)\, dt \qquad (5.9.19)$$

provided $\delta\mathbf{u}$ and $\delta\boldsymbol{\phi}$ vanish at $t = t_1$ and $t = t_2$:

$$\delta\mathbf{u}(\mathbf{x}, t_1) = \delta\mathbf{u}(\mathbf{x}, t_2) = \mathbf{0}, \qquad \delta\boldsymbol{\phi}(\mathbf{x}, t_1) = \delta\boldsymbol{\phi}(\mathbf{x}, t_2) = \mathbf{0}.$$

The proof of (5.9.19) is similar to that of the variational principle. In fact, Theorem 1 is a generalized form of Hamilton's Principle.

5.10   Conservation Laws

A very important consequence of the variational problem is that with each infinitesimal symmetry group of the Hamiltonian (Lagrangian), there is associated a conservation law. Roughly speaking, this is known as *Noether's theorem*. The importance of this theorem stems from the fact that, by finding all infinitesimal group symmetries of the Lagrangian (Hamiltonian), we can construct all conservation laws, in a systematic way. In fact, in Section 2.2, this idea was exploited by applying Galilean group symmetry to the expression of the energy balance. This then led to *all* balance laws of 3M continua. Conversely, to each conservation law, there corresponds a symmetry group of the Hamiltonian (Lagrangian).

For micropolar elasticity, Vukobrat [1989] discussed this problem. Pucci and Saccomandi [1990] determined the symmetry group and conservation laws by means of the Lie group and Noether's theorem. This method is quite general and applicable to variational problems and differential equations. The method is clearly presented and exploited through many examples of applications in an excellent book by Olver [1986]. Here we outline this method and apply it to micropolar elastostatics. Before we do that we give a summary of basic theorems, relevant to these applications:

Let $\mathbf{x} = (x^1, x^2, \ldots, x^p)$ denote a point in rectangular coordinates, in p-dimensional Euclidean space R_p, and let $\mathbf{w} = (w^1, w^2, \ldots, w^q)$ denote the continuously differentiable q functions of $\mathbf{x}$, in a bounded, closed, regular region W of the Euclidean space W. Consider a one-parameter, continous group of transformations

$$G: \quad \bar{\mathbf{x}} = \bar{\mathbf{x}}(\mathbf{x}, \epsilon), \qquad \bar{w}^\alpha = \bar{w}^\alpha(\bar{\mathbf{x}}), \qquad (5.10.1)$$

where $\epsilon = 0$ corresponds to identity transformations

$$\bar{\mathbf{x}}(\mathbf{x}, 0) = \mathbf{x}, \qquad \bar{w}^\alpha|_{\epsilon=0} = w^\alpha(\mathbf{x}). \qquad (5.10.2)$$

Under G, a function $f(\mathbf{x}, w^\alpha)$ transforms to $f(\bar{\mathbf{x}}, \bar{w}^\alpha)$. For sufficiently small values of ϵ, we can write

$$f(\bar{\mathbf{x}}, \bar{w}^\alpha) = f(\mathbf{x}, w^\alpha) + \left(\frac{df}{d\epsilon}\right)_{\epsilon=0} \epsilon + O(\epsilon^2), \qquad (5.10.3)$$

where

$$\left(\frac{df}{d\epsilon}\right)_{\epsilon=0} = X^i \left(\frac{\partial f}{\partial \bar{x}^i}\right)_{\epsilon=0} + W^\alpha \left(\frac{\partial f}{\partial w^\alpha}\right)_{\epsilon=0},$$

$$X^i = \left(\frac{d\bar{x}^i}{d\epsilon}\right)_{\epsilon=0}, \qquad W^\alpha = \left(\frac{dw^\alpha}{d\epsilon}\right)_{\epsilon=0}. \qquad (5.10.4)$$

Equivalently, we write

$$\bar{\mathbf{x}} = \mathbf{x} + \epsilon \mathbf{X}(\mathbf{x}, \mathbf{w}), \qquad \bar{w}^\alpha = w^\alpha(\mathbf{x}) + \epsilon W^\alpha(\mathbf{x}, \mathbf{w}), \qquad (5.10.5)$$

$$\left.\frac{df(\bar{\mathbf{x}}, \bar{w}^\alpha)}{d\epsilon}\right|_{\epsilon=0} = \mathbf{v}(f), \qquad (5.10.6)$$

$$\mathbf{v} = \sum_{i=1}^{p} X^i(\mathbf{x}, \mathbf{w}) \frac{\partial}{\partial x^i} + \sum_{\alpha=1}^{q} W^\alpha(\mathbf{x}, \mathbf{w}) \frac{\partial}{\partial w^\alpha}, \qquad (5.10.7)$$

where the "vector field" $\mathbf{v}$ is called the *infinitesimal generator* of the Lie group G.

From (5.10.3), it is clear that $f(\bar{\mathbf{x}}, \bar{w}^\alpha) = f(\mathbf{x}, w^\alpha)$ whenever $(df/d\epsilon)_{\epsilon=0} = 0$. This suggests:

Definition 1. A function $f(\mathbf{x}, w^\alpha)$ is said to be infinitesimally invariant under G if and only if

$$\left(\frac{df}{d\epsilon}\right)_{\epsilon=0} = 0 \qquad \text{or} \qquad \mathbf{v}(f) = \mathbf{0}. \tag{5.10.8}$$

This definition can be extended to a functional

$$\mathcal{L}[w] = \int_R L(\mathbf{x}, \mathbf{w}^{(n)})\, d\mathbf{x}, \tag{5.10.9}$$

where

$$\mathbf{w}^{(n)} = (\mathbf{w}, \mathbf{w}^{(1)}, \ldots, \mathbf{w}^{(n)}), \qquad \mathbf{w}^{(k)} \equiv \partial^k \mathbf{w}/\partial x^k. \tag{5.10.10}$$

Definition 2. The functional $\mathcal{L}[w]$ is said to be infinitesimally invariant under (5.10.5) if and only if

$$\left[\frac{d}{d\epsilon} \int_{\bar{R}} L(\bar{\mathbf{x}}, \bar{\mathbf{w}}^{(n)})\, d\bar{\mathbf{x}}\right]_{\epsilon=0} = 0, \tag{5.10.11}$$

where $\bar{R} \subset R$ and $\bar{\mathbf{w}}$ is a single-valued function over $\bar{R}$.

Theorem 1. A connected group of transformations G is a variational symmetry group of the functional (5.10.9) if and only if

$$\mathrm{pr}^{(n)}\mathbf{v}(L) + L \operatorname{div} \mathbf{X} = 0 \tag{5.10.12}$$

for all $(\mathbf{x}, \mathbf{w}^{(n)})$ and for every infinitesimal generator (5.10.7) of G.

Here, the symbol $\mathrm{pr}^{(n)}\mathbf{v}$ denotes the nth order prolongation of the vector field

$$\mathbf{v} = X^k \frac{\partial}{\partial x^k} + W^\alpha \frac{\partial}{\partial w^\alpha}, \tag{5.10.13}$$

where k is summed over 1 to p and α is summed over 1 to q. The operators $\mathrm{pr}^{(n)}\mathbf{v}$ are constructed as follows:

$$\mathrm{pr}^{(n)}\mathbf{v} = \mathbf{v} + \sum_{\alpha=1}^{q} \sum_{J} W_J^\alpha\left(\mathbf{x}, \mathbf{w}^{(n)}\right) \frac{\partial}{\partial w_J^\alpha}, \tag{5.10.14}$$

where the second summation is over all multi-indices $J = (j_1, \ldots, j_k)$, with $1 \leq j_\kappa \leq p$, $1 \leq k \leq n$. The coefficient functions W_J^α are given by

$$W_J^\alpha(x, w^{(n)}) = D_J\left(W^\alpha - \sum_{i=1}^{p} X^i w_i^\alpha\right) + \sum_{i=1}^{p} X^i w_{J,i}^\alpha,$$

$$w_i^\alpha = \partial w^\alpha/\partial x^i, \qquad w_{J,i}^\alpha = \partial w_J^\alpha/\partial x^i = \frac{\partial^{k+1} w^\alpha}{\partial x^i\, \partial x^{j_1} \ldots \partial x^{j_k}},$$

$$D_J = D_{j_1} D_{j_2} \ldots D_{j_k}. \tag{5.10.15}$$

For a function $P(\mathbf{x}, \mathbf{w}^{(n)})$ the total derivatives of P are defined by

$$D_i P = \frac{\partial P}{\partial x^i} + \sum_{\alpha=1}^{q} \sum_J w_{J,i}^\alpha \frac{\partial P}{\partial w_J^\alpha}, \tag{5.10.16}$$

where the sum is over all J's of order $0 \le \sharp J \le n$ and n is the highest-order derivative appearing in P. For these expressions and for the proof of the theorem, see Olver [1986, pp. 113 and pp. 257]. Here we give a proof of this theorem for the case $n = 1$.

$L(\mathbf{x}, \mathbf{w}^{(1)})$ is infinitesimally invariant if and only if (5.10.11) is valid for $n = 1$. When transferred to $\mathbf{x} \subset R$, (5.10.11) reads

$$\left[\frac{d}{d\epsilon} \int_R L(\bar{\mathbf{x}}, \bar{\mathbf{w}}^{(1)}) J\, d\mathbf{x} \right]_{\epsilon=0} = 0, \tag{5.10.17}$$

where J is the Jacobian determinant, given by

$$J(\mathbf{x}, \bar{\mathbf{x}}) = \det(\partial \bar{\mathbf{x}} / \partial \mathbf{x}). \tag{5.10.18}$$

Since the arbitrary region of integration R is independent of ϵ, we have

$$\left[\frac{d}{d\epsilon} L(\bar{\mathbf{x}}, \bar{\mathbf{w}}^{(1)}) \right]_{\epsilon=0} J_0 + L(\mathbf{x}, \mathbf{w}^{(1)}) \left[\frac{\partial J}{\partial \epsilon} \right]_{\epsilon=0} = 0. \tag{5.10.19}$$

But, we have

$$J_0 = 1, \qquad \frac{dJ}{d\epsilon} = X_{,k}^k,$$

$$\frac{dL(\bar{\mathbf{x}}, \bar{\mathbf{w}}^{(1)})}{d\epsilon} = \frac{\partial L}{\partial \bar{x}^k} \frac{d\bar{x}^k}{d\epsilon} + \frac{\partial L}{\partial \bar{w}^\alpha} \frac{d\bar{w}^\alpha}{d\epsilon} + \frac{\partial L}{\partial \bar{w}_{,k}^\alpha} \frac{d\bar{w}_{,k}^\alpha}{d\epsilon}, \tag{5.10.20}$$

$$\left(\frac{dw_{,k}^\alpha}{d\epsilon} \right)_{\epsilon=0} = W_{,k}^\alpha - w_{,l}^\alpha X_{,k}^l.$$

Consequently, (5.10.19) becomes

$$X^k \frac{\partial L}{\partial x^k} + W^\alpha \frac{\partial L}{\partial w^\alpha} + \left(W_{,k}^\alpha - X_{,k}^l w_{,l}^\alpha \right) \frac{\partial L}{\partial w_{,k}^\alpha} + L X_{,k}^k = 0. \tag{5.10.21}$$

But this is none other than

$$\mathrm{pr}^{(1)} \mathbf{v}(L) + L \mathrm{div} \mathbf{X} = 0, \tag{5.10.22}$$

which completes the proof of the theorem.

A very important theorem due to Noether [1918] tells us that, to every infinitesimal symmetry of the Lagrangian L, there corresponds a conservation law. Thus, if we can determine all infinitesimal symmetries of L, then we can construct the corresponding conservation laws. This is expressed by the following restricted form of Noether's theorem:

Theorem 2. Suppose G is a local one-parameter group of symmetries of the variational problem (5.10.9) with infinitesimal generators given by (5.10.7) and

$$Q_\alpha(\mathbf{x}, \mathbf{w}) = W^\alpha - \sum_{k=1}^p X^k w_{,k}^\alpha. \tag{5.10.23}$$

Then, there exist p-tuple $\mathbf{P}(\mathbf{x}, \mathbf{w}^{(m)}) = (P_1, \ldots, P_p)$ such that

$$\mathrm{div}\mathbf{P} = \mathbf{Q} \cdot \mathbf{E}(L) = \sum_{\nu=1}^q Q_\nu E_\nu(L). \tag{5.10.24}$$

This is a conservation law of the Euler–Lagrange equations $\mathbf{E}(L) = \mathbf{0}$.

We give the proof of this theorem for $n = 1$. In the expression (5.10.22), we use the Euler–Lagrange equation

$$\frac{\partial L}{\partial w^\alpha} - \frac{\partial}{\partial x^k}\left(\frac{\partial L}{\partial w_{,k}^\alpha}\right) = 0 \tag{5.10.25}$$

and the expression

$$(X^k L)_{,k} = X_{,k}^k L + X^k\left(\frac{\partial L}{\partial x^k} + \frac{\partial L}{\partial w^\alpha}w_{,k}^\alpha + \frac{\partial L}{\partial w_{,l}^\alpha}w_{,lk}^\alpha\right).$$

With this, (5.10.22) is expressed as

$$\mathrm{div}\,\mathbf{P} = 0, \tag{5.10.26}$$

where

$$P^k = \sum_{\alpha=1}^q W^\alpha \frac{\partial L}{\partial w_{,k}^\alpha} + X^k L - \sum_{\alpha=1}^q\sum_{l=1}^p X^l w_{,l}^\alpha \frac{\partial L}{\partial w_{,k}^\alpha} + B^k, \tag{5.10.27}$$

where B^k is an opportune analytic function. If $B^k \neq 0$, then the vector field $\mathbf{v}$ is an *infinitesimal "divergence"* symmetry of $\mathcal{L}$. If $B^k = 0$, then $\mathbf{v}$ is the generator of a variational symmetry group of the functional $\mathcal{L}$.

Theorem 3. If G is a variational symmetry group of the functional $\mathcal{L}[w]$, then G is a symmetry group of the Euler–Lagrange equations $E(L) = 0$.

However, it is *not* true that every symmetry group of the Euler–Lagrange equations is also a variational symmetry group of the original variational problem. Thus, we need to exclude those symmetry groups of $E(L) = 0$, by using (5.10.12), from the list of symmetries of the original variational problem.

The symmetry group of a system of differential equations

$$\Delta_\nu(\mathbf{x}, w^{(n)}) = 0, \qquad \nu = 1, \dots, l, \tag{5.10.28}$$

is similarly defined, under some mild conditions of *maximal rank*. This means that the Jacobian matrix

$$J_\Delta(\mathbf{x}, w^{(n)}) = \left(\frac{\partial \Delta_\nu}{\partial x^i}, \frac{\partial \Delta_\nu}{\partial w_j^\alpha}\right) \tag{5.10.29}$$

must be of rank l whenever (5.10.28) is satisfied.

Theorem 4. Suppose

$$\Delta_\nu(\mathbf{x}, w^{(n)}) = 0, \quad \nu = 1, \dots, l, \tag{5.10.30}$$

is a system of differential equations of maximal rank, defined over $M \subset R \times W$. If G is a local group of transformations acting on M, and

$$pr^{(n)}\mathbf{v}[\Delta_\nu(\mathbf{x}, w^{(n)})] = 0, \quad \nu = 1, \dots, l, \quad \text{whenever} \quad \Delta(\mathbf{x}, w^{(n)}) = 0, \tag{5.10.31}$$

for every infinitesimal generator $\mathbf{v}$ of G, then G is a symmetry group of the system. See Olver [1986], p. 106.

An equivalent condition to (5.10.31) is that there exist functions $Q_{\nu\mu}(\mathbf{x}, w^{(n)})$ such that

$$\mathrm{pr}^{(n)}\mathbf{v}[\Delta_\nu(\mathbf{x}, w^{(n)})] = \sum_{\mu=1}^{l} Q_{\nu\mu}(\mathbf{x}, w^{(n)})\Delta_\mu(\mathbf{x}, w^{(n)}). \tag{5.10.32}$$

The task of finding the symmetry group and conservation laws of a set of differential equations (5.10.30) requires the following steps:

(i) Calculate (5.10.31) to obtain the symmetry group of (5.10.30).

(ii) Using (5.10.12), discard those memebers that are not symmetries of the functional $\mathcal{L}[w]$ from which (5.10.30) are obtained as Euler–Lagrange equations.

(iii) Calculate Q_α given by (5.10.23).

(iv) Conservation laws are then given by (5.10.24).

In the following, we proceed to apply this process to elastostatic micropolar field equations (see Pucci and Saccomandi [1990]).

$$(\lambda + \mu)u_{l,lk} + (\mu + \kappa)u_{k,ll} + \kappa\epsilon_{klm}\phi_{m,l} = 0,$$
$$(\alpha + \beta)\phi_{l,lk} + \gamma\phi_{k,ll} + \kappa\epsilon_{klm}u_{m,l} - 2\kappa\phi_k = 0. \tag{5.10.33}$$

The $\Delta = 0$ system is linear and of second-order differential equations in the dependent variables $u_k(\mathbf{x})$ and $\phi_k(\mathbf{x})$, $k = 1, 2, 3$. The infinitesimal generator of the Lie group is expressed as

$$\mathbf{v} = X^k \frac{\partial}{\partial x_k} + U^\alpha \frac{\partial}{\partial u_\alpha} + \Phi^\alpha \frac{\partial}{\partial \phi_\alpha}. \tag{5.10.34}$$

The operator $\mathrm{pr}^{(2)}\mathbf{v}$ is of the form

$$\mathrm{pr}^{(2)}\mathbf{v} = \mathbf{v} + U^{\alpha i} \frac{\partial}{\partial u_{\alpha,i}} + \Phi^{\alpha i} \frac{\partial}{\partial \phi_{\alpha,i}} + U^{\alpha ij} \frac{\partial}{\partial u_{\alpha,ij}} + \Phi^{\alpha ij} \frac{\partial}{\partial \phi_{\alpha,ij}}, \tag{5.10.35}$$

with $U^{\alpha i}$, $U^{\alpha ij}$, etc., given by (5.10.15), i.e.,

$$U^{\alpha i} = D_i U^\alpha - u_{\alpha,j} D_i X^j,$$
$$U^{\alpha ij} = D_i D_j U^\alpha - u_{\alpha,ij} D_i X^k - u_{\alpha,ik} D_j X^k - u_{\alpha,k} D_i D_j X^k,$$
$$D_i = \frac{\partial}{\partial x_i} + u_{\alpha,i} \frac{\partial}{\partial u_\alpha} + \phi_{\alpha,i} \frac{\partial}{\partial \phi_\alpha}. \tag{5.10.36}$$

Expressions of $\Phi^{\alpha i}$ and $\Phi^{\alpha ij}$ are similar to these, with $\mathbf{u}$ replaced by ϕ and U by Φ. Applying (5.10.31) to Equations (5.10.33), the infinitesimal generators of $\mathbf{v}$ become

$$\left[(\lambda + \mu)U^{llk} + (\mu + \kappa)U^{kll} + \kappa\epsilon_{klm}\Phi^{ml}\right]_{\Delta=0} = 0, \tag{5.10.37}$$
$$\left[(\alpha + \beta)\Phi^{llk} + \gamma\Phi^{kll} + \kappa\epsilon_{klm}U^{ml} - 2\kappa\Phi^k\right]_{\Delta=0} = 0. \tag{5.10.38}$$

Substituting the expressions of $U^{kl}, U^{klm}, \Phi^{ml}$, etc., from (5.10.36) into (5.10.37) and (5.10.38) leads to the "defining equations" for the generators X^k, U^α, and Φ^α. However, these expressions are too long and unwieldy even for symbol manipulating programs. Instead, Pucci and Saccomandi make some simplifications, extracted from setting the coefficients of the second derivatives appearing in these equations to zero to arrive at

$$X^k = X^k(\mathbf{x}),$$
$$U^\alpha = a_j^\alpha(\mathbf{x})u_j + f^\alpha(\mathbf{x}),$$
$$\Phi^\alpha = b_j^\alpha(\mathbf{x})\phi_j + g^\alpha(\mathbf{x}).$$

With the use of these results then, infinitesimal generators of Lie groups which leave field equations of micropolar elastostatics invariant are found to be

$$X^i = d_i + \epsilon_{ijk}a_j x_k,$$
$$U^\alpha = cu_\alpha + \epsilon_{\alpha jk}a_j u_k + f^\alpha(\mathbf{x}),$$
$$\Phi^\alpha = c\phi_\alpha + \epsilon_{\alpha jk}a_j \phi_k + g^\alpha(\mathbf{x}), \tag{5.10.39}$$

where d_i, a_j, c are arbitrary parameters and $f^\alpha(\mathbf{x})$ and $g^\alpha(\mathbf{x})$ are the solutions of the systems (5.10.33) (f_α replacing u_k and g_α replacing ϕ_k).

From (5.10.39), it is clear that the symmetry algebra is generated by the following vector fields:

$$v_1 = \partial_x \quad \text{(translations)},$$
$$v_2 = \mathbf{x} \wedge \partial_x + \mathbf{u} \wedge \partial_u + \boldsymbol{\phi} \wedge \partial_\phi \quad \text{(rotations)},$$
$$v_3 = \mathbf{u}^T \cdot \partial_u + \boldsymbol{\phi}^T \cdot \partial_\phi \quad \text{(scaling)},$$
$$v_4 = f(\mathbf{x})^T \cdot \partial_u + g(\mathbf{x})^T \cdot \partial_\phi \quad \text{(addition of solutions)}, \quad (5.10.40)$$

where the superposed T denotes the transposed vector and

$$\partial_x = (\partial_{x1}, \partial_{x2}, \partial_{x3}), \quad \partial_u = (\partial_{u1}, \partial_{u2}, \partial_{u3}), \quad \partial_\phi = (\partial_{\phi1}, \partial_{\phi2}, \partial_{\phi3}).$$

With the symmetries of the field equations determined, we now proceed to step (ii) and use (5.10.12).

The Euler–Lagrange equations (5.10.33) are the result of the variational problem (5.10.11), where

$$\mathcal{L} = \int_R L \, dv, \qquad (5.10.41)$$

where L is the strain energy function given by

$$L = \tfrac{1}{2} \left[\lambda u_{k,k} u_{l,l} + (\mu + \kappa) u_{k,l} u_{k,l} + \mu u_{k,l} u_{l,k} + (2\mu + \kappa) \epsilon_{klm} u_{k,l} \phi_m \right.$$
$$\left. + (2\mu + \kappa) \phi_k \phi_k + \alpha \phi_{k,k} \phi_{l,l} + \beta \phi_{k,l} \phi_{l,k} + \gamma \phi_{k,l} \phi_{k,l} \right]. \qquad (5.10.42)$$

According to (5.10.12), only those members of (5.10.40) that do not violate (5.10.12) are the variational symmetry group of the functional (5.10.41). It may be verified that, excluding the scaling (v_3), all v_i are a symmetry group of the functional (5.10.41).

The flux vector $\mathbf{P}$ is given by

$$P_i = U^\alpha \frac{\partial L}{\partial u_{,k}^\alpha} + \Phi^\alpha \frac{\partial L}{\partial \phi_{\alpha,i}} + X^i L - X^j \left(u_{\alpha,j} \frac{\partial L}{\partial u_{\alpha,i}} + \phi_{\alpha,j} \frac{\partial L}{\partial \phi_{\alpha,i}} \right) - B_i. \qquad (5.10.43)$$

Translations: For the translation, we have the flux vector corresponding to the k-generator

$$P_i^k = u_{\alpha,k} \frac{\partial L}{\partial u_{\alpha,i}} + \phi_{\alpha,k} \frac{\partial L}{\partial \phi_{\alpha,i}} + \delta_{ik} L. \qquad (5.10.44)$$

Components of these vector fluxes lead to the extension of Eshelby's energy momentum tensor in micropolar elastostatics.

Rotations: Generators of the rotation group are also infinitesimal variational symmetries. The flux vector for the k-generator is given by

$$P_i^k = \epsilon_{kmj} \left(u_m \frac{\partial L}{\partial u_{j,i}} + \phi_m \frac{\partial L}{\partial \phi_{j,i}} \right)$$
$$- \epsilon_{kmj} x_j \left(u_{l,m} \frac{\partial L}{\partial u_{l,i}} + \phi_{l,m} \frac{\partial L}{\partial \phi_{l,i}} \right) + \epsilon_{kmi} X_m L. \qquad (5.10.45)$$

These are the consequence of the isotropy of the body.

Addition of Solutions: The vector field v_4 turns out to be a divergence symmetry. This is shown by setting

$$V_{ji}(h, z) = (\lambda + \mu)h_{r,r}\delta_{ij} + (\mu + \kappa)h_{j,i} + \kappa\epsilon_{ijk}z_k,$$
$$W_{ij}(z) = (\alpha + \beta)z_{r,r}\delta_{ij} + \gamma z_{j,i},$$

then the vector field **B** corresponding to v_4 is

$$B_i = u_j V_{ji}(f, g) + \phi_j W_{ij}(g).$$

The corresponding flux vector is

$$P_i = f_j V_{ij}(\mathbf{u}, \boldsymbol{\phi}) + g_i W_{ij}(\boldsymbol{\phi}) - \{u_j V_{ij}(f, g) + \phi_j W_{ij}(f, g)\}.$$

This is the result of Betti's reciprocal theorem and the linearity of the field equations (5.10.33).

The scale group generated by v_3 is neither a variational nor a divergence symmetry. Moreover, no linear combinations with other vector fields (5.10.40) generate such symmetry.

5.11 Plane Harmonic Waves

Micropolar effects become important in high-frequency and short wavelength regions of waves. In order to display these effects, and the new physical phenomena predicted, over and above classical elasticity, here we discuss the dispersion of plane harmonic waves in an isotropic micropolar elastic solid. This also provides a basis for experimental observations to determine the micropolar elastic constants and to compare with the predictions of the atomic lattice dynamics discussed in Section 5.12.

Disregarding thermal effects and body loads, the field equations (5.6.16) and (5.6.17) can be decomposed into scalar and vector wave equations by introducing scalar potentials (σ, τ) and vector potentials $\mathbf{U}, \boldsymbol{\Phi}$,

$$\mathbf{u} = \nabla\sigma + \nabla \times \mathbf{U}, \qquad \nabla \cdot \mathbf{U} = 0,$$
$$\boldsymbol{\phi} = \nabla\tau + \nabla \times \boldsymbol{\Phi}, \qquad \nabla \cdot \boldsymbol{\Phi} = 0. \tag{5.11.1}$$

Equations (5.6.16) and (5.6.17) are then satisfied if

$$c_1^2\nabla^2\sigma - \ddot{\sigma} = 0, \qquad c_3^2\nabla^2\tau - \omega_0^2\tau - \ddot{\tau} = 0, \tag{5.11.2}$$
$$c_2^2\nabla^2\mathbf{U} + \tfrac{1}{2}j\omega_0^2\nabla \times \boldsymbol{\Phi} - \ddot{\mathbf{U}} = \mathbf{0},$$
$$c_4^2\nabla^2\boldsymbol{\Phi} - \omega_0^2\boldsymbol{\Phi} + \tfrac{1}{2}\omega_0^2\nabla \times \mathbf{U} - \ddot{\boldsymbol{\Phi}} = \mathbf{0}, \tag{5.11.3}$$

where

$$c_1^2 = \frac{\lambda + 2\mu + \kappa}{\rho}, \quad c_2^2 = \frac{\mu + \kappa}{\rho}, \quad c_3^2 = \frac{\alpha + \beta + \gamma}{\rho j}, \quad c_4^2 = \frac{\gamma}{\rho j}, \quad \omega_0^2 = \frac{2\kappa}{\rho j}. \tag{5.11.4}$$

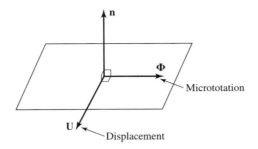

FIGURE 5.11.1. Coupled Transverse Micropolar Waves

For plane-harmonic waves, we write

$$(\sigma, \tau, \mathbf{U}, \boldsymbol{\Phi}) = (a, b, \mathbf{A}, \mathbf{B}) \exp[i(\mathbf{k} \cdot \mathbf{x} - \omega t)], \qquad (5.11.5)$$

where $\mathbf{k}$ is the wave-vector and ω is the circular frequency. With these, (5.11.2) give the dispersion relations of scalar waves:

$$\omega_r^2 = v_r^2 k^2, \quad r = 1, 3, \qquad v_1^2 = c_1^2, \qquad v_3^2 = \omega_0^2 k^{-2} + c_3^2. \qquad (5.11.6)$$

The vector wave equations (5.11.3) become

$$(\omega^2 - c_2^2 k^2)\mathbf{A} + i\frac{j\omega_0^2}{2}\mathbf{k} \times \mathbf{B} = \mathbf{0},$$

$$\frac{i}{2}\omega_0^2 \mathbf{k} \times \mathbf{A} + (\omega^2 - \omega_0^2 - c_4^2 k^2)\mathbf{B} = \mathbf{0}. \qquad (5.11.7)$$

From (5.11.7), we have

$$\mathbf{k} \cdot \mathbf{A} = 0, \qquad \mathbf{k} \cdot \mathbf{B} = 0. \qquad (5.11.8)$$

Consequently, $\mathbf{A}$ and $\mathbf{B}$ lie in a common plane whose unit normal is $\mathbf{k}/|\mathbf{k}|$, Figure 5.11.1.

Eliminating $\mathbf{A}$ and $\mathbf{B}$ from (5.11.7), we have

$$\nu\nu^* = k^2, \qquad \omega_0^2 \neq 0, \qquad (5.11.9)$$

where

$$\nu = \frac{c_2^2 k^2 - \omega^2}{j\omega_0^2/2}, \qquad \nu^* = \frac{c_4^2 k^2 + \omega_0^2 - \omega^2}{\omega_0^2/2}. \qquad (5.11.10)$$

Explicitly, $(5.11.9)_1$ reads

$$\omega^4 - 2p\omega^2 k^2 + qk^4 = 0, \qquad (5.11.11)$$

where

$$2p = \omega_0^2 k^{-2} + c_2^2 + c_4^2,$$

$$q = \omega_0^2\left(c_2^2 - j\frac{\omega_0^2}{4}\right)k^{-2} + c_2^2 c_4^2 = \frac{2\mu + \kappa}{2\rho}\left(\frac{\omega_0}{k}\right)^2 + c_2^2 c_4^2. \qquad (5.11.12)$$

The two roots of (5.11.11) for ω^2 are given by

$$\omega_r^2 = v_r^2 k^2, \qquad r = 2, 4, \tag{5.11.13}$$

where

$$v_2^2 = p - (p^2 - q)^{1/2}, \qquad v_4^2 = p + (p^2 - q)^{1/2}, \tag{5.11.14}$$

are the phase velocities of the vector waves.

For the discriminant, we have

$$\Delta = p^2 - q = \frac{1}{4}\left(\frac{\omega_0^2}{k^2} + c_4^2 - c_2^2\right)^2 + j\frac{\omega_0^4}{4k^2}. \tag{5.11.15}$$

Since $\Delta > 0$ for all k, both v_2^2 and v_4^2 are real and positive. Moreover,

$$v_4^2 > v_2^2 > 0 \tag{5.11.16}$$

for $k = 0$, $\omega_2 = 0$, and $\omega_4 = \omega_0$, so that the waves having frequency ω_4 have a cut-off frequency

$$\omega_{cr}^2 = \omega_0^2 = \frac{2\kappa}{\rho j}. \tag{5.11.17}$$

This is also the cut-off frequency of the longitudinal microrotation mode that has the frequency ω_3.

For some purposes, the roots k_r of the frequency equations as functions of ω prove to be useful. These are given by

$$\begin{aligned}
k_1^2 &= \omega^2/c_1^2, & k_3^2 &= (\omega^2 - \omega_0^2)/c_3^2, \\
k_2^2 &= r + \sqrt{r^2 - s}, & k_4^2 &= r - \sqrt{r^2 - s},
\end{aligned} \tag{5.11.18}$$

of which k_2^2 and k_4^2 are the roots of

$$k^4 - 2rk^2 + s = 0, \tag{5.11.19}$$

where

$$\begin{aligned}
r &= \frac{1}{2}\left(1 + \frac{c_2^2}{c_4^2}\right)\frac{\omega^2}{c_2^2} - \left(1 - \frac{j\omega_0^2}{4c_2^2}\right)\frac{\omega_0^2}{2c_4^2}, \\
s &= \frac{\omega^2}{c_2^2}\left(\frac{\omega^2}{c_4^2} - \frac{\omega_0^2}{c_4^2}\right),
\end{aligned} \tag{5.11.20}$$

provided $r^2 - s \geq 0$. This condition and positive k_2^2 and k_4^2 prevails when $\omega^2 \geq \omega_0^2$.

In summary, we have found that there are four different waves propagating with four different phase velocities

$$v_r = \omega_r/k, \qquad r = 1, 2, 3, 4. \tag{5.11.21}$$

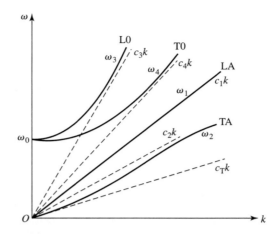

FIGURE 5.11.2. Dispersion of Micropolar Waves

(i) A longitudinal displacement wave propagates with the phase velocity v_1 in the direction $\mathbf{n} = \mathbf{k}/k$. This will be called the *Longitudinal Acoustic Branch* (LA).

(ii) A longitudinal microrotation wave propagates with the velocity v_3 in the direction of $\mathbf{n}$. This will be called the *Longitudinal Optic Branch* (LO).

(iii) A vector wave with phase velocity $v_2 = \omega_2/k$ propagates in a plane normal to $\mathbf{n}$. This will be called the *Transverse Acoustic Branch* (TA).

(iv) A vector wave coupled with TA propagates with the phase velocity $v_4 = \omega_4/k$, in a plane normal to $\mathbf{n}$. This will be called the *Transverse Optic Branch* (TO).

Dispersion relations are sketched in Figure 5.11.2. The longitudinal modes are uncoupled from each other and form the transverse modes. Excluding the LA-mode, all modes are dispersive. The asymptotes of LO, TO, and TA are shown by dotted lines. The asymptote of the LO-mode is c_3k and the asymptotes of the TA- and TO-modes are, respectively, c_2k and c_4k. The slope of the TA-mode at $k = 0$ is $c_T = [(\mu+\kappa/2)/\rho]^{1/2}$, which corresponds to the equivoluminal wave velocity in classical elasticity, since, in micropolar theory, $\mu+\kappa/2$ corresponds to the shear modulus G of classical elasticity. The existence of these four branches and the inequality (5.11.16) as $k \to \infty$ require that

$$c_1 \geq 0, \quad c_3 \geq 0, \quad \omega_0 \geq 0, \quad c_T \geq 0, \quad c_4 \geq c_2 \geq 0. \tag{5.11.22}$$

In terms of (5.11.4), these translate to

$$\lambda+2\mu+\kappa \geq 0, \quad \alpha+\beta+\gamma \geq 0, \quad \kappa \geq 0, \quad \mu+\kappa/2 \geq 0, \quad \gamma/j \geq \mu+\kappa \geq 0. \tag{5.11.23}$$

These results should be compared with the corresponding inequalities (5.3.12), based on the energy criterion.

It is also known that $\lambda \geq 0$ and α and β are estimated to be small, nonnegative quantities. On the basis of (5.11.23) and the lattice dynamical calculations, we can order c_i as follows:

$$c_3 \geq c_4 \geq c_1 \geq c_2. \tag{5.11.24}$$

Gauthier [1982] gave the following experimental values for the material he tested:

$$c_1 = 2.28 \,(\text{mm}/\mu\text{s}), \qquad c_2 = 0.94 \,(\text{mm}/\mu\text{s}),$$
$$c_4 = 2.48 \,(\text{mm}/\mu\text{s}), \qquad \omega_0 = 0.26 \,(\mu\text{s})^{-1}. \tag{5.11.25}$$

Excluding c_3, which he did not measure, these observations agree with (5.11.24).

According to the present results

$$\omega_3 \geq \omega_4 \geq \omega_1 \geq \omega_2, \tag{5.11.26}$$

which is in agreement with the lattice dynamical results (cf. Figure 5.12.3). This result supports (5.11.24). Of course, the present dispersion curves are valid only near the origin $k = 0$ when compared with those predicted by lattice dynamics.

The phase velocities of various modes follow from $v_r = \omega_r/k$ so that

$$v_3 \geq v_4 \geq v_1 \geq v_2. \tag{5.11.27}$$

5.12   Material Moduli

Linear, isotropic, micropolar, thermoelastic solids are characterized by six elastic constants $\lambda, \mu, \kappa, \alpha, \beta, \gamma$; one thermal expansion constant β_0; one heat conduction constant K; and a microinertia tensor which has a single component j. Compared to classical thermoelasticity, we have here five additional constants $\kappa, \alpha, \beta, \gamma$, and j which must be determined for each material. For anisotropic materials, as shown in Table 5.1, Section 5.2, the number of material contants is 196. Depending on the group symmetry, this number may be reduced for various classes of crystallographic groups. Contrasted with classical elasticity, the experimental task for the determination of the material constants of micropolar solids is major.

Furthermore, experiments with micropolar constants require much precision and elaborate instrumentation, since we are faced with the measurements of microscopic-level quantities in high-frequency, short-wavelength regions. At this range, many other physical phenomenon begin to interfere with observation, introducing distortions and errors. For example, internal damping, microslips, and polarization are some of these effects. Compared to theoretical work on the micropolar continuum theory, experimental work is meager.

In order to cast light on the problems and the direction of future pursuits, we consider two classes of materials:

Class I. Materials with periodic microstructure.

Class II. Materials with arbitrary microstructure.

Examples of Class I materials are all crystal classes with lattice structure, molecular crystals, composites with periodic structure, tall buildings and structures with grids, framework, and trusses. Here, the inner geometry of the material and the cohesive forces are known or can be estimated fairly accurately. Equations of motions of the microelements (lattice, cell) can be written in the form of difference equations. It is then a simple matter to make a continuum approximation, by expanding these equations in terms of an internal characteristic length (e.g., lattice parameter, beam-length). The situation here is similar to the passage to the continuum theory from lattice dynamics (cf. Born and Huang [1954], Maradudin et al. [1971], Aşkar [1985]). For grid-work, tall buildings, and trusses, this method leads to the identification of a micropolar continuum model, to simplify the structural calculations. For example, for an 80-story building, such calculations are shown to be very effective, accurate, and economical.

Analysis for Class I materials has been successful. It led to the determination of micropolar constants, in terms of force constants and geometrical orientations of the inner structure. We demonstrate this in the next section with two examples.

For Class II materials, the situation is somewhat more illusive since the inner structure is random. In this case, at the outset, it is necessary to decide on the isotropy group of the material. For isotropic media, several experiments are required for the determination of material constants. But, this is not all. Since these constants depend on the internal characteristic length l, and possibly, the internal characteristic time τ, it is necessary to select these quantities, appropriate to the material to be tested. These latter quantities enter into typical solutions that make the basis for experimental observations. The isotropy or the group symmetry of the material can be decided on the basis of additional observations (e.g., optical means, electron microscopy). Interference of unwanted effects (e.g., internal friction) will have to be checked in both classes.

In Section 5.13 we discuss the results of some attempts for this class of material. Composites with chopped fibers, bone, concrete, some man-made materials, and most real materials fall into this category.

Class I: Materials with Periodic Microstructures

The class of materials with microstructures may be divided into two subclasses:

(i) natural materials with lattice structure; and

(ii) man-made materials.

In class (i), we have all perfect crystals, molecular crystals, solid and liquid crystals. Examples for class (ii) are: fiber composites, layered structures, grid works, trusses.

The research work in area (i) is extensive. Condensed matter physicists and material scientists spend major parts of their time on the investigation of the material properties for this class of materials. Here, we discuss a simple model that connects with micropolar elastic solids.

The presence of polar effects in crystalline solids has been noticed through X-ray and neutron diffraction, specific heat measurements, and infrared and Raman spectroscopy, cf. Wagner and Hornig [1950], Balkanski and Teng [1969], Balkanski et al. [1968]. In large classes of materials (e.g., hydrocarbons such as benzene, aromatic crystals such as naphthalene; ionic crystals such as KNO_3, perovskite crystals such as $KTaO_3$, ice), frequencies of rotational modes, comparable to lattice translational modes, have been observed (cf. Musimovici [1967], Barett [1970], Akiyama et al. [1980], Rao and Chaplot [1980]). These modes are not predicted by means of classical elasticity.

Based on these observations, Aşkar [1972] employed the structure of KNO_3 to calculate the micropolar elastic moduli. This model was modified further by Inga Fisher-Hjalmar [1981a, 1981b] and Pouget et al. [1986a, 1986b], to obtain greater numerical accuracies for ferroelectric crystals. Ferroelectric crystals possess a peculiar mode associated with the ferroelectric property which is called the ferroelectric soft mode. This mode is connected with the phase transition as predicted by Cochran [1960] and Anderson [1960]. In the ferroelectric phase, these materials exhibit a strong electromechanical coupling. Thus, the piezoelectricity and acoustical wave propagations are important in this respect. Hence a wealth of experimental data has been obtained from Raman and Brillouin spectroscopies and inelastic neutron scattering (cf. Balkanski et al. [1968], Scott [1974]).

Thus, ferroelectric crystals represent an ideal lattice model to build a bridge between microcontinuum theories and lattice dynamics. In this way, the region of validity of the micropolar theory can be tested and material constants can be determined.

(i) KNO_3 Molecule

A unit cell of KNO_3 in its argonite structure is sketched in Figure 5.12.1. The complex ion NO_3 lies on the (x,y)-plane, while the simple ion K is

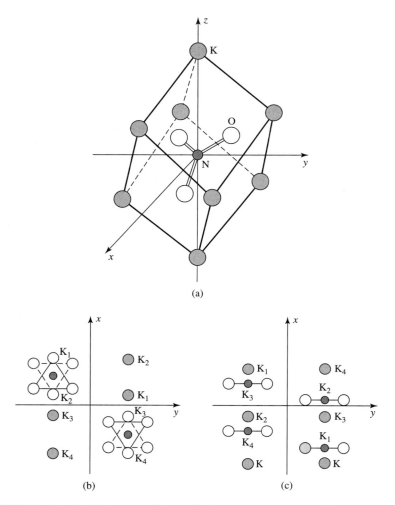

FIGURE 5.12.1. (a) Unit crystalline cell of potassium nitrate, (b) projection on (yz) plane, and (c) on (xy) plane; K denotes the potassium ion, N the nitrogen ion, and O the oxygen ion. (From Poujet et al [1986a])

on the (y, z)-plane. It is known that lattice motions (displacement and rotations) occur at much lower frequencies $(\sim 10^{13} \text{ s}^{-1})$ than the covalent modes, due to the relative displacements of N and O $(\sim 0.5 \times 10^{15} \text{ s}^{-1})$. Thus, at lattice frequencies, covalent modes may be considered as not excited. Based on this consideration, NO_3 may be modeled as a rigid ion. A simplified beam model for the lattice vibrations of KNO_3 may be considered as a mass M with rotation inertia J representing NO_3, attached to the mass m of the ion K, which possesses no rotatory inertia. The forces of attraction and repulsion between NO_3 and K ions consist of coulomb and

Born–Mayer types, which can be modeled by springs and bending coefficients.

Crystalline symmetry, with the symmetry group C_{3v} (trigonal symmetry), makes wave propagation, in three orthogonal directions (x, y, z), uncoupled, so that we can consider waves propagating along the x-axis. This
allows us to use a one-dimensional model along the x-axis of the crystal. In
the schematic diagram Figure 5.12.2, NO_3 and K ions are shown in alternat-
ing positions along the x-axis. Longitudinal and transverse displacements
of NO_3 at the location n_2 are marked as u_{n_2} and v_{n_2}, respectively. The
rotation of the NO_3 ion about the y-axis is denoted by θ_n. The K ion has
no intrinsic rotation, but it has longitudinal and transverse displacements
denoted by u_{n_1} and v_{n_1}, respectively. With the ion locations, indices n_2
and n_1 change, as shown in Figure 5.12.2. The total force acting on any
ion (and couple on NO_3) arises from the motions of the two nearest K ions
and the two NO_3 ions. Thus we have

$$\mathbf{F}_{n_1} = X_{\parallel}\mathbf{i}_1 + X_{\perp}\mathbf{i}_2, \qquad \mathbf{F}_{n_2} = Y_{\parallel}\mathbf{i}_1 + Y_{\perp}\mathbf{i}_2, \qquad (5.12.1)$$

where

$$
\begin{aligned}
X_{\parallel} &= k_{\parallel}^{11}(u_{n_1+1} + u_{n_1-1} - 2u_{n_1}) + k_{\parallel}^{21}(u_{n_2} - u_{n_1}) - k_{\parallel}^{12}(u_{n_1} - u_{n_2-1}), \\
X_{\perp} &= k_{\perp}^{11}(v_{n_1+1} + v_{n_1-1} - 2v_{n_1}) + k_{\perp}^{21}(v_{n_2} - v_{n_1}) - k_{\perp}^{12}(v_{n_1} - v_{n_2-1}) \\
&\quad + \gamma_{21}\theta_n - \gamma_{12}\theta_{n-1}, \qquad (5.12.2)
\end{aligned}
$$

where $k_{\parallel}^{ij}, k_{\perp}^{ij}$, and γ_{ij} are spring constants for horizontal and vertical displacements and rotations of the ion, respectively. $\mathbf{F}_{n_2}$ is obtained similarly,
except for additional contributions arising from the rotations of NO_3 ions.
Thus, we have

$$
\begin{aligned}
Y_{\parallel} &= k_{\parallel}^{22}(u_{n_2+1} + u_{n_2-1} - 2u_{n_2}) - k_{\parallel}^{21}(u_{n_2} - u_{n_1}) + k_{\parallel}^{12}(u_{n_1+1} - u_{n_2}), \\
Y_{\perp} &= k_{\perp}^{22}(v_{n_2+1} + v_{n_2-1} - 2v_{n_2}) - k_{\perp}^{21}(v_{n_2} - v_{n_1}) + k_{\perp}^{12}(v_{n_1+1} - v_{n_2}) \\
&\quad + \gamma_1(\theta_{n+1} + \theta_{n-1} - 2\theta_n) + \gamma_2(\theta_{n+1} - \theta_{n-1}) \\
&\quad - (\gamma_{21} - \gamma_{12})\theta_n. \qquad (5.12.3)
\end{aligned}
$$

Here, the terms containing γ_i come from the differential rotations of neigh-
boring ions to n_2 with respect to the NO_3 ion at the location n_2. For the
torque acting on the NO_3 ion at n_2, we have

$$
\begin{aligned}
T &= -\chi_1\theta_n + \chi_2(\theta_{n+1} + \theta_{n-1} - 2\theta_n) + \gamma_1(v_{n_2+1} + v_{n_2-1} - 2v_{n_2}) \\
&\quad - \gamma_2(v_{n_2+1} - v_{n_2-1}) - \gamma_{21}(v_{n_2} - v_{n_1}) - \gamma_{12}(v_{n_1+1} - v_{n_2}). \quad (5.12.4)
\end{aligned}
$$

The origins of various torque components (5.12.4) are clear from their compositions.

Equations of motion are

$$m_1\ddot{u}_{n_1} = X_{\parallel}, \quad m_2\ddot{u}_{n_2} = Y_{\parallel}, \quad m_1\ddot{v}_{n_1} = X_{\perp}, \quad m_2\ddot{v}_{n_2} = Y_{\perp},$$
$$I\ddot{\theta}_{n_2} = T. \qquad (5.12.5)$$

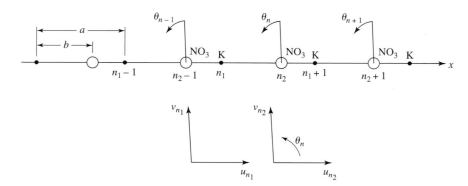

FIGURE 5.12.2. Diatomic Chain (a model for KNO_3)

It is useful to refer the motion to the inertial center of the crystalline cell. To this end, we set

$$u_{n_1} = U_n - \nu_2\zeta_n, \qquad u_{n_2} = U_n + \nu_1\zeta_n,$$
$$v_{n_1} = V_n - \nu_2\eta_n, \qquad v_{n_2} = V_n + \nu_1\eta_n, \qquad (5.12.6)$$

where

$$\nu_1 = m_1/m, \qquad \nu_2 = m_2/m,$$
$$m = m_1 + m_2, \qquad \lambda = m_1 m_2/m. \qquad (5.12.7)$$

From (5.12.5)–(5.12.7), it follows that

$$m\ddot{U}_n = K_\parallel^1(U_{n+1} + U_{n-1} - 2U_n) + K_\parallel^{12}(\zeta_{n+1} + \zeta_{n-1} - 2\zeta_n)$$
$$\quad - k_\parallel^{12}[\nu_1(\zeta_n - \zeta_{n-1}) + \nu_2(\zeta_{n+1} - \zeta_n)],$$

$$\lambda\ddot{\zeta}_n = K_\parallel^2(\zeta_{n+1} + \zeta_{n-1} - 2\zeta_n) - K_\parallel\zeta_n + K_\parallel^{12}(U_{n+1} + U_{n-1} - 2U_n)$$
$$\quad + k_\parallel^{12}[\nu_1(U_{n+1} - U_n) + \nu_2(U_n - U_{n-1})],$$

$$m\ddot{V}_n = K_\perp^1(V_{n+1} + V_{n-1} - 2V_n) + K_\perp^{12}(\eta_{n+1} + \eta_{n-1} - 2\eta_n)$$
$$\quad - k_\perp^{12}[\nu_1(\eta_n - \eta_{n-1}) + \nu_2(\eta_{n+1} - \eta_n)] + \tilde{\gamma}_1(\theta_{n+1} - \theta_{n-1})$$
$$\quad + \tilde{\gamma}_2(\theta_{n+1} + \theta_{n-1} - 2\theta_n),$$

$$\lambda\ddot{\eta}_n = K_\perp^2(\eta_{n+1} + \eta_{n-1} - 2\eta_n) - K_\perp\eta_n + K_\perp^{12}(V_{n+1} + V_{n-1} - 2V_n)$$
$$\quad + k_\perp^{12}[\nu_1(V_{n+1} - V_n) + \nu_2(V_n - V_{n-1})] - \tilde{\gamma}\theta_n$$
$$\quad + \tilde{\gamma}_1'(\theta_{n+1} - \theta_{n-1}) + \tilde{\gamma}_2'(\theta_{n+1} + \theta_{n-1} - 2\theta_n),$$

$$I\ddot{\theta}_n = -\chi_1\theta_n + \chi_2(\theta_{n+1} + \theta_{n-1} - 2\theta_n) - \tilde{\gamma}_1(V_{n+1} - V_{n-1})$$
$$\quad - \tilde{\gamma}_1'(\eta_{n+1} - \eta_{n-1}) - \tilde{\gamma}\eta_n + \tilde{\gamma}_2(V_{n+1} + V_{n-1} - 2V_n)$$
$$\quad + \tilde{\gamma}_2'(\eta_{n+1} + \eta_{n-1} - 2\eta_n), \qquad (5.12.8)$$

where

$$K_\parallel^1 = k_\parallel^{11} + k_\parallel^{22} + k_\parallel^{12}, \quad K_\parallel^{12} = \nu_1 k_\parallel^{22} - \nu_2 k_\parallel^{11},$$
$$K_\parallel^2 = \nu_2^2 k_\parallel^{11} + \nu_1^2 k_\parallel^{22} - \nu_1 \nu_2 k_\parallel^{12}, \quad k_\parallel = k_\parallel^{12} + k_\parallel^{21},$$
$$K_\perp^1 = k_\perp^{11} + k_\perp^{22} + k_\perp^{12}, \quad K_\perp^{12} = \nu_1 k_\perp^{22} - \nu_2 k_\perp^{11},$$
$$K_\perp^2 = \nu_2^2 k_\perp^{11} + \nu_1^2 k_\perp^{22} - \nu_1 \nu_2 k_\perp^{12}, \quad k_\perp = k_\perp^{12} + k_\perp^{21},$$
$$\tilde\gamma_1 = \gamma_2 + \frac{1}{2}\gamma_{12}, \quad \tilde\gamma_2 = \gamma_1 - \frac{1}{2}\gamma_{12},$$
$$\tilde\gamma_2' = \nu_1\gamma_1 + \nu_2\frac{\gamma_{12}}{2}, \quad \tilde\gamma_1' = \nu_1\gamma_2 - \frac{1}{2}\nu_2\gamma_{12}, \quad \tilde\gamma = \gamma_{21} - \gamma_{12}. \quad (5.12.9)$$

Thus, the axial and transverse motions of the inertial center are given by the first and third equations in (5.12.8), with accelerations $\ddot U_n$ and $\ddot V_n$. These two equations represent the longitudinal and transverse acoustic modes. The internal motions defined by ζ_n and η_n give the relative displacements of two adjacent ions. These lead to longitudinal and transverse optic modes. Finally, the last equation of (5.12.8) represents the rotational motions θ_n of the molecular group. This is known as the internal libration mode.

The plane harmonic wave solution of (5.12.8) was studied by Pouget et al.[4] [1986a]. This requires substituting

$$(U_n, \zeta_n, V_n, \eta_n, \theta_n) = (U, \zeta, V, \eta, \theta)\exp[i(nqa - \omega t)]$$

into (5.12.8) and studying the roots of the coefficients determinant for the frequency ω as a function of the wave number q, i.e., the dispersion relations $\omega = \omega(q)$. Omitting the detail of calculations, we sketch here the dispersion relations for the KNO$_3$ molecule. In Figure 5.12.3, frequencies of the longitudinal optic mode (LO), the logitudinal acoustic mode (LA), the transverse optic mode (TO), the transverse acoustic mode (TA), and the rotations mode (R) are plotted as functions of the wave number q. Relevant to the micropolar continuum theory, are the frequencies in the neighborhood of the origin $q = 0$. In fact, it is by means of the experimental values of frequencies near $q = 0$, that the micropolar material coefficients of KNO$_3$ are determined.

The long-wavelength approximation ($qa \ll 1$) may be made by expanding U_n, V_n, η_n, ζ_n, and θ_n into a Taylor series in the form

$$U_{n\pm1} = U \pm a\frac{\partial U}{\partial x} + \frac{a^2}{2}\frac{\partial^2 U}{\partial x^2} \pm \cdots, \quad U_n \equiv U.$$

Substituting this series into (5.12.8) and retaining the lowest order terms, we obtain

$$\rho\ddot U = C_\parallel^1 U_{xx} + C_\parallel^{12}\zeta_{xx} - D_\parallel\zeta_x,$$

---

[4]Several coefficients $(\tilde\gamma_1, \tilde\gamma_2, \tilde\gamma_1', \tilde\gamma_2')$ in (5.12.9) do not agree with those given by Pouget et al.

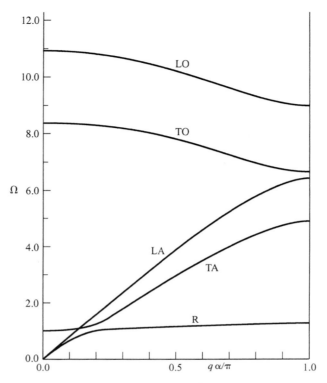

FIGURE 5.12.3. Dispersion curves for the KNO$_3$ longitudinal optic mode (LO), transverse optic mode (TO), longitudinal acoustic mode (LA). The transverse acoustic mode (TA) is coupled to the rotational mode (R). (From Poujet et al; [1986a])

$$\mu\ddot{\zeta} = C_{\parallel}^2 \zeta_{xx} - C_{\parallel}\zeta + C_{\parallel}^{12}U_{xx} + D_{\parallel}U_x,$$
$$\rho\ddot{V} = C_{\perp}^1 V_{xx} + C_{\perp}^{12}\eta_{xx} - D_{\perp}\eta_x + \epsilon_1\theta_x + \epsilon_2\theta_{xx},$$
$$\mu\ddot{\eta} = C_{\perp}^2 \eta_{xx} - C_{\perp}\eta + C_{\perp}^{12}V_{xx} + D_{\perp}V_x - \epsilon_0\theta + \epsilon_3\theta_x + \epsilon_4\theta_{xx},$$
$$J\ddot{\theta} = \kappa\theta_{xx} - \chi\theta - \epsilon_0\eta - \epsilon_1 V_x + \epsilon_2 V_{xx} - \epsilon_3\eta_x + \epsilon_4\eta_{xx}, \qquad (5.12.10)$$

where

$$
\begin{aligned}
\rho &= m/a^3, & \mu &= \lambda/a^3, \quad J = I/a^3, \\
C_{\parallel}^1 &= K_{\parallel}^1/a, & C_{\parallel}^{12} &= [K_{\parallel}^{12} + \tfrac{1}{2}(\nu_1 - \nu_2)k_{\parallel}^{12}]/a, \\
C_{\parallel}^2 &= K_{\parallel}^2/a, & C_{\parallel} &= K_{\parallel}/a^3, \quad D_{\parallel} = k_{\parallel}^{12}/a^2, \\
C_{\perp}^1 &= K_{\perp}^1/a, & C_{\perp}^{12} &= [K_{\perp}^{12} + \tfrac{1}{2}(\nu_1 - \nu_2)k_{\perp}^{12}]/a, \\
C_{\perp}^2 &= K_1^2/a, & C_{\perp} &= K_{\perp}/a^3, \quad D_{\perp} = k_{\perp}^{12}/a^2,
\end{aligned}
$$

$$\epsilon_0 = \tilde{\gamma}/a^3, \quad \epsilon_1 = 2\tilde{\gamma}_1/a^2, \quad \epsilon_2 = \tilde{\gamma}_2/a, \quad \epsilon_3 = 2\tilde{\gamma}_1'/a^2,$$
$$\epsilon_4 = \tilde{\gamma}_2'/a, \quad \kappa = \chi_2/a, \quad \chi = \chi_1/a^3.$$

$$(5.12.11)$$

At acoustic mode frequencies ($\omega < 10^{14}$ Hz), the ionic modes are not excited. Moreover, higher-order coupling terms and second gradient terms are negligible, as compared to first-order gradient terms. This means that, in the neighborhood of $q = 0$, we may neglect η_{xx} and ζ_{xx} in $(5.12.10)_{1,3}$, $\ddot{\zeta}, \zeta_{xx}, U_{xx}$ in $(5.12.10)_2$ and $\ddot{\eta}, \eta_{xx}, V_{xx}$, and θ_{xx} in $(5.12.10)_4$. With this, we can solve for

$$\zeta = -(D_{\parallel}/C_{\parallel})U_x, \quad \eta = (D_{\perp}/C_{\perp})V_x - (\epsilon_0/C_{\perp})\theta + (\epsilon_3/C_{\perp})\theta_x. \quad (5.12.12)$$

Using these in the first, third, and fifth of (5.12.10), we obtain

$$\rho\ddot{U} = [C_{\parallel}^1 + (D_{\parallel}^2/C_{\parallel})]U_{xx},$$
$$\rho\ddot{V} = \left(C_{\perp}^1 - \frac{D_{\perp}^2}{C_{\perp}}\right)V_{xx} + \left(\epsilon_1 + \frac{\epsilon_0 D_{\perp}}{C_{\perp}}\right)\theta_x + \left(\epsilon_2 - \frac{D_{\perp}\epsilon_3}{C_{\perp}}\right)\theta_{xx},$$
$$J\ddot{\theta} = \left(\kappa - \frac{\epsilon_3^2 + 2\epsilon_0\epsilon_4}{C_{\perp}}\right)\theta_{xx} + \left(\frac{\epsilon_0^2}{C_{\perp}} - \chi\right)\theta - \left(\epsilon_0\frac{D_{\perp}}{C_{\perp}} + \epsilon_1\right)V_x$$
$$+ \left(\epsilon_2 - \epsilon_3\frac{D_{\perp}}{C_{\perp}}\right)V_{xx}.$$

$$(5.12.13)$$

Further simplification is possible. The coupling terms with coefficients $\epsilon_2 - D_{\perp}\epsilon_3/C_{\perp}$ are also negligible. Alternatively, we can choose $\epsilon_2 - D_{\perp}\epsilon_3/C_{\perp} = 0$. With this, (5.12.13) agrees with the micropolar field equations

$$A_{1111}u_{,xx} - \rho\ddot{u} = 0,$$
$$A_{1212}v_{,xx} + (A_{1221} - A_{1212})\phi_{,x} - \rho\ddot{v} = 0,$$
$$B_{3131}\phi_{,xx} + (A_{1212} + A_{2121} - 2A_{1221})\phi + (A_{2112} - A_{1212})v_{,x} - \rho j_{33}\ddot{\phi} = 0,$$

$$(5.12.14)$$

for materials which are symmetrical about the ($x_1 = 0$)-plane. The two sets (5.12.13) and (5.12.14) become identical with

$$u = U, \quad v = V, \quad \theta = \phi, \quad \rho j_{33} = J, \quad A_{1111} = C_{\parallel}^1 + (D_{\parallel}^2)/C_{\parallel},$$
$$A_{1212} = C_{\perp}^1 - \frac{(D_{\perp})^2}{C_{\perp}}, \quad A_{1221} = \epsilon_1 + C_{\perp}^1 + \frac{\epsilon_0 D_{\perp} - (D_{\perp})^2}{C_{\perp}}$$
$$B_{3131} = \kappa - \frac{\epsilon_3^2 + 2\epsilon_0\epsilon_4}{C_{\perp}}, \quad A_{2121} = 2\epsilon_1 - C_{\perp}^1 - \chi + \frac{\epsilon_0^2 + 2\epsilon_0 D_{\perp} - (D_{\perp})^2}{C_{\perp}}.$$

$$(5.12.15)$$

Using Table II given in Pouget et al. [1986b], we calculate the micropolar constants for potassium nitrate

$$
\begin{aligned}
A_{1111} &= 1.31 \times 10^{10} \text{N/m}^2, & A_{1212} &= 0.68 \times 10^{10} \text{N/m}^2, \\
A_{1221} &= 0.72 \times 10^{10} \text{N/m}^2, & A_{2112} &= 0.71 \times 10^{10} \text{N/m}^2, \\
A_{2121} &= 0.76 \times 10^{10} \text{N/m}^2, & B_{3131} &= 1.43 \times 10^{-10} \text{N}, \\
J &= 2.15 \times 10^{-18} \text{ kg/m}, & \rho &= m/a^3 = 340 \text{ kg/m}^3. \quad (5.12.16)
\end{aligned}
$$

The model discussed for KNO_3 is representative of a large class of other crystalline solids. In fact, $BaTiO_3$, $NaNO_3$, and all perovskite crystals (e.g., $KTaO_3$), aromatic crystals (e.g., naphthalene), hydrogen-bonded crystals (e.g., ice), fall into this catagory. The corresponding micropolar moduli for these crystals can be calculated by means of the lattice parameters appropriate to these crystals, and/or for frequencies of various modes measured in the neighborhood of $q = 0$.

From (5.12.13) and (5.12.14), it is clear that the micropolar continuum model contains sufficient physics (beyond those of the classical continuum theory) for discussion of the micromotions of materials in the high-frequency region, for long wavelengths. In fact, Inga Fisher-Hjalmar [1981b] has shown that this model can be effective in the region up to one-third of the entire Brillouin zone around the origin.

(ii) Man-Made Structure-Grid Frameworks

Skeletons of tall buildings consist of steel and/or concrete grid structures in the form of frameworks which repeat in horizontal and vertical directions. In a way, such a structure is a macroscopic replica of the perfect crystal structures organized to meet the design and the optimum strength and stability requirements of buildings. For tall buildings, structural analysis requires solving a large number of algebraic equations which can even exhaust the capacity of larger computers, or at least pose major costs. For periodic structure, the micropolar theory offers simple analysis. The literature is fairly extensive in the study of micropolar modelization of grid structures, cf. Aşkar and Çakmak [1968], Wozniak [1966a,b], Tauchert [1970], Bazant and Christensen [1972], Sun and Yang [1973], Kanatani [1979], Noor and Nemeth [1980a,b], and Kim and Piziali [1987]. Here, we present the model explored by Bazant and Christensen for a two-dimensional grid-work constructed with bars of the same material with spans in two perpendicular directions. Moreover, we consider that, initially, the structure is unloaded and, in the loaded state, the state of stress is far below the buckling load. A two-dimensional grid framework is shown in Figure 5.12.5. A typical bar isolated from this structure, under axial load P, vertical shear V, and couples M_a and M_b, at its two ends, deforms as shown in Figure 5.12.4. Equations of equilibrium read

$$
V = -(M_a + M_b)/L,
$$

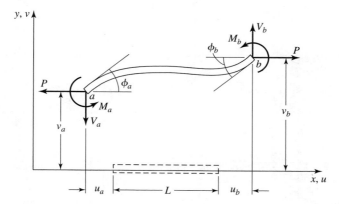

FIGURE 5.12.4. Deformation of a bar, initially lying along x-axis, under forces and couples

$$EIv_{xx} = -Vx - M_a. \qquad (5.12.17)$$

Upon solving this equation, under the boundary conditions

$$v_{a,x} = \phi_a, \qquad v_{b,x} = \phi_b, \qquad (5.12.18)$$

We find that the couples and forces can be expressed as

$$M_a = 2k \left[2\phi_a + \phi_b - \frac{3}{L}(v_b - v_a) \right],$$

$$M_b = 2k \left[\phi_a + 2\phi_b - \frac{3}{L}(v_b - v_a) \right],$$

$$V = -\frac{6k}{L} \left[\phi_a + \phi_b - \frac{2}{L}(v_b - v_a) \right],$$

$$P = c(u_b - u_a), \quad k = \frac{EI}{L}, \quad c = \frac{EA}{L}, \qquad (5.12.19)$$

where the last equation is the result of the elastic response of the bar to an axial force P. Using these results at a joint located at $x = i, y = j$, we express the equations of motion of the joint as

$$E_x(u_{i+1,j} - 2u_{i,j} + u_{i-1,j}) + \frac{6k_y}{L_y}(\phi_{i,j+1} - \phi_{i,j-1})$$

$$+ \frac{12k_y}{L_y^2}(u_{i,j+1} - 2u_{i,j} + u_{i,j-1}) - m\ddot{u}_i = 0,$$

$$E_y(v_{i,j+1} - 2v_{i,j} + v_{i,j-1}) - \frac{6k_x}{L_x}(\phi_{i+1,j} - \phi_{i-1,j})$$

$$+ \frac{12k_x}{L_x^2}(v_{i+1,j} - 2v_{i,j} + v_{i-1,j}) - m\ddot{v}_i = 0,$$

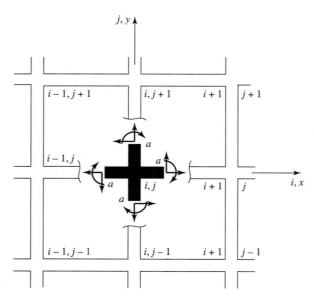

FIGURE 5.12.5. Forces and couples at a joint i, j of a grid structure

$$-2k_x(\phi_{i+1,j} + \phi_{i-1,j} + 4\phi_{i,j}) - 2k_y(\phi_{i,j+1} + \phi_{i,j-1} + 4\phi_{i,j})$$

$$-\frac{6k_y}{L_y}(u_{i,j+1} - u_{i,j-1}) + \frac{6k_x}{L_x}(v_{i+1,j} - v_{i-1,j}) - J\ddot{\phi}_{i,j} = 0, \quad (5.12.20)$$

where E_x, E_y, k_x, k_y, L_x, and L_y refer to the values of E, k, and L in the x- and y-directions. m is the mass of the black cross in Figure 5.12.5, and J is the rotatory inertia. Equations (5.12.20) also agree with those obtained by Demiray and Eringen [1977], for the orthogonally reinforced fiber composites, when the effect of the matrix is eliminated by setting its shear modules $G = 0$.

We pass to the continuum, by using expressions of the form

$$u_{i\pm1,j} = u_{i,j} \pm L_x \frac{\partial u_{i,j}}{\partial x} + \tfrac{1}{2}L_x^2 \frac{\partial^2 u_{i,j}}{\partial x^2} + \cdots,$$

$$u_{i,j\pm1} = u_{i,j} \pm L_y \frac{\partial u_{i,j}}{\partial y} + \tfrac{1}{2}L_y^2 \frac{\partial^2 u_{i,j}}{\partial y^2} + \cdots,$$

where we also set $u_{i,j} = u, v_{i,j} = v$ and $\phi_{i,j} = \phi$.

Equations (5.12.20) give the continuum equations

$$(E_x L_x/L_y)u_{,xx} + (12k_y/L_x L_y)u_{,yy} + (12k_y/L_x L_y)\phi_{,y} - (m/L_x L_y)\ddot{u} = 0,$$

$$(12k_x/L_x L_y)v_{,xx} + (E_y L_y/L_x)v_{,yy} - (12k_x/L_x L_y)\phi_{,x} - (m/L_x L_y)\ddot{v} = 0,$$

$$-(2k_x L_x/L_y)\phi_{,xx} - (2k_y L_y/L_x)\phi_{,yy} - [8(k_x + k_y)/L_x L_y]\phi$$

$$-(12k_y/L_x L_y)u_{,y} + (12k_x/L_x L_y)v_{,x} - (J/L_x L_y)\ddot{\phi} = 0.$$

$$(5.12.21)$$

In two-dimensions, the field equations (5.5.10) and (5.5.11) of the orthotropic micropolar elastic solids reduce to

$$A_{1111}u_{,xx} + A_{2121}u_{,yy} + (A_{2112} + A_{1122})v_{,xy} + (A_{2121} - A_{2112})\phi_{,y} - \rho\ddot{u} = 0,$$
$$A_{1212}v_{,xx} + A_{2222}v_{,yy} + (A_{1221} + A_{2211})u_{,xy} + (A_{1221} - A_{1212})\phi_{,x} - \rho\ddot{v} = 0,$$
$$B_{3131}\phi_{,xx} + B_{3232}\phi_{,yy} + (2A_{1221} - A_{2121} - A_{1212})\phi + (A_{1221} - A_{2121})u_{,y}$$
$$+ (A_{1212} - A_{2112})v_{,x} - \rho j\ddot{\phi} = 0. \tag{5.12.22}$$

Comparing (5.12.21) and (5.12.22), and remembering the symmetry regulations $A_{ijkl} = A_{klij}$, we arrive at the micropolar elastic constants

$$A_{1111} = E_x L_x/L_y, \qquad A_{2121} = 12k_y/L_x L_y, \qquad A_{2112} = 0, \quad (5.12.23)$$
$$A_{2222} = E_y L_y/L_x, \qquad A_{1212} = 12k_x/L_x L_y, \qquad\qquad\qquad (5.12.24)$$
$$B_{3131} = -2k_x L_x/L_y, \qquad B_{3232} = -2k_y L_y/L_x, \qquad\qquad\quad (5.12.25)$$
$$\rho = m/L_x L_y, \qquad j = J/m. \qquad\qquad\qquad\qquad\qquad (5.12.26)$$

The coupling terms $u_{,xy}$ and $v_{,xy}$ appearing in (5.12.22) will have to be discarded, since the state of stress in the grid framework is not fully two-dimensional in the classical sense. This can be achieved by setting $A_{1122} = 0$. However, note that this restriction does not follow from the material symmetry regulations.

It is remarkable that the continuum approximation of the gird framework provides the values of all micropolar elastic constants.

5.13   Experimental Attempts

The mathematical determination of micropolar elastic constants, in terms of microscopic parameters, faces major difficulties when the internal geometry of solids is not periodic. The situation here is akin to suspensions in fluids, however, requiring further micromechanical considerations involving the microrotations. While some progress has been made on this problem for classical elastic solids, presently, the micromechanics of micropolar elasticity and fluids remain untouched. From previous discussions on materials with periodic inner structures it is clear that micropolar elastic moduli depend on an *internal characteristic length l* and *internal characteristic time* τ. For isotropic materials, these may be selected as

$$l = \gamma/G, \quad \tau = (\rho j/G)^{1/2}, \tag{5.13.1}$$

where $G \equiv \mu + \kappa/2$ is the shear modulus. Consequently, micropolar effects will be observable when l is of the order of magnitude of the *external characteristic length L* and τ is of the order of the *external characteristic time T*. In static experiments, a crack tip radius, a plate thickness, or the average grain size can serve as the external characteristic length, and in dynamic experiments, as the wavelength λ. Similarly, $1/\tau$ must be in the range of high enough frequencies ω, for the observation of the effects of micromotions.

The early experimental efforts of Schijve [1966], Ellis and Smith [1967], and Gauthier and Jahsman [1975] did not disclose any couple stress or micropolar effects in composite materials. Experiments carried out by Perkins and Thomson [1973] on foam material suggested couple stress behavior for the foam. However, the viscoelastic nature of the foam may have clouded the issue. The dynamical experiments of Jahsman and Gauthier [1980] and Gauthier [1982] led to determination of some of the micropolar moduli $(\lambda, \mu, \kappa, \gamma, j)$ for an epoxy polymer carrying particulate aluminum shots. An account on these experimental results is given below.

Yang and Lakes [1982] published some experimental results on bone, suggesting that animal bone may be modeled as a micropolar solid. A brief discussion on their experiments is also included here.

A. Experimental Results of Gauthier and Jahsman

The experimental work of Gauthier [1982] involves wave propagation experiments on a fabricated micropolar material which consists of an elastic epoxy matrix with uniformly distributed "rigid" aluminum shots.

As can be seen from the lattice dynamical discussion, a fundamental difference between classical elasticity and micropolar elasticity is in the existence of the optical branches, and the coupling between shear waves and microrotational modes (cf. Sections 5.11 and 5.12). This coupling leads to a frequency spectrum for microrotational waves which is missing from classical elasticity solutions. For this mode, the frequency ω is given by

$$\frac{\omega^2}{c_4^2} = \frac{2\kappa}{\gamma} + \left(\frac{n\pi}{h}\right)^2 + k^2, \qquad n = 1, 3, 5, \tag{5.13.2}$$

where k is the wave number, h is the plate thickness, and $c_4^2 \equiv \gamma/\rho j$.

For long wavelengths, k^2 may also be dropped. Using (5.13.2) for $n = 1$ and $n = 3$, Gauthier determined

$$\frac{\kappa}{\rho j} = \frac{9\omega_1^2 - \omega_3^2}{16}, \qquad c_4^2 = \frac{\gamma}{\rho j} = \frac{1}{8}\left(\frac{h}{\pi}\right)^2 (\omega_3^2 - \omega_1^2). \tag{5.13.3}$$

For $j = 0.196$ mm^2, $h = 38.1$ mm, and $\omega_1 = \frac{1}{3}$, $\omega_3 = \frac{2}{3}$, we have

$$\frac{\kappa}{\rho} = 0.0067 \text{mm } (\mu s)^{-1}, \qquad c_4 = 2.48 \text{mm } (\mu s)^{-1}. \tag{5.13.4}$$

Then, using micropolar moduli

$$\lambda = \frac{E\nu}{(1 + \nu)(1 - 2\nu)}, \qquad \mu + \frac{\kappa}{2} = G = \frac{E}{2(1 + \nu)},$$

with $\nu = 0.4$, calculations give

$$\lambda = 7.59\text{G Pa}, \quad \mu = 1.89\text{G Pa}, \quad \gamma = 2.63\text{kN},$$
$$\frac{\kappa}{\mu} = 0.00788, \quad \frac{\gamma}{\mu j} = 7.11. \tag{5.13.5}$$

The use of a characteristic length based on bending

$$a = \left[(1 - \nu^2)\frac{\gamma}{E}\right]^{1/2} \simeq 0.76\text{mm}$$

turns out to be very close to the radius 0.7 mm of aluminum shots! Determination of the remaining constants α and β was not pursued by Gauthier. Further experiments are needed to determine these constants. For the phase velocities of plane harmonic waves (Section 5.11), we obtain

$$c_4 = (\gamma/\rho j)^{1/2} \simeq 2.48 \text{ mm}/\mu\text{s},$$
$$c_1 = [(\lambda + 2\mu + \kappa)/\rho]^{1/2} \simeq 2.28 \text{ mm}/\mu\text{s},$$
$$c_2 = [(\mu + \kappa)/\rho]^{1/2} \simeq 0.94 \text{ mm}/\mu\text{s}.$$

B. Experiments with Animal Bones

Yang and Lakes [1982] published some experimental observations on the elastic properties of human compact bone. They state that: "Experimental data are fitted more accurately by an exact solution to the bending problem in micropolar theory, than by an upper bound solution in indeterminate couple stress theory." Experimental data are compared with the theoretical curve displaying J/d^2 as a function of d^2, where J is the flexural rigidity of a long circular cylinder of diameter $d = 2a$. Using different diameters, the material constants E, N, and a are calculated for $\beta/\gamma = 1$. Here, E is Young's modulus, N is the coupling parameter, and a is the internal characteristic length, defined by

$$E \equiv \frac{(2\mu + \kappa)(3\lambda + 2\mu + \kappa)}{2\lambda + 2\mu + \kappa}, \quad N \equiv \frac{\kappa}{2(\mu + \kappa)}, \quad a \simeq \left[\frac{\gamma(\mu + \kappa)}{\kappa(2\mu + \kappa)}\right]^{1/2}. \tag{5.13.6}$$

The interpolated results are listed in Table (5.4).

Characteristic lengths obtained in this work are comparable to the size of osteons (ca. 0.25 mm diameter). This is interpreted to signify the dominant role played by the osteon as a structural element of bone. The nonclassical behavior of the bone is attributed to the presence of osteon in human compact bone.

Yang and Lakes' experimental results, summarized in Table 5.4, are subject to certain criticisms. In their calculations

$$0.50 \leq N \leq 1, \quad \beta/\gamma = 1.$$

TABLE 5.4. Micropolar elastic moduli of human compact bone.

Specimen	$E(GN/m^2)$	N^2	a	β/γ	N^2/a^2	Residual
Bone No. 2	19.1	0.50	0.232	1.0	0.16	0.75
Bone No. 3	12.0	0.75	0.971	1.0	0.84	7.37
Bone No. 4	12.4	1.00	0.629	1.0	0.63	1.63
Bone No. 5	12.0	1.00	0.605	1.0	0.61	0.22
Bone No. 6	14.1	0.65	0.489	1.0	0.39	0.56
Mean 2, 4, 5, 6	14.40	0.79	0.49	1.0	0.45	—
Standard deviation	3.26	0.25	0.18	0	0.22	—

With the lowest value of $N = 0.5$, we get $\kappa/G = 2$ ($G \equiv \mu + \kappa/2$), which is too high, since $\kappa \ll G$. For $N = 1$, $G = 0$, which is clearly not possible. Moreover, for $N = 1$, E also vanishes, indicating a serious problem.

5.14 Displacement Potentials

Field equations of linear, isotropic, micropolar, elastic solids are simplified or uncoupled by the introduction of appropriate scalar and vector potentials. Here, we discuss two different methods for the isothermal case, in three- and two-dimensional cases.

I. Displacement Potentials in Three Dimensions

For the isothermal case, the field equations (5.6.16) and (5.6.17) may be expressed in the form

$$c_1^2 \nabla \nabla \cdot \mathbf{u} - c_2^2 \nabla \times \nabla \times \mathbf{u} + j \frac{\omega_0^2}{2} \nabla \times \boldsymbol{\phi} + \mathbf{f} - \ddot{\mathbf{u}} = \mathbf{0},$$

$$c_3^2 \nabla \nabla \cdot \boldsymbol{\phi} - c_4^2 \nabla \times \nabla \times \boldsymbol{\phi} + \frac{\omega_0^2}{2} \nabla \times \mathbf{u} - \omega_0^2 \boldsymbol{\phi} + \frac{1}{j} - \ddot{\boldsymbol{\phi}} = \mathbf{0}, \quad (5.14.1)$$

where c_r are given by

$$c_1^2 = \frac{\lambda + 2\mu + \kappa}{\rho}, \quad c_2^2 = \frac{\mu + \kappa}{\rho}, \quad c_3^2 = \frac{\alpha + \beta + \gamma}{\rho j}, \quad c_4^2 = \frac{\gamma}{\rho j}, \quad \omega_0^2 = \frac{2\kappa}{\rho j}.$$

$$(5.14.2)$$

Extending the concept of Lamé potentials (Eringen and Şuhubi [1975, p. 350]) to micropolar elasticity, we decompose $\mathbf{u}, \boldsymbol{\phi}, \mathbf{f}$, and $1/j$ into scalar and vector potentials as follows (Parfitt and Eringen [1969]):

$$\mathbf{u} = \nabla \sigma + \nabla \times \mathbf{U}, \qquad \nabla \cdot \mathbf{U} = 0,$$

$$\boldsymbol{\phi} = \nabla \tau + \nabla \times \boldsymbol{\Phi}, \qquad \nabla \cdot \boldsymbol{\Phi} = 0, \qquad (5.14.3)$$

$$\mathbf{f} = \nabla g + \nabla \times \mathbf{F}, \qquad \nabla \cdot \mathbf{F} = 0,$$
$$\frac{1}{j} = \nabla h + \nabla \times \mathbf{L}, \qquad \nabla \cdot \mathbf{L} = 0. \qquad (5.14.4)$$

Substituting these into (5.14.1), we obtain

$$\nabla(c_1^2 \nabla^2 \sigma - \ddot{\sigma} + g) + \nabla \times (c_2^2 \nabla^2 \mathbf{U} + \tfrac{1}{2} j \omega_0^2 \nabla \times \boldsymbol{\Phi} - \ddot{\mathbf{U}} + \mathbf{F}) = \mathbf{0},$$
$$\nabla(c_3^2 \nabla^2 \tau - \omega_0^2 \tau - \ddot{\tau} + h)$$
$$+ \nabla \times (c_4^2 \nabla^2 \boldsymbol{\Phi} - \omega_0^2 \boldsymbol{\Phi} + \tfrac{1}{2} \omega_0^2 \nabla \times \mathbf{U} - \ddot{\boldsymbol{\Phi}} + \mathbf{L}) = \mathbf{0}. \quad (5.14.5)$$

These equations are of the form

$$\nabla a + \nabla \times \mathbf{A} = \mathbf{B} = \mathbf{0}, \quad \nabla \cdot \mathbf{A} = 0.$$

We can write (cf. Morse and Feshback [1953, p. 53])

$$a = -\nabla \cdot \mathbf{C}, \quad \mathbf{A} = \nabla \times \mathbf{C},$$

where C is defined by

$$\mathbf{C} = \int_V \frac{\mathbf{B}}{4\pi R} \, dv(\mathbf{x}'), \quad R^2 = (\mathbf{x} - \mathbf{x}') \cdot (\mathbf{x} - \mathbf{x}').$$

In the present case, $\mathbf{B} \equiv \mathbf{0}$, hence $\mathbf{C} = \mathbf{0}$. Consequently, $a = 0, \mathbf{A} = \mathbf{0}$. Thus, we have proved that:

The necessary and sufficient conditions for (5.14.5) to be satisfied are:

$$c_1^2 \nabla^2 \sigma - \ddot{\sigma} + g = 0, \qquad (5.14.6)$$
$$c_3^2 \nabla^2 \tau - \omega_0^2 \tau - \ddot{\tau} + h = 0, \qquad (5.14.7)$$
$$c_2^2 \nabla^2 \mathbf{U} + \tfrac{1}{2} j \omega_0^2 \nabla \times \boldsymbol{\Phi} - \ddot{\mathbf{U}} + \mathbf{F} = \mathbf{0}, \qquad (5.14.8)$$
$$c_4^2 \nabla^2 \boldsymbol{\Phi} - \omega_0^2 \boldsymbol{\Phi} + \tfrac{1}{2} \omega_0^2 \nabla \times \mathbf{U} - \ddot{\boldsymbol{\Phi}} + \mathbf{L} = \mathbf{0}. \qquad (5.14.9)$$

From (5.14.4), it follows that

$$\nabla^2(g, h) = \nabla \cdot (\mathbf{f}, 1/j),$$
$$\nabla^2(\mathbf{F}, \mathbf{L}) = -\nabla \times \left(\mathbf{f}, \frac{1}{j}\right). \qquad (5.14.10)$$

Particular solutions of these equations are

$$\{g, h\} = -\frac{1}{4\pi} \int_V \{\mathbf{f}(\boldsymbol{\xi}), \mathbf{l}(\boldsymbol{\xi})/j\} \cdot \nabla_{\mathbf{x}}[R^{-1}(\boldsymbol{\xi}, \mathbf{x})] \, dv(\boldsymbol{\xi}),$$
$$\{\mathbf{F}, \mathbf{L}\} = -\frac{1}{4\pi} \int_V \{\mathbf{f}(\boldsymbol{\xi}), \mathbf{l}(\boldsymbol{\xi})/j\} \times \nabla_{\mathbf{x}}[R^{-1}(\boldsymbol{\xi}, \mathbf{x})] \, dv(\boldsymbol{\xi}), \quad (5.14.11)$$

where

$$R(\boldsymbol{\xi}, \mathbf{x}) = [(\mathbf{x} - \boldsymbol{\xi}) \cdot (\mathbf{x} - \boldsymbol{\xi})]^{1/2}. \qquad (5.14.12)$$

Consequently, upon the solution of (5.14.6)–(5.14.9), we would obtain the potentials $\sigma, \tau, \mathbf{U}$, and $\boldsymbol{\Phi}$ in terms of body force and body couple distributions. Expressions (5.14.3) then give $\mathbf{u}$ and ϕ.

For harmonic time-dependence, the Fourier transform technique proves to be useful. Denoting the Fourier transform by a superposed bar, i.e.,

$$\bar{f}(\mathbf{x}, \omega) = \frac{1}{\sqrt{2\pi}} \int_{-\infty}^{\infty} e^{i\omega t} f(\mathbf{x}, t)\, dt. \qquad (5.14.13)$$

Equations (5.14.6)–(5.14.9) may be expressed as

$$\nabla^2 \bar{\sigma} + k_1^2 \bar{\sigma} + \bar{g}/c_1^2 = 0, \qquad (5.14.14)$$

$$\nabla^2 \bar{\tau} + k_3^2 \bar{\tau} + \bar{h}/c_3^2 = 0, \qquad (5.14.15)$$

$$(\nabla^2 + k_2^2)(\nabla^2 + k_4^2)\bar{\mathbf{U}} + \left(\nabla^2 + \frac{c_3^2}{c_4^2}k_3^2\right)\frac{\bar{\mathbf{F}}}{c_2^2} - \frac{1}{2}\frac{j\omega_0^2}{c_2^2 c_4^2}\nabla \times \bar{\mathbf{L}} = \mathbf{0}, \qquad (5.14.16)$$

$$(\nabla^2 + k_2^2)(\nabla^2 + k_4^2)\bar{\boldsymbol{\Phi}} - \frac{1}{2}\frac{\omega_0^2}{c_2^2 c_4^2}\nabla \times \bar{\mathbf{F}} + \left(\nabla^2 + \frac{\omega^2}{c_2^2}\right)\frac{\bar{\mathbf{L}}}{c_4^2} = \mathbf{0}, \qquad (5.14.17)$$

where k_r are given by (5.11.18). The last two equations are obtained by eliminating $\bar{\mathbf{U}}$ and $\bar{\boldsymbol{\Phi}}$ from (5.14.8) and (5.14.9). Consequently, $\bar{\mathbf{U}}$ is related to $\bar{\boldsymbol{\Phi}}$ by

$$c_2^2 \nabla^2 \bar{\mathbf{U}} + \omega^2 \bar{\mathbf{U}} + \tfrac{1}{2} j\omega_0^2 \nabla \times \bar{\boldsymbol{\Phi}} + \bar{\mathbf{F}} = \mathbf{0}. \qquad (5.14.18)$$

Note that, when body loads are absent, $\bar{\mathbf{U}}$ and $\bar{\boldsymbol{\Phi}}$ satisfy

$$(\nabla^2 + k_2^2)(\nabla^2 + k_4^2)\{\bar{\mathbf{U}}, \bar{\boldsymbol{\Phi}}\} = \mathbf{0},$$
$$c_2^2 \nabla^2 \bar{\mathbf{U}} + \omega^2 \bar{\mathbf{U}} + \tfrac{1}{2} j\omega_0^2 \nabla \times \bar{\boldsymbol{\Phi}} = \mathbf{0}. \qquad (5.14.19)$$

II. Displacement Potentials in Two Dimensions

When displacement and rotation fields depend on two space variables x_1, x_2, and time t, the field equations can be decomposed into two independent sets.

A. The Plain-Strain

$$u_3 = \phi_1 = \phi_2 = 0, \quad f_3 = l_1 = l_2 = 0, \quad \partial/\partial x_3 = 0.$$

In this case, the field equations (5.14.1) reduce to

$$(c_1^2 - c_2^2)\nabla\nabla \cdot \mathbf{u} + c_2^2 \nabla^2 \mathbf{u} + j\frac{\omega_0^2}{2}\nabla \times (\phi_3 \mathbf{e}_3) - \ddot{\mathbf{u}} + \mathbf{f} = \mathbf{0},$$

$$c_4^2 \nabla^2 \phi_3 - \omega_0^2 \phi_3 + \frac{\omega_0^2}{2}(\nabla \times \mathbf{u}) \cdot \mathbf{e}_3 - \ddot{\phi}_3 + \frac{l}{j} = 0. \qquad (5.14.20)$$

The displacement potential $\mathbf{U} = U\mathbf{e}_3$ and $\mathbf{F} = F\mathbf{e}_3$, so that

$$u_1 = \sigma_{,1} + U_{,2}, \quad u_2 = \sigma_{,2} - U_{,1}, \quad \phi_3 = \phi,$$
$$f_1 = g_{,1} + F_{,2}, \quad f_2 = g_{,2} - F_{,1}, \quad l_3 = l. \tag{5.14.21}$$

Substituting these into (5.14.20), we obtain

$$c_1^2 \nabla^2 \sigma - \ddot{\sigma} + g = 0,$$
$$c_2^2 \nabla^2 U + j\frac{\omega_0^2}{2}\phi - \ddot{U} + F = 0,$$
$$c_4^2 \nabla^2 \phi - \omega_0^2 \phi - \frac{\omega_0^2}{2}\nabla^2 U - \ddot{\phi} + \frac{l}{j} = 0. \tag{5.14.22}$$

The body load potentials g and F are determined by solving

$$\nabla^2 g = \nabla \cdot \mathbf{f}, \quad \nabla^2 F = -(\nabla \times \mathbf{f}) \cdot \mathbf{e}_3, \tag{5.14.23}$$

in two dimensions. Hence,

$$g = -\frac{1}{2\pi} \int_{\partial v} \mathbf{f} \cdot \nabla_{\mathbf{x}} \ln[R(\boldsymbol{\xi}, \mathbf{x})] \, dv(\boldsymbol{\xi}),$$
$$F = -\frac{1}{2\pi} \int_{\partial v} \mathbf{f} \times \nabla_{\mathbf{x}} \ln[R(\boldsymbol{\xi}, \mathbf{x})] \cdot \mathbf{e}_3 \, dv(\boldsymbol{\xi}). \tag{5.14.24}$$

The stress and couple stress tensors have the following components:

$$\mathbf{t} = \begin{bmatrix} t_{11} & t_{12} & 0 \\ t_{21} & t_{22} & 0 \\ 0 & 0 & t_{33} \end{bmatrix}, \quad \mathbf{m} = \begin{bmatrix} 0 & 0 & m_{13} \\ 0 & 0 & m_{23} \\ m_{31} & m_{32} & 0 \end{bmatrix}, \tag{5.14.25}$$

where t_{kl} and m_{kl} are given by

$$t_{11} = \lambda \nabla^2 \sigma + (2\mu + \kappa)(\sigma_{,11} + U_{,12}),$$
$$t_{22} = \lambda \nabla^2 \sigma + (2\mu + \kappa)(\sigma_{,22} - U_{,12}),$$
$$t_{12} = (2\mu + \kappa)\sigma_{,12} + \mu U_{,22} - (\mu + \kappa)U_{,11} - \kappa\phi,$$
$$t_{21} = (2\mu + \kappa)\sigma_{,12} - \mu U_{,11} + (\mu + \kappa)U_{,22} + \kappa\phi,$$
$$m_{k3} = \gamma \phi_{,k}, \quad m_{3k} = \beta \phi_{,k}. \tag{5.14.26}$$

For harmonic waves (in the Fourier domain), by elimination of $\bar{U}$ and $\bar{\phi}$ among (5.14.22), we obtain

$$\nabla^2 \bar{\sigma} + (\omega/c_1)^2 \bar{\sigma} + (\bar{g}/c_1^2) = 0,$$
$$(\nabla^2 + k_2^2)(\nabla^2 + k_4^2)\begin{Bmatrix} \bar{U} \\ \bar{\phi} \end{Bmatrix} + \begin{Bmatrix} \bar{P} \\ \bar{Q} \end{Bmatrix} = 0,$$
$$c_2^2 \nabla^2 \bar{U} + \omega^2 \bar{U} + j\frac{\omega_0^2}{2}\bar{\phi} + \bar{F} = 0. \tag{5.14.27}$$

where

$$\bar{P} \equiv \left(\nabla^2 + \frac{c_3^2}{c_4^2}k_3^2\right)\frac{\bar{F}}{c_2^2} - \frac{\omega_0^2\bar{l}}{2c_2^2c_4^2},$$

$$\bar{Q} \equiv \frac{\omega_0^2}{2c_2^2c_4^2}\nabla^2\bar{F} + \left(\nabla^2 + \frac{\omega^2}{c_2^2}\right)\frac{\bar{l}}{jc_4^2}. \tag{5.14.28}$$

In this case, it is simpler to disregard $(5.14.27)_2$ for $\bar{\phi}$ and determine $\bar{\phi}$ by using $(5.14.27)_3$.

B. Antiplane Strain

$$u_1 = u_2 = \phi_3 = 0, \quad u_3 = w(x_1, x_2), \quad \partial/\partial x_3 = 0,$$
$$f_1 = f_2 = l_3 = 0, \quad f_3 = f.$$

The field equations are

$$c_2^2\nabla^2 w + \frac{j\omega_0^2}{2}(\phi_{2,1} - \phi_{1,2}) - \ddot{w} + f = 0,$$
$$(c_3^2 - c_4^2)\phi_{l,lk} + c_4^2\nabla^2\phi_k + \tfrac{1}{2}\omega_0^2\epsilon_{kl3}w_{,l} - \omega_0^2\phi_k - \ddot{\phi}_k + \frac{l_k}{j} = 0. \tag{5.14.29}$$

Introducing potentials

$$\phi_1 = \tau_{,1} + \Phi_{,2}, \qquad \phi_2 = \tau_{,2} - \Phi_{,1},$$
$$l_1/j = h_{,1} + L_{,2}, \qquad l_2/j = h_{,2} - L_{,1}, \tag{5.14.30}$$

We obtain

$$c_3^2\nabla^2\tau - \omega_0^2\tau - \ddot{\tau} + h = 0,$$
$$c_2^2\nabla^2 w - \ddot{w} - j\frac{\omega_0^2}{2}\nabla^2\Phi + f = 0,$$
$$c_4^2\nabla^2\Phi - \omega_0^2\Phi - \ddot{\Phi} + \frac{\omega_0^2}{2}w + L = 0. \tag{5.14.31}$$

The couple potentials h and L are determined in a similar way to (5.14.11)

$$h = -\frac{1}{2\pi}\int_{\partial v}\frac{\mathbf{l}(\boldsymbol{\xi})}{j}\cdot\nabla_{\mathbf{x}}\ln[R(\boldsymbol{\xi},\mathbf{x})]\,dv(\boldsymbol{\xi}),$$
$$L = -\frac{1}{2\pi}\int_{\partial v}\frac{\mathbf{l}(\boldsymbol{\xi})}{j}\times\nabla_{\mathbf{x}}[\ln R(\boldsymbol{\xi},\mathbf{x})]\,dv(\boldsymbol{\xi}), \tag{5.14.32}$$

The stress and couple-stress fields have the following components:

$$\mathbf{t} = \begin{bmatrix} 0 & 0 & t_{13} \\ 0 & 0 & t_{23} \\ t_{31} & t_{32} & 0 \end{bmatrix}, \qquad \mathbf{m} = \begin{bmatrix} m_{11} & m_{12} & 0 \\ m_{21} & m_{22} & 0 \\ 0 & 0 & m_{33} \end{bmatrix}, \tag{5.14.33}$$

where t_{kl} and m_{kl} are given by

$$t_{\alpha 3} = (\mu + \kappa)w_{,\alpha} - \kappa\delta_{\alpha 2}(\tau_{,1} + \Phi_{,2}) + \kappa\delta_{\alpha 1}(\tau_{,2} - \Phi_{,1}),$$
$$t_{3\alpha} = \mu w_{,\alpha} + \kappa\delta_{\alpha 2}(\tau_{,1} + \Phi_{,2}) - \kappa\delta_{\alpha 1}(\tau_{,2} - \Phi_{,1}),$$
$$m_{11} = \alpha\nabla^2\tau + (\beta + \gamma)(\tau_{,11} + \Phi_{,12}),$$
$$m_{22} = \alpha\nabla^2\tau + (\beta + \gamma)(\tau_{,22} - \Phi_{,12}),$$
$$m_{12} = (\beta + \gamma)\tau_{,12} - \gamma\Phi_{,11} + \beta\Phi_{,22},$$
$$m_{21} = (\beta + \gamma)\tau_{,12} - \beta\Phi_{,11} + \gamma\Phi_{,22},$$
$$m_{33} = \alpha\nabla^2\tau, \qquad w = U_{2,1} - U_{1,2}. \tag{5.14.34}$$

Eliminating w or Φ among (5.14.31), we obtain

$$\nabla^2\bar{\tau} + \frac{\omega^2 - \omega_0^2}{c_3^2}\bar{\tau} + \frac{\bar{h}}{c_3^2} = 0,$$
$$(\nabla^2 + k_2^2)(\nabla^2 + k_4^2)\left\{\begin{matrix}\bar{w}\\\bar{\Phi}\end{matrix}\right\} + \left\{\begin{matrix}\bar{P}_A\\\bar{Q}_A\end{matrix}\right\} = 0,$$
$$c_4^2\nabla^2\bar{\Phi} + (\omega^2 - \omega_0^2)\bar{\Phi} + \tfrac{1}{2}\omega_0^2\bar{w} + \bar{L} = 0, \tag{5.14.35}$$

where

$$\bar{P}_A = \left(\nabla^2 + \frac{c_3^2}{c_4^2}k_3^2\right)\frac{\bar{f}}{c_2^2} + \frac{j\omega_0^2}{2c_2^2c_4^2}\nabla^2\bar{L},$$
$$\bar{Q}_A = -\frac{\omega_0^2}{2c_2^2c_4^2}\bar{f} + \left(\nabla^2 + \frac{\omega^2}{c_2^2}\right)\frac{\bar{L}}{c_4^2}. \tag{5.14.36}$$

Here too, it is simpler to employ equations for $\bar{\tau}$ and $\bar{\Phi}$ from (5.14.35) and to determine $\bar{w}$ from $(5.14.35)_3$.

III. Sandru's Representation[5]

The field equations can be expressed in a matrix form, by introducing the following operators:

$$X_i \equiv \partial/\partial x_i, \qquad T \equiv \partial/\partial t, \qquad\qquad Q \equiv X_1^2 + X_2^2 + X_3^2,$$
$$Q_1 \equiv (\lambda + 2\mu + \kappa)Q - \rho T^2, \qquad\qquad Q_2 \equiv (\mu + \kappa)Q - \rho T^2,$$
$$Q_3 \equiv (\alpha + \beta + \gamma)Q - \rho j T^2 - 2\kappa, \qquad Q_4 \equiv \gamma Q - \rho j T^2 - 2\kappa, \tag{5.14.37}$$

---

[5]Sandru [1966], see also Eringen [1968a, p. 719]. This representation is a generalization of the Somigliana representation to micropolar elasticity (cf. Eringen and Şuhubi [1975, p. 854]).

and the 6×6 matrix $\mathbf{L} = \|L_{ij}\|$ $(i, j = 1, 2, \ldots, 6)$ is given in partitioned form, in terms of 3×3 matrices $L_{\alpha\beta}$ $(\alpha, \beta = 1, 2)$ as

$$
\mathbf{L} = \begin{bmatrix} \mathbf{L}_{11} & \vdots & \mathbf{L}_{12} \\ \cdots & \cdots & \cdots \\ \mathbf{L}_{21} & \vdots & \mathbf{L}_{22} \end{bmatrix},
$$

$$\mathbf{L}_{11} = [Q_2\delta_{ij} + (\lambda + \mu)X_iX_j], \quad \mathbf{L}_{12} = \mathbf{L}_{21} = -\kappa\epsilon_{ijk}X_k,$$
$$\mathbf{L}_{22} = [Q_4\delta_{ij} + (\alpha + \beta)X_iX_j], \quad i, j = 1, 2, 3. \tag{5.14.38}$$

Equations (5.14.1) are expressed in the matrix form

$$
\mathbf{L} \begin{bmatrix} u_1 \\ u_2 \\ u_3 \\ \phi_1 \\ \phi_2 \\ \phi_3 \end{bmatrix} = -\rho \begin{bmatrix} f_1 \\ f_2 \\ f_3 \\ jl_1 \\ jl_2 \\ jl_3 \end{bmatrix}. \tag{5.14.39}
$$

The inverse L_{ij}^{-1} of the matrix L_{ij} is given formally, by

$$L_{ij}^{-1} = \frac{N_{ij}}{Q_1Q_3(Q_2Q_4 + \kappa^2Q)}, \tag{5.14.40}$$

where a 6×6 matrix N_{ij} is obtained in the partitioned form, in terms of 3×3 matrices $\mathbf{N}_{\alpha\beta}, \alpha, \beta = 1, 2$,

$$
\mathbf{N} = \begin{bmatrix} \mathbf{N}_{11} & \vdots & \mathbf{N}_{12} \\ \cdots & \cdots & \cdots \\ \mathbf{N}_{21} & \vdots & \mathbf{N}_{22} \end{bmatrix},
$$

$$\mathbf{N}_{11} \therefore Q_3\{Q_1Q_4\delta_{ij} - [(\lambda + \mu)Q_4 - \kappa^2]\}X_iX_j,$$
$$\mathbf{N}_{12} = \mathbf{N}_{21} \therefore Q_1Q_3\kappa\epsilon_{ijk}X_k,$$
$$\mathbf{N}_{22} \therefore Q_1\{Q_2Q_3\delta_{ij} - [(\alpha + \beta)Q_2 - \kappa^2]\}X_iX_j,$$
$$i, j = 1, 2, 3. \tag{5.14.41}$$

Consider now

$$
\begin{bmatrix} u_1 \\ u_2 \\ u_3 \\ \phi_1 \\ \phi_2 \\ \phi_3 \end{bmatrix} = \mathbf{N} \begin{bmatrix} F_1 \\ F_2 \\ F_3 \\ F_1^* \\ F_2^* \\ F_3^* \end{bmatrix}. \tag{5.14.42}
$$

If $\boldsymbol{\Phi}_1 = Q_3\mathbf{F}, \boldsymbol{\Phi}_2 = Q_1\mathbf{F}$, through Equations (5.14.41) and (5.14.42), we find that

$$\mathbf{u}(\mathbf{x},t) = \Box_1\Box_4\boldsymbol{\Phi}_1 - [(\lambda + \mu)\Box_4 - \kappa^2]\nabla\nabla\cdot\boldsymbol{\Phi}_1 - \kappa\Box_3\nabla\times\boldsymbol{\Phi}_2,$$
$$\phi(\mathbf{x},t) = \Box_2\Box_3\boldsymbol{\Phi}_2 - [(\alpha + \beta)\Box_2 - \kappa^2]\nabla\nabla\cdot\boldsymbol{\Phi}_2 - \kappa\Box_1\nabla\times\boldsymbol{\Phi}_1,$$

$$(5.14.43)$$

where $\Box_i$ are the wave operators

$$\Box_1 \equiv (\lambda + 2\mu + \kappa)\nabla^2 - \rho\partial^2/\partial t^2, \qquad \Box_2 \equiv (\mu + \kappa)\nabla^2 - \rho\partial^2/\partial t^2,$$
$$\Box_3 \equiv (\alpha + \beta + \gamma)\nabla^2 - \rho j\partial^2/\partial t^2 - 2\kappa, \quad \Box_4 \equiv \gamma\nabla^2 - \rho j\partial^2/\partial t^2 - 2\kappa.$$

$$(5.14.44)$$

From (5.14.39), (5.14.42), and (5.14.44), it follows that $\boldsymbol{\Phi}_1$ and $\boldsymbol{\Phi}_2$ satisfy the following uncoupled equations:

$$\Box_1(\Box_2\Box_4 + \kappa^2\nabla^2)\boldsymbol{\Phi}_1 = -\rho\mathbf{f},$$
$$\Box_3(\Box_2\Box_4 + \kappa^2\nabla^2)\boldsymbol{\Phi}_2 = -\rho j\mathbf{l}. \qquad (5.14.45)$$

In the case $\kappa = 0$, the first of these gives a representation known in classical elasticity.

In the **static case**, we set $T = 0$, and (5.14.43) and (5.14.44) give a generalization of the *Galerkin* representation for micropolar elasticity, namely

$$\mathbf{u}(\mathbf{x}) = (\lambda + 2\mu + \kappa)\nabla^2(\gamma\nabla^2 - 2\kappa)\boldsymbol{\Phi}_1 - [\gamma(\lambda + \mu)\nabla^2$$
$$- \kappa(2\lambda + 2\mu + \kappa)]\nabla\nabla\cdot\boldsymbol{\Phi}_1 - \kappa[(\alpha + \beta + \gamma)\nabla^2 - 2\kappa]\nabla\times\boldsymbol{\Phi}_2,$$
$$\phi(\mathbf{x}) = (\mu + \kappa)\nabla^2[(\alpha + \beta + \gamma)\nabla^2 - 2\kappa]\boldsymbol{\Phi}_2 - [(\mu + \kappa)(\alpha + \beta)\nabla^2$$
$$- \kappa^2]\nabla\nabla\cdot\boldsymbol{\Phi}_2 - \kappa(\lambda + 2\mu + \kappa)\nabla^2(\nabla\times\boldsymbol{\Phi}_1), \qquad (5.14.46)$$

where now $\boldsymbol{\Phi}_1$ and $\boldsymbol{\Phi}_2$ satisfy the equations

$$(\lambda + 2\mu + \kappa)\nabla^4[(\mu + \kappa)\gamma\nabla^2 - \kappa(2\mu + \kappa)]\boldsymbol{\Phi}_1 = -\rho\mathbf{f},$$
$$[(\alpha + \beta + \gamma)\nabla^2 - 2\kappa]\,[(\mu + \kappa)\gamma\nabla^4 - \kappa(2\mu + \kappa)\nabla^2]\,\boldsymbol{\Phi}_2 = -\rho j\mathbf{l}.$$

$$(5.14.47)$$

We decompose $\mathbf{f}$ and $\mathbf{l}$ into irrotational and solenoidal parts

$$\rho\mathbf{f} = \nabla\pi_0 + \nabla\times\boldsymbol{\Pi}, \quad \nabla\cdot\boldsymbol{\Pi} = 0,$$
$$\rho j\mathbf{l} = \nabla\pi_0^* + \nabla\times\boldsymbol{\Pi}^*, \quad \nabla\cdot\overset{*}{\boldsymbol{\Pi}} = 0, \qquad (5.14.48)$$

and let

$$(\Box_2\Box_4 + \kappa^2\nabla^2)\boldsymbol{\Phi}_1 = \nabla\Lambda_0, \qquad \Box_1\boldsymbol{\Phi}_1 = \nabla\times\boldsymbol{\Lambda}, \qquad \nabla\cdot\boldsymbol{\Lambda} = 0,$$
$$(\Box_2\Box_4 + \kappa^2\nabla^2)\boldsymbol{\Phi}_2 = \nabla\Lambda_0^*, \qquad \Box_3\boldsymbol{\Phi}_2 = \nabla\times\boldsymbol{\Lambda}^*, \qquad \nabla\cdot\boldsymbol{\Lambda}^* = 0.$$

$$(5.14.49)$$

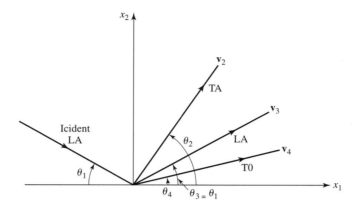

FIGURE 5.15.1. Reflection of Micropolar Waves

From (5.14.45) and (5.14.49), it follows that

$$\square_1 \Lambda_0 = -\pi_0, \qquad (\square_2\square_4 + \kappa^2\nabla^2)\boldsymbol{\Lambda} = -\boldsymbol{\Pi},$$
$$\square_3 \Lambda_0^* = -\pi_0^*, \qquad (\square_2\square_4 + \kappa^2\nabla^2)\boldsymbol{\Lambda}^* = -\boldsymbol{\Pi}^*, \qquad (5.14.50)$$

and we obtain

$$\mathbf{u} = \nabla\Lambda_0 + \nabla \times (\square_4\boldsymbol{\Lambda}) - \kappa\nabla \times (\nabla \times \boldsymbol{\Lambda}^*),$$
$$\boldsymbol{\phi} = \nabla\Lambda_0^* - \kappa\nabla \times (\nabla \times \boldsymbol{\Lambda}) + \nabla \times (\square_2\boldsymbol{\Lambda}^*). \qquad (5.14.51)$$

Thus, $\mathbf{u}$ and $\boldsymbol{\phi}$ are fully determined by $\Lambda_0, \Lambda_0^*, \boldsymbol{\Lambda}$, and $\boldsymbol{\Lambda}^*$.

5.15   Micropolar Waves in Half-Space

If $x_3 = 0$ is the plane of an incident plane harmonic wave, then the reflected waves from the boundary surface $x_2 = 0$ can be shown to remain in this plane, Figure 5.15.1. The boundary $x_2 = 0$ being free of tractions, we must have $\mathbf{t}_{(\mathbf{n})} = \mathbf{m}_{(\mathbf{n})} = \mathbf{0}$. Since $u_3 = \phi_1 = \phi_2 = 0$, we see that the problem is a plane strain problem.

The solution vectors appropriate to this problem satisfy (5.14.27):

$$\sigma_\alpha = a_\alpha \exp[i(k_1\mathbf{n}_\alpha \cdot \mathbf{x} - \omega t)],$$
$$\tau_\alpha = b_\alpha \exp[i(k_3\mathbf{n}_\alpha \cdot \mathbf{x} - \omega t)], \quad \alpha = 1, 3,$$
$$U_\beta = A_\beta \exp[i(k_\beta\mathbf{n}_\beta \cdot \mathbf{x} - \omega t)],$$
$$\phi_\beta = A_\beta \nu_\beta k_\beta^2 \exp[i(k_\beta\mathbf{n}_\beta \cdot \mathbf{x} - \omega t)], \quad \beta = 2, 4, \qquad (5.15.1)$$

where $\mathbf{n}_\alpha$ and $\mathbf{n}_\beta$ are unit vectors ($\alpha = 1, 3; \quad \beta = 2, 4$) and a_α, b_α, and A_β are constants. Repeated indices are not summed. For $\alpha = 1, 3$, we have two solutions for σ and two solutions for τ. Similarly, for $\beta = 2, 4$, we have two solutions for U and two for ϕ. These solutions enable us to represent the incident and reflected waves. Expressions k_α and k_β are given by (5.11.18), and that of ν_β by (5.11.10), namely

$$\nu_\beta = \frac{c_2^2 k^2 - \omega^2}{j\omega_0^2/2}. \qquad (5.15.2)$$

For vanishing boundary loads, from (5.14.26), we have

$$t_{22} = \lambda(\sigma_{,11} + \sigma_{,22}) + (2\mu + \kappa)(\sigma_{,22} - U_{,12}) = 0,$$
$$t_{21} = (2\mu + \kappa)\sigma_{,12} - \mu U_{,11} + (\mu + \kappa)U_{,22} + \kappa\phi = 0,$$
$$m_{23} = \gamma\phi_{,2} = 0, \quad x_2 = 0. \qquad (5.15.3)$$

Substituting from (5.15.1) into (5.15.3), we obtain

$$t_{22} = a_\alpha S_1^\alpha E_\alpha + S_2^\beta A_\beta E_\beta = 0,$$
$$t_{21} = a_\alpha T_1^\alpha E_\alpha + T_2^\beta A_\beta E_\beta = 0,$$
$$m_{23} = iA_\beta M^\beta E_\beta = 0 \quad \text{at } x_2 = 0, \qquad (5.15.4)$$

where

$$S_1^\alpha = -k_1^2[\lambda + (2\mu + \kappa)n_{\alpha 2}^2], \qquad S_2^\beta = (2\mu + \kappa)k_\beta^2 n_{\beta 1}n_{\beta 2},$$
$$T_1^\alpha = -k_1^2(2\mu + \kappa)n_{\alpha 1}n_{\alpha 2}, \qquad T_2^\beta = k_\beta^2[\mu n_{\beta 1}^2 - (\mu + \kappa)n_{\beta 2}^2 + \kappa\nu_\beta],$$
$$M^\beta = \gamma k_\beta^3 \nu_\beta n_{\beta 2},$$
$$E_\alpha = \exp[i(k_1\mathbf{n}_\alpha \cdot \mathbf{x} - \omega t)], \qquad E_\beta = \exp[i(k_\beta\mathbf{n}_\beta \cdot \mathbf{x} - \omega t)]. \qquad (5.15.5)$$

A. Incident LA-Wave

The incident LA-wave is represented by σ_1 and reflected waves by σ_3, U_2, and ϕ_4

$$\sigma_1 = a_1 \exp[i(k_1\mathbf{n}_1 \cdot \mathbf{x} - \omega t)],$$
$$\sigma_3 = a_3 \exp[i(k_1\mathbf{n}_3 \cdot \mathbf{x} - \omega t)],$$
$$U_2 = A_2 \exp[i(k_2\mathbf{n}_2 \cdot \mathbf{x} - \omega t)],$$
$$\phi_4 = A_4\nu_4 k_4^2 \exp[i(k_4\mathbf{n}_4 \cdot \mathbf{x} - \omega t)]. \qquad (5.15.6)$$

From the boundary conditions (5.15.4), it follows that

$$k_1 n_{11} = k_1 n_{31} = k_2 n_{21} = k_4 n_{41},$$
$$k_1 n_{12} = k_1 n_{32} = k_2 n_{22} = k_4 n_{42}, \qquad (5.15.7)$$

and

$$S_1^1 a_1 + S_1^3 a_3 + S_2^2 A_2 + S_2^4 A_4 = 0,$$
$$T_1^2 a_1 + T_1^3 a_3 + T_2^2 A_2 + T_2^4 A_4 = 0,$$
$$M^2 A_2 + M^4 A_4 = 0. \qquad (5.15.8)$$

Since the incident wave is in the $(x_3 = 0)$-plane, we have $n_{13} = 0$, and (5.15.6) yields

$$n_{23} = n_{33} = n_{43} = 0,$$

which verifies the fact that all reflected waves are in the $(x_3 = 0)$-plane. Writing

$$\mathbf{n}_1 = (\cos\theta_1, -\sin\theta_1), \qquad \mathbf{n}_3 = (\cos\theta_3, \sin\theta_3),$$
$$\mathbf{n}_2 = (\cos\theta_2, \sin\theta_2), \qquad \mathbf{n}_4 = (\cos\theta_4, \sin\theta_4), \qquad (5.15.9)$$

for the x_1- and x_2-components of $\mathbf{n}_\alpha$ and $\mathbf{n}_\beta$, from $\nu_r = \omega/k_r$ and (5.15.7), there follows:

$$\cos\theta_1 = \cos\theta_3, \quad \cos\theta_2 = (v_2/v_1)\cos\theta_1, \quad \cos\theta_4 = (v_4/v_1)\cos\theta_1.$$
$$(5.15.10)$$

The first of these shows that the incidence and reflection angles of the LA-waves are equal.

The solution of the system (5.15.8) gives the amplitude ratios of reflected waves to that of the incident wave

$$Da_3/a_1 = [S_2^2 - (M^2/M^4)S_2^4]T_1^1 - [T_2^2 - (M^2/M^4)T_2^4]S_1^1,$$
$$DA_2/a_1 = -S_1^3 T_1^1 + S_1^1 T_1^3,$$
$$DA_4/a_1 = (M^2/M^4)(S_1^3 T_1^1 - T_1^3 S_1^1),$$
$$D = -[S_2^2 - (M^2/M^4)S_2^4]T_1^3 + [T_2^2 - (M^2/M^4)T_2^4]S_1^3,$$
$$(5.15.11)$$

where, upon using (5.15.5) and (5.15.9), we obtain

$$S_1^1 = -k_1^2[\lambda + (2\mu + \kappa)\sin^2\theta_1], \quad T_1^1 = k_1^2(2\mu + \kappa)\sin\theta_1\cos\theta_1,$$
$$S_2^2 = k_2^2(2\mu + \kappa)\sin\theta_2\cos\theta_2, \quad T_2^2 = k_2^2[\mu\cos^2\theta_2 - (\mu + \kappa)\sin^2\theta_2 + \kappa\nu_2],$$
$$S_1^3 = -k_1^2[\lambda + (2\mu + \kappa)\sin^2\theta_3], \quad T_1^3 = -k_1^2(2\mu + \kappa)\sin\theta_3\cos\theta_3,$$
$$S_2^4 = (2\mu + \kappa)k_4^2\sin\theta_4\cos\theta_4, \quad T_2^4 = k_4^2[\mu\cos^2\theta_4 - (\mu + \kappa)\sin^2\theta_4 + \kappa\nu_4],$$

$$M^2/M^4 = \frac{k_2^3\nu_2\sin\theta_2}{k_4^3\nu_4\sin\theta_4}. \qquad (5.15.12)$$

An examination of (5.15.11) indicates that the incident LA-wave is reflected from the boundary of the half-space, producing three different waves:

(a) an LA-branch propagating in the $\mathbf{n}_3$-direction;

(b) a TA-branch propagating in $\mathbf{n}_2$-direction; and

(c) a TO-branch propagating in $\mathbf{n}_4$-direction.

(see Figure 5.15.1).

The following special cases are of interest:

(i) **Normal Incidence.** In this case, $\theta_1 = \theta_3 = 90$ deg , and (5.15.11) shows that

$$a_3/a_1 = -1, \quad A_2/a_1 = A_4/a_1 = 0. \tag{5.15.13}$$

Consequently, the reflected LA-wave is normal to the boundary, and the TA-wave and TO-wave disappear.

(ii) **Grazing Reflection of the TA-Wave.** For a grazing TA-wave, we must have $\theta_2 = 0$. This gives

$$\cos\theta_1 = v_1/v_2.$$

Since $v_1 > v_2$, the grazing TA-wave is not possible.

(iii) **Grazing Reflection of the TO-Wave.** This requires that $\theta_4 = 0$, and

$$\cos\theta_1 = v_1/v_4, \quad \cos\theta_2 = v_2/v_4. \tag{5.15.14}$$

Since $v_4 \geq v_1 \geq v_2$, this is possible. For this case, (5.15.11) gives

$$a_3/a_1 = -1, \quad A_2/a_1 = 0,$$
$$A_4/a_1 = -\frac{2(2\mu + \kappa)}{\mu + \kappa\nu_4}[1 - (v_1/v_4)^2]^{1/2}(v_4/v_1). \tag{5.15.15}$$

B. Reflection of Coupled TA- and TO-Waves

The incident waves are characterized by

$$U_1 = A_1 \exp[i(k_2\mathbf{n}_1 \cdot \mathbf{x} - \omega_2 t)],$$
$$\phi_1 = A_1\nu_2 k_2^2 \exp[i(k_2\mathbf{n}_1 \cdot \mathbf{x} - \omega t)]. \tag{5.15.16}$$

For the reflected waves, we have

$$\sigma = a_3 \exp[i(k_1\mathbf{n}_3 \cdot \mathbf{x} - \omega_1 t)],$$
$$U_\beta = A_\beta \exp[i(k_\beta\mathbf{n}_\beta \cdot \mathbf{x} - \omega_\beta t)],$$
$$\phi_\beta = A_\beta\nu_\beta k_\beta^2 \exp[i(k_\beta\mathbf{n}_\beta \cdot \mathbf{x} - \omega t)], \quad \beta = 2, 4. \tag{5.15.17}$$

The boundary conditions require that the incident and reflected waves will be on the same ($x_3 = 0$)-plane and

$$k_2 n_{11} = k_2 n_{21} = k_1 n_{31} = k_4 n_{41}. \tag{5.15.18}$$

In terms of the angles θ_i, for the normal $\mathbf{n}_i$, this reads

$$\cos\theta_1 = \cos\theta_2, \quad \cos\theta_3 = (v_1/v_2)\cos\theta_2, \quad \cos\theta_4 = (v_4/v_2)\cos\theta_2. \tag{5.15.19}$$

For $\cos\theta_3 \leq 1$ and $\cos\theta_4 \leq 1$, the reflected waves consist of:

(a) an LA-mode propagating with phase velocity v_3 at angle θ_3;

(b) a TA-mode propagating with phase velocity v_2 at angle θ_2; and

(c) a TO-mode propagating with phase velocity v_4 at angle θ_4;

provided that $\omega > \omega_{cr}$. When $\omega < \omega_{cr}$, the reflected TO-wave degenerates into a vibration. The amplitude ratios can be calculated by solving three equations arising from the boundary conditions (5.15.4) (see Parfitt and Eringen [1969]).

Grazing Reflected LA-Wave: This requires that $\theta_3 = 0$ and

$$\cos \theta_2 = v_2/v_1 \leq 1. \tag{5.15.20}$$

Grazing Reflected TO-Wave: In this case, we have $\theta_4 = 0$ and

$$\cos \theta_2 = v_2/v_4 \leq 1, \quad \cos \theta_3 = v_1/v_4 \leq 1. \tag{5.15.21}$$

C. Reflection of an LO-Wave

The reflection of an LO-wave can be studied in a similar fashion (see Parfitt and Eringen [1969]). This requires that $\omega > \omega_{cr}$. For $\omega < \omega_{cr}$, the incident wave degenerates into a vibratory motion of the medium and there is no reflection.

The incident LO-wave with velocity v_3 is reflected as one LA-wave with velocity v_1, and two coupled TA- and TO-waves with velocities v_2 and v_4. Corresponding angles are related to each other by

$$\cos \theta_1 = \cos \theta_3, \quad \cos \theta_2 = (v_2/v_3) \cos \theta_1, \quad \cos \theta_4 = (v_4/v_3) \cos \theta_1, \tag{5.15.22}$$

where θ_1 is the angle of incidence of the LO-wave. The grazing incidence of the LO-wave ($\theta_1 = 0$) is possible with reflection angles $\cos \theta_2 = v_2/v_3 \leq 1$, $\cos \theta_4 = v_4/v_3 \leq 1$, for the TA- and TO-waves. Grazing reflections of TA- and TO-waves are not possible.

5.16   Micropolar Surface Waves

In this section we investigate the propagation of surface waves in an isotropic micropolar half-space.[6] Since the incident and reflected waves propagate in the same plane, the problem is a plane-strain problem. We consider the waves in the $(z = 0)$-plane over the half-plane $y \geq 0$. For vanishing body

---

[6]Eringen and Şuhubi [1964]. See also Eringen [1968a].

loads and monochromatic waves, the field equations are given by (5.14.27), namely,

$$\nabla^2 \bar{\sigma} + (\omega/c_1)^2 \bar{\sigma} = 0,$$

$$(\nabla^2 + k_2^2)(\nabla^2 + k_4^2) \left\{ \begin{matrix} \bar{U} \\ \bar{\phi} \end{matrix} \right\} = 0,$$

$$c_2^2 \nabla^2 \bar{U} + \omega^2 \bar{U} + j\frac{\omega_0^2}{2} \bar{\phi} = 0. \tag{5.16.1}$$

The boundary conditions require that the waves decay with increasing y and that at the boundary $y = 0$, we must have $t_{yy} = t_{yx} = m_{yz} = 0$. From (5.14.26) we have

$$t_{yy} = \lambda \nabla^2 \bar{\sigma} + (2\mu + \kappa)(\bar{\sigma}_{,yy} - \bar{U}_{,xy}) = 0,$$

$$t_{yx} = (2\mu + \kappa)\bar{\sigma}_{,xy} - \mu \bar{U}_{,xx} + (\mu + \kappa)\bar{U}_{,yy} + \kappa\bar{\phi} = 0,$$

$$m_{yz} = \gamma\bar{\phi}_{,y} = 0 \quad \text{at } y = 0. \tag{5.16.2}$$

General solutions of (5.16.1) are given by

$$\bar{\sigma} = a \exp(-\delta y + ikx),$$

$$\bar{U} = [A \exp(-\lambda_2 y) + B \exp(-\lambda_4 y)] \exp(ikx),$$

$$\bar{\phi} = [\nu_2 A \exp(-\lambda_2 y) + \nu_4 B \exp(-\lambda_4 y)] \exp(ikx), \tag{5.16.3}$$

where

$$\delta^2 = k^2 - (\omega/c_1)^2, \quad \lambda_2^2 = k^2 - k_2^2, \quad \lambda_4^2 = k^2 - k_4^2,$$

$$\nu_\beta = \frac{\rho}{\kappa}[c_2^2(k^2 - \lambda_\beta^2) - \omega^2] \quad (\kappa \neq 0), \quad \beta = 2, 4. \tag{5.16.4}$$

Here k_β denotes the wave numbers given by (5.11.18). Substituting (5.16.3) into the boundary conditions expressed by (5.16.2), we obtain

$$\bar{t}_{yy} = [(\lambda + 2\mu + \kappa)\delta^2 - \lambda k^2]a + ik\lambda_2(2\mu + \kappa)A$$
$$+ ik\lambda_4(2\mu + \kappa)B = 0,$$

$$\bar{t}_{yx} = -ik(2\mu + \kappa)\delta a + [\mu k^2 + (\mu + \kappa)\lambda_2^2 + \kappa\nu_2]A$$
$$+ [\mu k^2 + (\mu + \kappa)\lambda_4^2 + \kappa\nu_4]B = 0,$$

$$\bar{m}_{yz} = -\gamma(\nu_2\lambda_2 A + \nu_4\lambda_4 B) = 0. \tag{5.16.5}$$

Setting the determinant of the coefficients $a, A,$ and B equal to zero, we obtain the dispersion relations

$$(\lambda_4 - \lambda_2)\{ 4 (\lambda_2\delta/k^2)(1 - \kappa/2\rho c_2^2)^2(\lambda_4^2 + \lambda_4\lambda_2)$$

$$- \left(2 - \frac{v^2}{c_2^2} - \frac{\kappa}{\rho c_2^2}\right)^2 [\lambda_4^2 + \lambda_4\lambda_2 + \lambda_2^2 - k^2(1 - v^2/c_2^2)]\} = 0.$$

$$\tag{5.16.6}$$

Hence we have, for the phase velocity v of the micropolar surface waves, either

$$4(1 - v^2/c_1^2)^{1/2}\lambda_2/k = [2 - (v/c_2)^2 - (\kappa/\rho c_2^2)]^2[\lambda_2^2 + \lambda_2\lambda_4 + \lambda_4^2$$
$$- k^2(1 - v^2/c_2^2)](1 - \kappa/2\rho c_2^2)^{-2}(\lambda_4^2 + \lambda_2\lambda_4)^{-1}$$

$$(5.16.7)$$

or

$$\lambda_4 = \lambda_2. \qquad (5.16.8)$$

In the limit $\kappa \to 0$, from (5.11.18), it follows that

$$k_2^2 = \omega^2/c_2^2, \quad k_4^2 = \omega^2/c_4^2. \qquad (5.16.9)$$

With these, (5.16.7) gives the well-known equations of Rayleigh surface wave velocity (cf. Eringen and Şuhubi [1975, p. 519]).

$$4(1 - v^2/c_1^2)^{1/2}(1 - v^2/c_2^2)^{1/2} = (2 - v^2/c_2^2)^2. \qquad (5.16.10)$$

Equations (5.16.4) and (5.16.8) gives $k_2^2 = k_4^2$. Using (5.11.18), we find that

$$\omega^2/c_2^2 = (\omega_0^2/c_4^2)\{(1 + m)\theta - 2 \pm 2[(m\theta - 1)(\theta - 1)]^{1/2}\}(1 - m)^{-2}, \quad (5.16.11)$$

where

$$\theta \equiv \frac{2\mu + \kappa}{2\mu + 2\kappa}, \quad m \equiv (c_2/c_4)^2. \qquad (5.16.12)$$

Since $\theta < 1$ and $m < 1$, it can be seen that the right-hand side of (5.6.11) is negative. Thus, this branch does not correspond to any surface wave.

For $\theta = 0.996$ and $m = c_2^2/c_4^2 = 0.1418$, corresponding to Gauthier's experimental results, the roots v/c_2 of surface waves are calculated from (5.16.8) as a function of k. These are listed in Table (5.5)[7] and displayed as a graph on Figure 5.16.1.

5.17   Micropolar Waves in Plates

Consider a micropolar elastic plate bounded by surfaces $y = \pm h$, and free of applied loads. We wish to determine the frequency spectrum of the plate for a monochromatic wave propagating along the x-axis in the $(z = 0)$-plane.

---

[7]Courtesy of Professor E.S. Şuhubi.

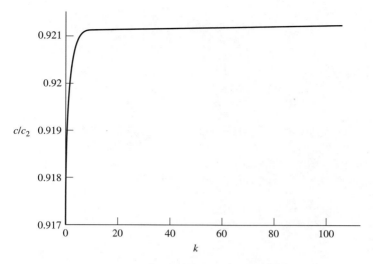

FIGURE 5.16.1. Surface Wave Velocity Versus k

A. Plane Strain

For the plane strain, the displacement potentials must satisfy (5.14.27), i.e.,

$$\nabla^2 \bar{\sigma} + (\omega^2/c_1^2)\bar{\sigma} = 0,$$
$$(\nabla^2 + k_2^2)(\nabla^2 + k_4^2)\bar{U} = 0,$$
$$c_2^2 \nabla^2 \bar{U} + \omega^2 \bar{U} + \frac{\kappa}{\rho}\bar{\phi} = 0, \tag{5.17.1}$$

where k_2^2 and k_4^2 are given by (5.11.18) and $j\omega_0^2/2 = \kappa/\rho$. A superposed bar denotes the Fourier transform.

The boundary conditions can be read from (5.14.26), namely

$$\bar{t}_{yy} = \lambda \nabla^2 \bar{\sigma} + (2\mu + \kappa)(\bar{\sigma}_{,yy} - \bar{U}_{,xy}) = 0,$$
$$\bar{t}_{yx} = (2\mu + \kappa)\bar{\sigma}_{,xy} - \mu\bar{U}_{,xx} + (\mu + \kappa)\bar{U}_{,yy} + \kappa\bar{\phi} = 0,$$
$$m_{yz} = \gamma\bar{\phi}_{,y} = 0 \quad \text{at } y = \pm h. \tag{5.17.2}$$

The general solution of (5.17.1) is given by

$$\bar{\sigma} = [a_1 \sinh(\delta y) + a_2 \cosh(\delta y)]e^{ikx},$$
$$\bar{U} = [A_1 \sinh(\lambda_2 y) + A_2 \cosh(\lambda_2 y) + A_3 \sinh(\lambda_4 y) + A_4 \cosh(\lambda_4 y)]e^{ikx},$$
$$\bar{\phi} = \nu_2[A_1 \sinh(\lambda_2 y) + A_2 \cosh(\lambda_2 y)]$$
$$+ \nu_4[A_3 \sinh(\lambda_4 y) + A_4 \cosh(\lambda_4 y)]e^{ikx}, \tag{5.17.3}$$

where

$$\delta^2 = k^2 - (\omega/c_1)^2, \quad \lambda_\beta^2 = k^2 - k_\beta^2,$$
$$\nu_\beta = \frac{\rho}{\kappa}[c_2^2(k^2 - \lambda_\beta^2) - \omega^2] \quad (\kappa \neq 0), \quad \beta = 2, 4. \tag{5.17.4}$$

TABLE 5.5. Surface wave velocity.

k	v/c_2	k	v/c_2
0.001	0.917842	1	0.918948
0.01	0.917843	2	0.920053
0.05	0.917846	3	0.920563
0.1	0.917859	4	0.920804
0.2	0.917909	5	0.920931
0.3	0.917988	6	0.921004
0.4	0.918093	7	0.921050
0.5	0.918216	8	0.921081
0.6	0.918353	9	0.921103
0.7	0.918499	10	0.921118
0.8	0.918650	50	0.921183
0.9	0.918800	100	0.921185
		>142	0.921186

The problem is divided naturally into symmetric and antisymmetric vibrations.

(a) Symmetric Vibrations

For the vibrations symmetric with respect to $(y = 0)$-plane, $a_1 = A_2 = A_4 = 0$. With these, upon substituting (5.17.3) into boundary conditions (5.17.2), we obtain three linear equations for the three unknowns a_2, A_1, and A_3. The determinant of these coefficients must vanish leading to the dispersion relations

$$(2\mu + \kappa)^2 k^2 \delta \lambda_2 \lambda_4 (\nu_2 - \nu_4) \tanh(\delta h) + (\rho c_2^2 k^2)^2 \left[2 - (v/c_2)^2 - \frac{\kappa}{\kappa + \mu} \right]^2$$
$$\times [\nu_4 \lambda_4 \tanh(\lambda_2 h) - \nu_2 \lambda_2 \tanh(\lambda_4 h)] = 0. \tag{5.17.5}$$

This equation is satisfied if (Nowacki [1970])[8]

$$\frac{\tanh(\delta h)}{\tanh(\lambda_2 h)} = \frac{k^2 \left[2 - \left(\frac{v}{c_2}\right)^2 - \frac{\kappa}{\mu + \kappa} \right]^2 \left[\nu_4 - \nu_2 \frac{\lambda_2 \tanh(\lambda_4 h)}{\lambda_4 \tanh(\lambda_2 h)} \right]}{\left(2 - \frac{\kappa}{\mu + \kappa} \right)^2 \delta \lambda_2 (\nu_4 - \nu_2)}, \tag{5.17.6}$$

provided $\lambda_2 \neq \lambda_4$ or

$$\lambda_2 = \lambda_4. \tag{5.17.7}$$

---

[8]K.M. Rao [1988] also analyzed the longitudinal waves. However, his results appear not to be correct, since his wave number k is purely imaginary (for real p and q) which corresponds to δ. Equation (5.17.7) appears not to have been noticed by Nowacki.

In the limit as $\kappa \to 0$, from (5.11.18) we have

$$k_2^2 = \omega^2/c_2^2, \qquad k_4^2 = \omega^2/c_4^2, \qquad (5.17.8)$$

and (5.17.6) gives the Rayleigh–Lamb equation of the classical elasticity

$$\frac{\tanh[kh(1 - v^2/c_1^2)]^{1/2}}{\tanh[kh(1 - v^2/c_2^2)]^{1/2}} = \frac{[2 - (v/c_2)^2]^2}{4[1 - (v/c_1)^2]^{1/2}[1 - (v/c_2)^2]^{1/2}}. \qquad (5.17.9)$$

Equation (5.17.7) is valid while (5.17.6) is not satisfied necessarily. In this case, we obtain again (5.16.11). As discussed before, (5.17.7) does not correspond to any wave.

(i) *Long-wave approximation.* If we assume that the wavelength $2\pi/k$ is very large as compared to the thickness $2h$ of the plate, then kh is very small. Replacing $\tanh z \simeq z$ for small z, we obtain

$$v^2 = \frac{\omega^2}{k^2} = \frac{(2\mu + \kappa)(2\lambda + 2\mu + \kappa)}{(\lambda + 2\mu + \kappa)\rho}. \qquad (5.17.10)$$

For $\kappa = 0$ this gives the phase velocity of extensional waves in classical elasticity (cf. Eringen and Şuhubi [1975, p. 561]).

(ii) *Short-wave approximation.* For short wavelengths, we set $\tanh z \simeq 1$ in (5.17.6). This leads to

$$4(1 - v^2/c_1^2)^{1/2}\lambda_2/k$$
$$= \left[2 - (v/c_2)^2 - \frac{\kappa}{\rho c_2^2}\right]^2 [\lambda^2 + \lambda_2\lambda_4 + \lambda_4^2 - k^2(1 - v^2/c_2^2)]$$
$$\times (1 - \kappa/2\rho c_2^2)^{-2}(\lambda_4^2 + \lambda_2\lambda_4)^{-1}. \qquad (5.17.11)$$

This is the dispersion relation for surface waves (see (5.16.7)). For $\kappa = 0$, (5.17.11) gives the classical Rayleigh surface wave velocity.

(b) Antisymmetric Vibrations

For antisymmetric motions about $y = 0$, $a_2 = A_1 = A_3 = 0$. Substituting $\bar{\sigma}, \bar{U}$, and $\bar{\phi}$ into boundary conditions (5.17.2), we obtain a system of three linear, homogeneous equations for the unknown coefficients a_1, A_2, and A_4. Setting the determinant of these coefficients equal to zero, we obtain the dispersion relations

$$\left(\frac{\nu_4\lambda_4}{\tanh(\lambda_2 h)} - \frac{\nu_2\lambda_2}{\tanh(\lambda_4 h)}\right)\tanh(\delta h) = \frac{\left(2 - \frac{\kappa}{\mu+\kappa}\right)^2 (\nu_4 - \nu_2)\lambda_2\lambda_4\delta}{k^2\left[2 - (v/c_2)^2 - \frac{\kappa}{\mu+\kappa}\right]^2},$$
$$(5.17.12)$$

provided $\lambda_2 \neq \lambda_4$, or

$$\lambda_2 = \lambda_4. \qquad (5.17.13)$$

In the special case $\kappa = 0$, (5.17.12) goes into the dispersion relations of the antisymmetric waves of plates in classical elasticity

$$\frac{\tanh[kh(1 - v^2/c_1^2)]^{1/2}}{\tanh[kh(1 - v^2/c_2^2)]^{1/2}} = \frac{4(1 - v^2/c_1^2)^{1/2}(1 - v^2/c_2^2)^{1/2}}{(2 - v^2/c_2^2)^2}, \qquad (5.17.14)$$

cf. Eringen and Şuhubi [1975, p. 562].

Equation (5.17.13) has the same form as (5.16.11) which gives no real waves. For small wavelengths, (5.17.12) reduces to (5.17.11), the dispersion relations for surface waves. For long wavelengths, we expand the hyperbolic functions and obtain

$$\left(1 - \frac{\delta^2 h^2}{3}\right)\left[\frac{\nu_4}{\lambda_2^2\left(1 - \frac{\lambda_2^2 h^2}{3}\right)} - \frac{\nu_2}{\lambda_4^2\left(1 - \frac{\lambda_4^2 h^2}{3}\right)}\right] = \frac{\left(2 - \frac{\kappa}{\mu + \kappa}\right)^2 (\nu_4 - \nu_2)}{k^2\left[2 - (v/c_2)^2 - \frac{\kappa}{\mu + \kappa}\right]^2}. \qquad (5.17.15)$$

B. Antiplane Strain

Antiplane strain requires solving the field equations (5.14.35) with vanishing body loads $h = f = L = 0$:

$$\nabla^2 \bar{\tau} + \frac{\omega^2 - \omega_0^2}{c_3^2}\bar{\tau} = 0,$$

$$(\nabla^2 + k_2^2)(\nabla^2 + k_4^2)\left\{\begin{matrix} \bar{w} \\ \bar{\Phi} \end{matrix}\right\} = 0,$$

$$c_4^2 \nabla^2 \bar{\Phi} + (\omega^2 - \omega_0^2)\bar{\Phi} + \tfrac{1}{2}\omega_0^2 \bar{w} = 0, \qquad (5.17.16)$$

subject to the boundary conditions

$$\bar{t}_{yz} = (\mu + \kappa)\bar{w}_{,y} - \kappa(\bar{\tau}_{,x} + \bar{\Phi}_{,y}) = 0,$$
$$\bar{m}_{yy} = \alpha\nabla^2\bar{\tau} + (\beta + \gamma)(\bar{\tau}_{,yy} - \bar{\Phi}_{,xy}) = 0,$$
$$\bar{m}_{yx} = (\beta + \gamma)\bar{\tau}_{,xy} - \beta\bar{\Phi}_{,xx} + \gamma\bar{\Phi}_{,yy} = 0 \quad \text{at } y = \pm h, \quad (5.17.17)$$

where, as usual, quantities carrying a overbar denote the Fourier transforms with respect to time.

The general solution of (5.17.16) for monochromatic waves, advancing in the x-direction, is given by

$$\bar{\tau} = [a_1 \sinh(\delta y) + a_2 \cosh(\delta y)]e^{ikx},$$
$$\bar{\Phi} = [A_1 \sinh(\lambda_2 y) + A_2 \cosh(\lambda_2 y) + A_3 \sinh(\lambda_4 y)$$
$$\qquad + A_4 \cosh(\lambda_4 y)]e^{ikx},$$
$$\bar{w} = \tilde{\nu}_2[A_1 \sinh(\lambda_2 y) + A_2 \cosh(\lambda_2 y)] + \tilde{\nu}_4[A_3 \sinh(\lambda_4 y)$$
$$\qquad + A_4 \cosh(\lambda_4 y)]e^{ikx}, \qquad (5.17.18)$$

where

$$\delta^2 = k^2 + \frac{\omega_0^2 - \omega^2}{c_3^2}, \qquad \lambda_\beta^2 = k^2 - k_\beta^2,$$

$$\tilde{\nu}_\beta = \frac{c_4^2(k^2 - \lambda_\beta^2) + \omega_0^2 - \omega^2}{\omega_0^2/2} \qquad (\beta = 2, 4). \tag{5.17.19}$$

As usual, k_2 and k_4 are given by (5.11.18) and $\omega_0^2 = 2\kappa/\rho j$.

(a) Symmetric Vibrations

In this case $a_1 = A_2 = A_4 = 0$. Boundary conditions (5.17.17) lead to a system of three homogeneous, linear equations for a_2, A_1, and A_3 whose coefficient determinant must vanish. This leads to, either

$$\frac{\tanh(\delta h)}{\tanh(\lambda_4 h)} = \left\{ \kappa k^2 (\beta + \gamma)(\lambda_2 m_4 - \lambda_4 m_2) + [(\alpha + \beta + \gamma)\delta^2 - \alpha k^2] \right.$$

$$\times \left. \left[(n_2 m_4 - n_4 m_2) \frac{\tanh(\lambda_2 h)}{\tanh(\lambda_4 h)} \right] \right\} [(\beta + \gamma)^2 k^2 \delta(n_2 \lambda_4$$

$$- n_4 \lambda_2)]^{-1}, \tag{5.17.20}$$

where

$$m_\alpha = \beta k^2 + \gamma \lambda_\alpha, \quad n_\alpha = (\mu + \kappa)\tilde{\nu}_\alpha \lambda_\alpha - \kappa \lambda_\alpha, \quad \alpha = 2, 4,$$

provided $\lambda_2 \neq \lambda_4$, or

$$\lambda_2 = \lambda_4. \tag{5.17.21}$$

In the special case $\kappa \to 0$, (5.17.20) reduces to

$$\frac{\tanh[kh(1 - v^2/c_3^2)]^{1/2}}{\tanh[kh(1 - v^2/c_4^2)]^{1/2}} = \frac{(2 - v^2/c_0^2)^2}{4(1 - v^2/c_3^2)^{1/2}(1 - v^2/c_4^2)^{1/2}}, \tag{5.17.22}$$

where

$$c_0^2 = \frac{\beta + \gamma}{2\rho j}. \tag{5.17.23}$$

In classical elasticity, there is no phase velocity corresponding to (5.17.22). This is because no LO- or TO-wave is predicted.

For small wavelengths as compared to the thickness of the plate, we take $\tanh x \simeq 1$, reducing (5.17.20) to

$$\kappa k^2 (\beta + \gamma)(\lambda_2 m_4 - \lambda_4 m_2) + [(\alpha + \beta + \gamma)\delta^2 - \alpha k^2][(n_2 m_4 - n_4 m_2)]$$

$$= [(\beta + \gamma)^2 k^2 \delta(n_2 \lambda_4 - n_4 \lambda_2). \tag{5.17.24}$$

This indicates the possibility of surface waves of the TO-type. In classical elasticity, this corresponds to Love waves in layered media, under the assumption of certain inequalities among the material constants (cf. Ewing et al. [1957, p. 154]).

The second root (5.17.21) leads to the imaginary frequency equation represented by (5.16.11) obtained before, and corresponds to no real waves.

(b) Antisymmetric Vibrations

For the antisymmetric motions $a_2 = A_1 = A_3 = 0$. The analysis leads to the dispersion relations, either

$$\frac{\tanh(\delta h)}{\tanh(\lambda_4 h)} = [(\beta + \gamma)^2 k^2 \delta (n_2 \lambda_4 - n_4 \lambda_2)]$$

$$\left\{ \kappa k^2 (\beta + \gamma) \left(\lambda_2 m_4 - \lambda_4 m_2 \frac{\tanh(\lambda_4 h)}{\tanh(\lambda_2 h)} \right) \right.$$

$$\left. + [(\alpha + \beta + \gamma)\delta^2 - \alpha k^2] \left(n_2 m_4 - n_4 m_2 \frac{\tanh(\lambda_4 h)}{\tanh(\lambda_2 h)} \right) \right\}^{-1},$$

$$(5.17.25)$$

or

$$\lambda_2 = \lambda_4. \tag{5.17.26}$$

In the case $\kappa \to 0$, (5.17.25) reduces to

$$\frac{\tanh[k(1 - v^2/c_3^2)]^{1/2}}{\tanh[k(1 - v^2/c_4^2)]^{1/2}} = \frac{4(1 - v^2/c_3^2)^{1/2}(1 - v^2/c_4^2)^{1/2}}{(2 - v^2/c_0^2)^2}. \tag{5.17.27}$$

The small wavelength limit of (5.17.25) is identical to (5.17.24).

The second branch (5.17.26) gives complex frequency (5.16.11) which does not correspond to any real waves.

5.18   Fundamental Solutions

The displacements and rotations in a micropolar solid due to a concentrated force and couple are known as the fundamental solutions. This problem has been treated by several authors.[9] Here we employ Sandru's representation (5.14.45) to determine the potentials Φ_1 and Φ_2 for the unbounded solid. Expressions (5.14.43) will then give $\mathbf{u}(\mathbf{x}, t)$ and $\phi(\mathbf{x}, t)$.

We employ the Fourier transform with respect to time, denoted by letters carrying an overbar, e.g.,

$$\bar{f}(\mathbf{x}, \omega) = (2\pi)^{-1/2} \int_{-\infty}^{\infty} f(\mathbf{x}, t) e^{i\omega t} \, dt. \tag{5.18.1}$$

The Fourier transforms of (5.14.45) are

$$(\nabla^2 + k_1^2)(\nabla^2 + k_2^2)(\nabla^2 + k_4^2)\bar{\Phi}_1 = \bar{\mathbf{F}}(\mathbf{x}, \omega),$$
$$(\nabla^2 + k_3^2)(\nabla^2 + k_2^2)(\nabla^2 + k_4^2)\bar{\Phi}_2 = \bar{\mathbf{L}}(\mathbf{x}, \omega), \tag{5.18.2}$$

---

[9]Sandru [1966], Nowacki [1970], Ieşan [1971], Bhargava and Ghosh [1975], Hirashima [1976], and Dragoş [1984].

where k_i are given by (5.11.18) and

$$\bar{\mathbf{F}} = -\frac{\rho \bar{\mathbf{f}}}{(\lambda + 2\mu + \kappa)(\mu + \kappa)\gamma}, \quad \bar{\mathbf{L}} = -\frac{\rho j \bar{\mathbf{l}}}{(\alpha + \beta + \gamma)(\mu + \kappa)\gamma}. \quad (5.18.3)$$

For a solid of infinite extent in all directions, the three-dimensional Fourier transform with respect to space variables $\mathbf{x}$, may be used to find the solution of the system (5.18.2), i.e.,

$$\tilde{f}(\boldsymbol{\xi}, t) = (2\pi)^{-3/2} \int_{-\infty}^{\infty} f(\mathbf{x}, t) \exp(i\boldsymbol{\xi} \cdot \mathbf{x}) \, d^3\mathbf{x}. \quad (5.18.4)$$

Applying (5.18.4)–(5.18.2) we obtain

$$\tilde{\bar{\boldsymbol{\Phi}}}_1(\boldsymbol{\xi}, \omega) = \left(\frac{A_1}{k_1^2 - \xi^2} + \frac{A_2}{k_2^2 - \xi^2} + \frac{A_4}{k_4^2 - \xi^2} \right) \tilde{\bar{\mathbf{F}}}(\xi, \omega),$$

$$\tilde{\bar{\boldsymbol{\Phi}}}_2(\boldsymbol{\xi}, \omega) = \left(\frac{B_1}{k_3^2 - \xi^2} + \frac{B_2}{k_2^2 - \xi^2} + \frac{B_4}{k_4^2 - \xi^2} \right) \tilde{\bar{\mathbf{L}}}(\xi, \omega), \quad (5.18.5)$$

where

$$A_1 = [(k_2^2 - k_1^2)(k_4^2 - k_1^2)]^{-1}, \quad A_2 = [(k_1^2 - k_2^2)(k_4^2 - k_2^2)]^{-1},$$
$$A_4 = [(k_1^2 - k_4^2)(k_2^2 - k_4^2)]^{-1},$$
$$B_1 = [(k_2^2 - k_3^2)(k_4^2 - k_3^2)]^{-1}, \quad B_2 = [(k_3^2 - k_2^2)(k_4^2 - k_2^2)]^{-1},$$
$$B_3 = [(k_3^2 - k_4^2)(k_2^2 - k_4^2)]^{-1}. \quad (5.18.6)$$

The inverse transforms of (5.18.5) are obtained by means of the convolution theorem:

$$\boldsymbol{\Phi}_1(\mathbf{x}, t) = \frac{1}{4\pi} \int_{-\infty}^{\infty} \frac{\tilde{\bar{\mathbf{F}}}(\xi, \omega)}{R(\boldsymbol{\xi}, \mathbf{x})} \sum_{r=1,2,4} A_r \exp(ik_r R - i\omega t) \, d^3\boldsymbol{\xi} \, d\omega,$$

$$\boldsymbol{\Phi}_2(\mathbf{x}, t) = \frac{1}{4\pi} \int_{-\infty}^{\infty} \frac{\tilde{\bar{\mathbf{L}}}(\xi, \omega)}{R(\boldsymbol{\xi}, \mathbf{x})} \sum_{r=2,3,4} B_r \exp(ik_r R - i\omega t) \, d^3\boldsymbol{\xi} \, d\omega, \quad (5.18.7)$$

where

$$R(\boldsymbol{\xi}, \mathbf{x}) \equiv [(\mathbf{x} - \boldsymbol{\xi}) \cdot (\mathbf{x} - \boldsymbol{\xi})]^{1/2}. \quad (5.18.8)$$

This completes the formal solution. However, calculations of the fourfold integrals in (5.18.7) often present major difficulties. Here we treat four simple cases.

I. Concentrated Harmonic Force

For a concentrated harmonic force $\mathbf{f}$ acting alone at $\mathbf{x} = \mathbf{x}_0$, we substitute

$$\mathbf{f} = \mathbf{f}_0 \delta(\mathbf{x} - \mathbf{x}_0) e^{i\omega t}, \quad \mathbf{l} = \mathbf{0}, \quad (5.18.9)$$

where $\mathbf{f}_0$ is a constant vector. Expression (5.18.7) gives

$$\boldsymbol{\Phi}_1 = -\frac{\rho \mathbf{f}_0}{4\pi(\lambda + 2\mu + \kappa)(\mu + \kappa)\gamma} e^{-i\omega t} \sum_{r=1,2,4} \frac{A_r}{R} e^{ik_r R},$$

$$\boldsymbol{\Phi}_2 = 0, \tag{5.18.10}$$

where

$$R = R(\mathbf{x}_0, \mathbf{x}). \tag{5.18.11}$$

The displacement and rotation fields are calculated from (5.14.43):

$$\mathbf{u}(\mathbf{x}, t) = (\lambda + 2\mu + \kappa)\gamma \left(\nabla^2 + \frac{\omega^2}{c_1^2}\right)\left(\nabla^2 + \frac{\omega^2 - \omega_0^2}{c_4^2}\right)\boldsymbol{\Phi}_1$$

$$- (\lambda + \mu)\gamma \left(\nabla^2 + \frac{\omega^2 - \omega_0^2}{c_4^2} - \frac{j\omega_0^2}{4c_2^2 c_4^2}\right)\nabla\nabla \cdot \boldsymbol{\Phi}_1,$$

$$\boldsymbol{\phi}(\mathbf{x}, t) = -\kappa(\lambda + 2\mu + \kappa)\left(\nabla^2 + \frac{\omega^2}{c_1^2}\right)\nabla \times \boldsymbol{\Phi}_1. \tag{5.18.12}$$

II. Concentrated Harmonic Couple

In the case of a concentrated harmonic couple $\mathbf{l}$ acting alone at $\mathbf{x} = \mathbf{x}_0$, we substitute

$$\mathbf{f} = \mathbf{0}, \quad \mathbf{l} = \mathbf{l}_0 \delta(\mathbf{x} - x_0) e^{-i\omega t} \tag{5.18.13}$$

into (5.18.3) and (5.18.7) to obtain

$$\boldsymbol{\Phi}_1 = 0,$$

$$\boldsymbol{\Phi}_2 = -\frac{\rho j \mathbf{l}_0}{4\pi(\alpha + \beta + \gamma)(\mu + \kappa)\gamma} e^{-i\omega t} \sum_{r=2,3,4} \frac{B_r}{R} e^{ik_r R}, \tag{5.18.14}$$

where R is given by (5.18.11).

The displacement and microrotation fields are obtained from (5.14.43)

$$\mathbf{u}(\mathbf{x}, t) = -\kappa(\alpha + \beta + \gamma)\left(\nabla^2 + \frac{\omega^2 - \omega_0^2}{c_3^2}\right)\nabla \times \boldsymbol{\Phi}_2,$$

$$\boldsymbol{\phi}(\mathbf{x}, t) = (\alpha + \beta + \gamma)(\mu + \kappa)\left(\nabla^2 + \frac{\omega^2}{c_2^2}\right)\left(\nabla^2 + \frac{\omega^2 - \omega_0^2}{c_3^2}\right)\boldsymbol{\Phi}_2$$

$$- \left[(\alpha + \beta)(\mu + \kappa)\left(\nabla^2 + \frac{\omega^2}{c_2^2}\right) - \kappa^2\right]\nabla\nabla \cdot \boldsymbol{\Phi}_2. \tag{5.18.15}$$

III. Concentrated Static Force

Let $\mathbf{F}$ be a concentrated force acting at the origin of coordinates $\mathbf{x} = \mathbf{0}$ and $\mathbf{l} = \mathbf{0}$:

$$\rho \mathbf{f} = \mathbf{P} \delta(\mathbf{x}), \tag{5.18.16}$$

where $\mathbf{P}$ is a constant vector and $\delta(\mathbf{x})$ is the three-dimensional Dirac delta function. From (5.14.48), we have

$$\nabla^2 \pi_0 = \nabla \cdot (\rho \mathbf{f}), \quad \nabla^2 \mathbf{\Pi} = -\nabla \times (\rho \mathbf{f}).$$

The solutions of these equations are

$$\pi_0(\mathbf{x}) = -\frac{1}{4\pi} \int_{-\infty}^{\infty} \rho \mathbf{f}(\boldsymbol{\xi}) \cdot \nabla_{\mathbf{x}}(1/R) \, d^3\boldsymbol{\xi},$$

$$\mathbf{\Pi}(\mathbf{x}) = -\frac{1}{4\pi} \int_{-\infty}^{\infty} \rho \mathbf{f}(\boldsymbol{\xi}) \times \nabla_{\mathbf{x}}(1/R) \, d^3\boldsymbol{\xi},$$

$$R \equiv R(\mathbf{x}, \boldsymbol{\xi}) = [(\mathbf{x} - \boldsymbol{\xi}) \cdot (\mathbf{x} - \boldsymbol{\xi})]^{1/2}. \tag{5.18.17}$$

Using (5.18.16) these give

$$\pi_0 = -\frac{1}{4\pi} \mathbf{P} \cdot \nabla(1/r), \quad \mathbf{\Pi} = -\frac{1}{4\pi} \mathbf{P} \times \nabla(1/r),$$

$$r \equiv (\mathbf{x} \cdot \mathbf{x})^{1/2}. \tag{5.18.18}$$

Carrying these into (5.14.50), we have, for the static case,

$$(\lambda + 2\mu + \kappa)\nabla^2 \Lambda_0 = \frac{1}{4\pi} \mathbf{P} \cdot \nabla(1/r),$$

$$\left[\gamma \left(\mu + \kappa \right) \nabla^4 - \kappa \left(2\mu + \kappa \right) \nabla^2 \right] \mathbf{\Lambda} = \frac{1}{4\pi} \mathbf{P} \times \nabla(1/r). \tag{5.18.19}$$

These equations possess the solutions

$$\Lambda_0 = \frac{1}{8\pi(\lambda + 2\mu + \kappa)} \mathbf{P} \cdot \frac{\mathbf{x}}{r},$$

$$\mathbf{\Lambda} = \frac{1}{8\pi\kappa(2\mu + \kappa)} \nabla \times (\mathbf{P}r) + \frac{l^2}{4\pi\kappa(2\mu + \kappa)} \nabla \times \left[(\mathbf{P}/r) \left(1 - e^{-r/l} \right) \right],$$

$$l^2 \equiv \frac{\gamma(\mu + \kappa)}{\kappa(2\mu + \kappa)}. \tag{5.18.20}$$

With $\mathbf{\Lambda}^* = \mathbf{0}$, we have the solutions

$$\mathbf{u} = \frac{2\lambda + 6\mu + 3\kappa}{8\pi(2\mu + \kappa)(\lambda + 2\mu + \kappa)} \frac{\mathbf{P}}{r} + \frac{2\lambda + 2\mu + \kappa}{8\pi(2\mu + \kappa)(\lambda + 2\mu + \kappa)} \frac{\mathbf{P} \cdot \mathbf{x}}{r^3} \mathbf{x}$$

$$+ \frac{\gamma}{4\pi(2\mu + \kappa)^2} \nabla \times \left\{ \nabla \times \left[(\mathbf{P}/r) \left(e^{-R/l} - 1 \right) \right] \right\},$$

$$\phi = \frac{1}{4\pi(2\mu + \kappa)} \nabla \times \left[(\mathbf{P}/r) \left(1 - e^{-R/l} \right) \right]. \tag{5.18.21}$$

IV. Concentrated Static Couple

If $\mathbf{f} = \mathbf{0}$, but a concentrated couple $\mathbf{l}$ acts at the origin $\mathbf{x} = \mathbf{0}$ of coordinates in an unbounded space, then

$$\rho j \mathbf{l} = \mathbf{M}\delta(\mathbf{x}), \tag{5.18.22}$$

where $\mathbf{M}$ is a constant vector. In this case, (5.14.50) lead to

$$[(\alpha + \beta + \gamma)\nabla^2 - 2\kappa]\Lambda_0^* = -\pi_0^*,$$
$$[\gamma(\mu + \kappa)\nabla^4 - \kappa(2\mu + \kappa)\nabla^2]\boldsymbol{\Lambda}^* = -\boldsymbol{\Pi}^*, \tag{5.18.23}$$

where

$$\pi_0^* = -\frac{1}{4\pi}\mathbf{M}\cdot\nabla(1/r), \quad \boldsymbol{\Pi}^* = -\frac{1}{4\pi}\mathbf{M}\times\nabla(1/r), \tag{5.18.24}$$

and (5.14.50) have the solutions

$$\Lambda_0^* = \frac{1}{8\pi\kappa}\mathbf{M}\cdot\nabla\left[(1/r)\left(e^{-r/h} - 1\right)\right],$$
$$\boldsymbol{\Lambda}^* = \frac{1}{8\pi(2\mu + \kappa)\kappa}\nabla\times(\mathbf{M}r) + \frac{l^2}{4\pi(2\mu + \kappa)\kappa}\nabla\times\left[(\mathbf{M}/r)\left(1 - e^{-r/l}\right)\right],$$
$$h^2 \equiv (\alpha + \beta + \gamma)/2\kappa. \tag{5.18.25}$$

Substituting these into (5.14.51) with $\Lambda_0 = 0$ and $\boldsymbol{\Lambda} = \mathbf{0}$, we obtain

$$\mathbf{u} = \frac{1}{4\pi(2\mu + \kappa)}\nabla\times\left[(\mathbf{M}/r)\left(1 - e^{-r/l}\right)\right],$$
$$\boldsymbol{\phi} = \frac{1}{8\pi\kappa}\nabla\left\{\mathbf{M}\cdot\nabla\left[\frac{1}{r}\left(e^{-R/h} - 1\right)\right]\right\} + \frac{\mu + \kappa}{4\pi(2\mu + \kappa)\kappa}\nabla$$
$$\times\left\{\nabla\times\left[(\mathbf{M}/r)\left(1 - e^{-r/l}\right)\right]\right\}. \tag{5.18.26}$$

Using these in constitutive equations, the stress and couple stress fields are determined.

Remark 1. We note that the concentrated couple is a fundamental problem in micropolar elasticity. It is *not* deduced as a limiting solution of two equal parallel forces directed in the opposite sense, as in classical elasticity. Parallel to the practices in elasticity, concentrated couples may also be used to obtain higher-order couple singularities.

A state of stress due to higher-order singularities may be obtained by introducing multipole singularities. For example, a double force singularity is obtained by superposing a force of intensity $\rho\mathbf{f} = -P/\Delta\xi_1$ at the point $(\xi_1 + \Delta\xi_1/2, \xi_2, \xi_3)$ parallel to the x_1-axis with another force of the same intensity at the point $(\xi_1 - \Delta\xi_1/2, \xi_2, \xi_3)$:

$$\bar{u}_j = \frac{-P}{\Delta\xi_1}\left[\bar{u}_j\left(\mathbf{x}; \xi_1 + \frac{\Delta\xi_1}{2}, \xi_2, \xi_3\right) - \bar{u}_j\left(\mathbf{x}; \xi_1 - \frac{\Delta\xi_1}{2}, \xi_2, \xi_3\right)\right].$$

In the limit $\Delta\xi_1 \to 0$, we obtain

$$\bar{u}_j = -P\frac{\partial}{\partial\xi_1}\bar{u}_j(\mathbf{x},\boldsymbol{\xi}). \tag{5.18.27}$$

Thus, we can obtain any order singularities by substituting

$$\rho\bar{\mathbf{f}} = \mathbf{P}\frac{\partial^k}{\partial\xi_{i_1}\cdots\partial\xi_{i_k}}, \qquad \rho j\bar{\mathbf{l}} = \mathbf{M}\frac{\partial^l}{\partial\xi_{i_1}\cdots\partial\xi_{i_l}}, \tag{5.18.28}$$

for multipole force $\mathbf{P}$ and multicouple $\mathbf{M}$ singularites. For example, the action of three double couples gives a center of microrotation.

In micropolar elasticity, concentrated couple being a fundamental concept, higher-order singularities due to multipole couples can be constructed by superposition of multipole couple singularities. Consequently, the concept of disclination is as fundamental as the dislocation.

Remark 2. By means of the reciprocal theorem, integral representations can be provided for the initial value problems. The approach here is similar to that discussed by Eringen and Şuhubi [1975, Section 5.11]. These expressions are rather lengthy and we do not reproduce them here. The reciprocal theorem for the static case is given in Section 5.25.

5.19 Problems of Sphere and Spherical Cavity

The solution of the dynamical problems concerning the sphere and spherical cavity[10] requires the determination of two scalar displacement potentials, whose Fourier transforms $\bar{\sigma}$ and $\bar{\tau}$ must satisfy the scalar Helmholtz equation

$$(\nabla^2 + k^2)\psi = 0; \quad \psi = (\bar{\sigma},\bar{\tau}), \quad k^2 = (k_1^2, k_3^2), \tag{5.19.1}$$

and two vector potentials, whose Fourier transform $\bar{\mathbf{U}}$ and $\bar{\boldsymbol{\Phi}}$ must satisfy the vector Helmholtz equation (see Section 5.14).

$$(\nabla^2 + k^2)\boldsymbol{\Psi} = 0; \quad \boldsymbol{\Psi} = (\bar{\mathbf{U}},\bar{\boldsymbol{\Phi}}), \quad k^2 = (k_2^2, k_4^2). \tag{5.19.2}$$

Vector potentials are subject to the gauge condition

$$\nabla \cdot \boldsymbol{\Psi} = 0. \tag{5.19.3}$$

In spherical coordinates (r,θ,ϕ), the general solution of (5.19.1) is of the form

$$\psi_{ml}^{\nu}(kr) = f_l(kr)Y_{ml}^{\nu}(\theta,\phi), \tag{5.19.4}$$

---

[10]Singh [1975].

where f_l is a linear combination of the spherical Bessel functions of the first and second kinds and

$$Y_{ml}^{c,s}(\theta,\phi) = P_l^m(\cos\theta)(\cos m\phi, \sin m\phi). \tag{5.19.5}$$

Here P_l^m is the associated Legendre function of the first kind. Superscripts c and s denote $\cos m\phi$ and $\sin m\phi$, respectively. Three independent solutions of the vector Helmholtz equation (5.19.2) are (cf. Morse and Feshbach [1953, p. 1823])

$$\begin{aligned}
\mathfrak{M}_{ml}^\nu(kr) &= \nabla \times [\mathbf{e}_r r\psi_{ml}^\nu(kr)], \\
k\mathfrak{N}_{ml}^\nu(kr) &= \nabla \times \mathfrak{M}_{ml}^\nu(kr), \\
k\mathfrak{L}_{ml}^\nu(kr) &= \nabla\psi_{ml}^\nu(kr).
\end{aligned} \tag{5.19.6}$$

It may be noted that

$$\nabla \cdot \mathfrak{L} = -k\mathfrak{L}. \tag{5.19.7}$$

Displacement potentials may now be expressed as

$$\bar{\sigma} = \sum_{\nu=c,s}\sum_{l=0}^{\infty}\sum_{m=0}^{l} \sigma_{ml}^\nu \psi_{ml}^\nu(k_1 r),$$

$$\bar{\tau} = \sum_{\nu=c,s}\sum_{l=0}^{\infty}\sum_{m=0}^{l} \tau_{ml}^\nu \psi_{ml}^\nu(k_3 r),$$

$$\begin{aligned}
\bar{\mathbf{U}} = \sum_{\nu=c,s}\sum_{l=1}^{\infty}\sum_{m=0}^{l} &[a_{ml}^\nu \mathfrak{M}_{ml}^\nu(k_2 r) + b_{ml}^\nu \mathfrak{M}_{ml}^\nu(k_4 r) \\
&+ c_{ml}^\nu \mathfrak{N}_{ml}^\nu(k_2 r) + d_{ml}^\nu \mathfrak{N}_{ml}^\nu(k_4 r)],
\end{aligned}$$

$$\begin{aligned}
\bar{\boldsymbol{\Phi}} = \sum_{\nu=c,s}\sum_{l=1}^{\infty}\sum_{m=0}^{\infty} &[\alpha_{ml}^\nu \mathfrak{M}_{ml}^\nu(k_2 r) + \beta_{ml}^\nu \mathfrak{M}_{ml}^\nu(k_4 r) \\
&+ \gamma_{ml}^\nu \mathfrak{N}_{ml}^\nu(k_2 r) + \delta_{ml}^\nu \mathfrak{N}_{ml}^\nu(k_4 r)].
\end{aligned} \tag{5.19.8}$$

Note that the gauge condition (5.19.3) is satisfied since

$$\nabla \cdot (\mathfrak{M}_{ml}^\nu, \mathfrak{N}_{ml}^\nu) = 0.$$

It remains to verify whether (5.14.8) and (5.14.9), which led to (5.19.2), are satisfied. Substituting $\bar{\mathbf{U}}$ and $\bar{\boldsymbol{\Phi}}$ into the Fourier transforms of (5.14.8) and (5.14.9) (with $\mathbf{F} = \mathbf{0}$, $\mathbf{L} = \mathbf{0}$), we find that these equations are satisfied if

$$\alpha_{ml}^\nu/c_{ml}^\nu = \gamma_{ml}^\nu/a_{ml}^\nu = \frac{2k_2}{j\omega_0^2}(c_2^2 - v_2^2) \equiv \delta_1,$$

$$\beta_{ml}^\nu/d_{ml}^\nu = \delta_{ml}^\nu/b_{ml}^\nu = \frac{2k_4}{j\omega_0^2}(c_2^2 - v_4^2) \equiv \delta_2. \tag{5.19.9}$$

Substituting (5.19.8) and (5.19.9) into (5.14.3), we obtain displacement and microrotation fields

$$\mathbf{u} = \sigma_{00}^c k_1 \mathfrak{L}_{00}^c(k_1 r) + \sum_{\nu=c,s} \sum_{l=1}^{\infty} \sum_{m=0}^{l} \sigma_{ml}^{\nu} k_1 \mathfrak{L}_{ml}^{\nu}(k_1 r)$$

$$+ a_{ml}^{\nu} k_2 \mathfrak{N}_{ml}^{\nu}(k_2 r) + b_{ml}^{\nu} k_4 \mathfrak{N}_{ml}^{\nu}(k_4 r)$$

$$+ c_{ml}^{\nu} k_2 \mathfrak{M}_{ml}^{\nu}(k_2 r) + d_{ml}^{\nu} k_4 \mathfrak{M}_{ml}^{\nu}(k_4 r),$$

$$\boldsymbol{\phi} = \tau_{00}^c k_3 \mathfrak{L}_{00}^c(k_3 r) + \sum_{\nu=c,s} \sum_{l=1}^{\infty} \sum_{m=0}^{l} \tau_{ml}^{\nu} k_3 \mathfrak{L}_{ml}^{\nu}(k_3 r)$$

$$+ c_{ml}^{\nu} \delta_1 k_2 \mathfrak{N}_{ml}^{\nu}(k_2 r) + d_{ml}^{\nu} \delta_2 k_4 \mathfrak{N}_{ml}^{\nu}(k_4 r)$$

$$+ a_{ml}^{\nu} \delta_1 k_2 \mathfrak{M}_{ml}^{\nu}(k_2 r) + b_{ml}^{\nu} \delta_2 k_4 \mathfrak{M}_{ml}^{\nu}(k_4 r). \qquad (5.19.10)$$

This constitutes the general solution of the field equations for linear, isotropic micropolar elastic solids in spherical coordinates. For each triplet (ν, m, l) $(l \neq 0)$. There are (6×2) twelve arbitrary constants, since f_l is a linear combination of two spherical Bessel functions.

On a spherical surface, there are six boundary conditions. Consequently, a spherical shell requires twelve constants (six on each surface). If the sphere has no cavity, then at the center $r = 0$ of the sphere, regularity of the displacement and the traction fields requires that the coefficient of one set of Bessel's functions (which are singular) must vanish. This eliminates six constants. For a spherical cavity in infinite medium, the boundary conditions at $r = a$ at $r = \infty$ (e.g., outgoing waves) require twelve constants. Thus, the number of arbitrary constants in (5.19.10) is just what is needed to satisfy the boundary conditions.

Traction boundary conditions on a spherical surface require computing the radial components $\mathbf{t}_{(\mathbf{r})}$ and $\mathbf{m}_{(\mathbf{r})}$ of the surface tractions and couples. They are given by

$$\mathbf{t}_{(\mathbf{r})} = \mathbf{e}_r \cdot \mathbf{t} = \lambda \mathbf{e}_r (\nabla \cdot \mathbf{u}) + (2\mu + \kappa) \frac{\partial \mathbf{u}}{\partial r} + \mu \mathbf{e}_r \times (\nabla \times \mathbf{u}) + \kappa \mathbf{e}_r \times \boldsymbol{\phi},$$

$$\mathbf{m}_{(\mathbf{r})} = \mathbf{e}_r \cdot \mathbf{m} = \alpha \mathbf{e}_r (\nabla \cdot \boldsymbol{\phi}) + (\beta + \gamma) \frac{\partial \boldsymbol{\phi}}{\partial r} + \beta \mathbf{e}_r \times (\nabla \times \boldsymbol{\phi}). \qquad (5.19.11)$$

Calculations of these are straightforward, but lengthy. We copy these results from the paper of Singh [1975] who studied the problem of spherical cavity

$$\mathbf{t}_{(\mathbf{r})} = S_{00}^c(r) \mathfrak{P}_{00}^c(\theta, \phi) + \sum_{\nu=c,s} \sum_{l=1}^{\infty} \sum_{m=0}^{l} [S_{ml}^{\nu}(r) \mathfrak{P}_{ml}^{\nu}(\theta, \phi)$$

$$+ A_{ml}^{\nu}(r) \sqrt{l(l+1)} \mathfrak{B}_{ml}^{\nu}(\theta, \phi) + \mathfrak{B}_{ml}^{\nu}(r) \sqrt{l(l+1)} \mathfrak{C}_{ml}^{\nu}(\theta, \phi)],$$

$$\mathbf{m}_{(\mathbf{r})} = T_{00}^c(r)\mathfrak{P}_{00}^c(\theta,\phi) + \sum_{\nu=c,s}\sum_{l=1}^{\infty}\sum_{m=0}^{l}[T_{ml}^\nu(r)\mathfrak{P}_{ml}^\nu(\theta,\phi)$$

$$+ C_{ml}^\nu(r)\sqrt{l(l+1)}\mathfrak{B}_{ml}^\nu(\theta,\phi) + D_{ml}^\nu(r)\sqrt{l(l+1)}\mathfrak{C}_{ml}^\nu(\theta,\phi)],$$

$$(5.19.12)$$

where

$$\mathfrak{P}_{ml} = \mathbf{e}_r Y_{ml}, \quad \sqrt{l(l+1)}\mathfrak{B}_{ml} = \left(\mathbf{e}_\theta\frac{\partial}{\partial\theta} + \mathbf{e}_\phi\frac{1}{\sin\theta}\frac{\partial}{\partial\phi}\right)Y_{ml},$$

$$\sqrt{l(l+1)}\mathfrak{C}_{ml} = \left(\mathbf{e}_\theta\frac{1}{\sin\theta}\frac{\partial}{\partial\phi} - \mathbf{e}_\phi\frac{\partial}{\partial\theta}\right)Y_{ml}, \qquad (5.19.13)$$

and

$$S_{00}(r) = (2\mu+\kappa)k_1^2 F_{0,4}(k_1 r)\sigma_{00},$$
$$T_{00}(r) = (\beta+\gamma)k_3^2 F_{0,5}(k_3 r)\tau_{00},$$

$$\begin{pmatrix} S_{ml}(r) \\ A_{ml}(r) \\ D_{ml}(r) \end{pmatrix} = \begin{pmatrix} G_{11}(r) & G_{12}(r) & G_{13}(r) \\ G_{21}(r) & G_{22}(r) & G_{23}(r) \\ G_{31}(r) & G_{32}(r) & G_{33}(r) \end{pmatrix} \begin{pmatrix} \sigma_{ml} \\ a_{ml} \\ b_{ml} \end{pmatrix},$$

$$\begin{pmatrix} T_{ml}(r) \\ C_{ml}(r) \\ B_{ml}(r) \end{pmatrix} = \begin{pmatrix} H_{11}(r) & H_{12}(r) & H_{13}(r) \\ H_{21}(r) & H_{22}(r) & H_{23}(r) \\ H_{31}(r) & H_{32}(r) & H_{33}(r) \end{pmatrix} \begin{pmatrix} \tau_{ml} \\ c_{ml} \\ d_{ml} \end{pmatrix}, \quad (5.19.14)$$

$$G_{11}(r) = (2\mu+\kappa)k_1^2 F_{l,4}(k_1 r),$$
$$G_{12}(r) = (2\mu+\kappa)l(l+1)k_2^2 F_{l,1}(k_2 r),$$
$$G_{21}(r) = (2\mu+\kappa)k_1^2 F_{l,1}(k_1 r),$$
$$G_{22}(r) = (1/2)k_2[(2\mu+\kappa)k_2 F_{l,3}(k_2 r) + (2\delta_1 - k_2)\kappa f_l(k_2 r)],$$
$$G_{31}(r) = 0,$$
$$G_{32}(r) = (1/2)\delta_1 k_2^3 r[(\beta+\gamma)F_{l,1}(k_2 r) - (\beta-\gamma)F_{l,2}(k_2 r)],$$
$$H_{11}(r) = (\beta+\gamma)k_3^2 F_{l,5}(k_3 r),$$
$$H_{12}(r) = (\beta+\gamma)l(l+1)\delta_1 k_2^2 F_{l,1}(k_2 r),$$
$$H_{21}(r) = (\beta+\gamma)k_3^2 F_{l,1}(k_3 r),$$
$$H_{22}(r) = (1/2)\delta_1 k_2^2[(\beta+\gamma)F_{l,3}(k_2 r) + (\beta-\gamma)f_l(k_2 r)],$$
$$H_{31}(r) = -(\kappa/r)f_1(k_3 r),$$
$$H_{32}(r) = (1/2)k_2^2 r[(2\mu+\kappa)k_2 F_{l,1}(k_2 r) - (2\delta_1 - k_2)\kappa F_{l,2}(k_2 r)].$$

A_{i3} is obtained from A_{i2} and B_{i3} is obtained from B_{i2} on changing k_2 to k_4 and δ_1 to δ_2. In the foregoing expression

$$F_{l,1}(x) = \frac{l-1}{x^2} f_l(x) - \frac{1}{x} f_{l+1}(x),$$

$$F_{l,2}(x) = \frac{l+1}{x^2} f_l(x) - \frac{1}{x} f_{l+1}(x),$$

$$F_{l,3}(x) = \left[\frac{2}{x^2}(l^2-1) - 1\right] f_l(x) + \frac{2}{x} f_{l+1}(x),$$

$$F_{l,4}(x) = \left[\frac{l(l-1)}{x^2} - \frac{\lambda+2\mu+\kappa}{2\mu+\kappa}\right] f_l(x) + \frac{2}{x} f_{l+1}(x),$$

$$F_{l,5}(x) = \left[\frac{l(l-1)}{x^2} - \frac{\alpha+\beta+\gamma}{\beta+\gamma}\right] f_l(x) + \frac{2}{x} f_{l+1}(x). \quad (5.19.15)$$

Spherical Cavity

Consider a spherical cavity of radius $r = a$, in a micropolar elastic medium of infinite extent. The motion vanishes at infinity and the waves are an outgoing type, so that $f_l = h_l^{(2)}$, the spherical Hankel function. The surface tractions and couples are presented on the surface as

$$\mathbf{t}_{(\mathbf{r})} = \mathbf{P}(\theta,\phi), \quad \mathbf{m}_{(\mathbf{r})} = \mathbf{M}(\theta,\phi) \quad \text{at } r=a. \quad (5.19.16)$$

We express $\mathbf{P}$ and $\mathbf{M}$ in terms of the vector spherical harmonics

$$\mathbf{P}(\theta,\phi) = p_{00}^c \mathfrak{P}_{00}^c(\theta,\phi) + \sum_{\nu=c,s}\sum_{l=1}^{\infty}\sum_{m=0}^{l} [p_{ml}^\nu \mathfrak{P}_{ml}^\nu(\theta,\phi)$$
$$+ q_{ml}^\nu \sqrt{l(l+1)} \mathfrak{B}_{ml}^\nu(\theta,\phi) + r_{ml}^\nu \sqrt{l(l+1)} \mathfrak{C}_{ml}^\nu(\theta,\phi)],$$

$$\mathbf{M}(\theta,\phi) = \xi_{00}^c \mathfrak{P}_{00}^c(\theta,\phi) + \sum_{\nu=c,s}\sum_{l=1}^{\infty}\sum_{m=0}^{l} [\xi_{ml}^\nu \mathfrak{P}_{ml}^\nu(\theta,\phi)$$
$$+ \eta_{ml}^\nu \sqrt{l(l+1)} \mathfrak{B}_{ml}^\nu(\theta,\phi) + \zeta_{ml}^\nu \sqrt{l(l+1)} \mathfrak{C}_{ml}^\nu(\theta,\phi)].$$
$$(5.19.17)$$

Given $\mathbf{P}$ and $\mathbf{M}$, the coefficients $p_{ml}^\nu, \xi_{ml}^\nu, \ldots$ are calculated with the help of the orthogonality relations for the vector spherical harmonics (cf. Morse and Feshbach [1953, p. 1900]). Equation (5.19.12) now yields

$$S_{ml}^\nu(a) = p_{ml}^\nu, \quad A_{ml}^\nu(a) = q_{ml}^\nu, \quad D_{ml}^\nu(a) = \zeta_{ml}^\nu, \quad (5.19.18)$$
$$T_{ml}^\nu(a) = \xi_{ml}^\nu, \quad C_{ml}^\nu = \eta_{ml}^\nu, \quad B_{ml}^\nu = r_{ml}^\nu. \quad (5.19.19)$$

Substituting $r = a$ in (5.19.14), (5.19.18) gives six simultaneous equations and six unknowns. An examination of these equations shows that (5.19.18) are uncoupled from (5.19.19), so that we have two independent third-order systems (three equations). Equation (5.19.16) determines $\sigma_{ml}^{\nu}, a_{ml}^{\nu}$, and b_{ml}^{ν} and it gives $\tau_{ml}^{\nu} = c_{ml}^{\nu} = d_{ml}^{\nu} = 0$. For this problem

$$(\nabla \times \mathbf{u})_r = \nabla \cdot \boldsymbol{\phi} = \phi_r = 0. \tag{5.19.20}$$

Consequently $\mathbf{u}$ can be expressed only in terms of $\mathfrak{P}$ and $\mathfrak{B}$ vectors, and $\boldsymbol{\phi}$ can be expressed in terms of $\mathfrak{M}$ vectors. This motion is called *spheroidal motion*.

Equation (5.19.19), on the other hand, determines $c_{ml}^{\nu}, d_{ml}^{\nu}$, and τ_{ml}^{ν} with $\sigma_{ml}^{\nu} = a_{ml}^{\nu} = b_{ml}^{\nu} = 0$. In this case $\mathbf{u}$ is expressed in terms of $\mathfrak{M}$ vectors and microrotation $\boldsymbol{\phi}$ is expressed in terms of $\mathfrak{L}$ and $\mathfrak{N}$ vectors. Corresponding motion is a *torsional* (or *toroidal*) *motion* and it satisfies the condition

$$\nabla \cdot \mathbf{u} = u_r = (\nabla \times \boldsymbol{\phi})_r = \mathbf{0}. \tag{5.19.21}$$

The spheroidal and toridal motions reduce to their classical counterparts for the elastic medium in elastic limit ($\kappa \to 0$).

From (5.19.14), (5.19.17), and (5.19.19), we obtain

$$\sigma_{00} = p_{00}/[(2\mu + \kappa)k_1^2 F_{0,4}(k_1 a)],$$
$$\tau_{00} = \xi_{00}/[(\beta + \gamma)k_3^2 F_{0,5}(k_3 a)],$$
$$(\sigma_{ml}, a_{ml}, b_{ml})^T = G^{-1}(a)[p_{ml}, q_{ml}, \zeta_{ml}]^T,$$
$$(\tau_{ml}, c_{ml}, d_{ml})^T = H^{-1}(a)[\xi_{ml}, \eta_{ml}, r_{ml}]^T, \quad l \geq 1, \tag{5.19.22}$$

where G^{-1} and H^{-1} are the inverse matrices of G and H, respectively. Finally, we must remember to obtain the inverse Fourier transforms of the solutions.

The solution of the sphere problem is similar to the spherical cavity problem. In fact, the solution of the sphere problem is obtained by replacing $h_l^{(2)}$ with the spherical Bessel function of the first kind j_l.

A. Radial Motions

From the general solution given above, several special solutions can be extracted, e.g., pressure on spherical cavity and spheres, breathing modes of a sphere, etc. However, it is simpler to treat these one-dimensional motions by using the field equations. In this case

$$\mathbf{u} = \mathbf{e}_r u(r, t), \quad \boldsymbol{\phi} = \mathbf{0}, \tag{5.19.23}$$

and the only equation to be satisfied is

$$\frac{\partial}{\partial r}\left(\frac{\partial}{\partial r} + \frac{2}{r}\right)u = \frac{1}{c_1^2}\frac{\partial^2 u}{\partial t^2}, \quad c_1^2 \equiv (\lambda + 2\mu + \kappa)/\rho. \tag{5.19.24}$$

In this case, we have $\mathbf{m} = \mathbf{0}$ and $\mathbf{t}_{(\mathbf{r})}$ reads

$$\mathbf{t}_{(\mathbf{r})} = \mathbf{e}_r\left[\lambda\left(\frac{\partial}{\partial r} + \frac{2}{r}\right)u + (2\mu + \kappa)\frac{\partial u}{\partial r}\right]. \tag{5.19.25}$$

For the static problem, we need to satisfy

$$\frac{d}{dr}\left(\frac{d}{dr}+\frac{2}{r}\right)u = 0. \tag{5.19.26}$$

From these equations, it is clear that radial solutions (with $\phi = 0$) are identical to the classical elasticity solutions with μ replaced by $\mu_e = \mu + \kappa/2$.

B. Radial Solution for Microrotation

In micropolar elastic bodies, twist waves and/or twist vibrations exist that have no counterpart in classical elasticity. In the case $\mathbf{u} = \mathbf{0}$, $\phi = \phi(r,t)$, the field equations read

$$c_3^2\left(\phi_{,rr} + \frac{2}{r}\phi_{,r} - \frac{2}{r^2}\phi\right) - \omega_0^2\phi - \ddot{\phi} = 0, \tag{5.19.27}$$

where

$$c_3^2 = \frac{\alpha + \beta + \gamma}{\rho j}, \quad \omega_0^2 = 2\kappa/\rho j. \tag{5.19.28}$$

Equation (5.19.27) is a Klein–Gordon equation in two dimensions. Two solutions of this equation for impulsive boundary conditions were considered by Maiti [1987]

(i) $\phi = M_1\delta(t)$,   (ii) $m_{rr} = (\alpha+\beta+\gamma)\phi_{,r} + 2\alpha r^{-1}\phi = M_2\delta(t)$. (5.19.29)

By means of the Laplace transform, the solutions of (i) and (ii) are obtained, as indicated by ϕ_1 and ϕ_2,

$$\bar{\phi}_1 = \frac{M_1}{R^2}\frac{1+kaR}{1+ka}\exp[-ka(R-1)], \tag{5.19.30}$$

$$\bar{\phi}_2 = -\frac{M_2 a}{R^2}\frac{(1+kaR)\exp[-ka(R-1)]}{(\alpha+\beta+\gamma)k^2a^2 + 2(\beta+\gamma)(1+ka)}, \tag{5.19.31}$$

where a superposed bar denotes the Laplace transform with parameter p and

$$ka = (a\omega_0/c_3)[(p/\omega_0)^2 + 1]^{1/2}, \quad R = r/a. \tag{5.19.32}$$

The inverse transforms of these expressions are obtained by recalling that if $\bar{f}(p)$ is the Laplace transform of $f(t)$, then $\bar{f}(p/\omega_0)$ is the Laplace transform of $\omega_0 f(\omega_0 t)$, and $\bar{f}(\sqrt{p^2+1})$ is the Laplace transform of

$$f(t) - \int_0^t f(\sqrt{t^2-\xi^2})J_1(\xi)\,d\xi,$$

cf. Magnus and Oberhettinger [1949, p. 123]. Using these, we obtain

$$R^2\phi_1(R,t)/M_1\omega_0 = f_1(\omega_0 t) - \int_0^{\omega_0 t} f_1(\sqrt{\omega_0^2 t^2 - \xi^2})J_1(\xi)\,d\xi, \tag{5.19.33}$$

$$-2\kappa c_3 R^2\phi_2(R,t)/M_2\omega_0^2 = f_2(\omega_0 t) - \int_0^{\omega_0 t} f_2(\sqrt{\omega_0^2 t^2 - \xi^2})J_1(\xi)\,d\xi, \tag{5.19.34}$$

where

$$f_1(t) = R\delta(t) - \frac{(R-1)c_3}{a\omega_0}\exp(-c_3\tau/a\omega_0)H(\tau),$$

$$f_2(t) = \left[\frac{1-(a\omega_0/c_3)\sigma R}{(a\omega_0 d/c_3)}\sin(\tau d) + R\cos(\tau d)\right]\exp(-\sigma\tau)H(\tau),$$

$$\tau = t - \frac{a\omega_0}{c_3}(R-1), \quad \sigma = \frac{\beta+\gamma}{\alpha+\beta+\gamma}\frac{c_3}{a\omega_0},$$

$$d^2 = \frac{(\beta+\gamma)(2\alpha+\beta+\gamma)c_3^2}{(\alpha+\beta+\gamma)^2 a^2\omega_0^2}. \tag{5.19.35}$$

Here δt is the Dirac delta measure and $H(\tau)$ is the Heaviside unit function.

Maiti gave plots of the microrotation for the second case. However, his equations contain some errors.[11]

5.20   Axisymmetric Problems

For axisymmetric problems, it is convenient to employ cylindrical coordinates (r,θ,z). The fields are independent of θ, so that displacement vector **u** and microrotation vector $\boldsymbol{\phi}$ are of the form

$$\mathbf{u}(r,z) = u(r,z)\mathbf{e}_r + w(r,z)\mathbf{e}_z, \qquad \boldsymbol{\phi}(r,z) = \phi(r,z)\mathbf{e}_\theta. \tag{5.20.1}$$

The stress and couple stress possess the following nonvanishing components

$$t_{rr} = \lambda e + (2\mu+\kappa)u_{,r}, \qquad t_{rz} = \mu u_{,z} + (\mu+\kappa)w_{,r} + \kappa\phi,$$

$$t_{\theta\theta} = \lambda e + (2\mu+\kappa)\frac{u}{r}, \qquad t_{zr} = (\mu+\kappa)u_{,z} + \mu w_{,r} - \kappa\phi,$$

$$t_{zz} = \lambda e + (2\mu+\kappa)w_{,z}, \qquad e \equiv \frac{1}{r}(ru)_{,r} + w_{,z}, \tag{5.20.2}$$

$$m_{\theta r} = \beta\phi_{,r} - \frac{\gamma}{r}\phi, \qquad m_{r\theta} = -\frac{\beta}{r}\phi + \gamma\phi_{,r},$$

$$m_{z\theta} = \gamma\phi_{,z}, \qquad m_{\theta z} = \beta\phi_{,z}.$$

In terms of potentials for **u**, we have

$$u = \sigma_{,r} - U_{,z}, \quad w = \sigma_{,z} + \frac{1}{r}(rU)_{,r}. \tag{5.20.3}$$

The Fourier transforms of the field equations (5.14.27) (with $\bar{P} = \bar{Q} = \bar{g} = 0$), with respect to time, read

$$\nabla_2^2\bar\sigma + k_1^2\bar\sigma = 0,$$
$$(\nabla_3^2 + k_2^2)(\nabla_3^2 + k_4^2)\bar U = 0,$$
$$c_2^2\nabla_3^2\bar U + \omega^2\bar U + \tfrac{1}{2}j\omega_0^2\bar\phi = 0, \tag{5.20.4}$$

---

[11]Maiti's expressions for ϕ_2 and d are not correct. After we pointed out our disagreement, he concurred with my solutions.

where k_α^2 are given by (5.11.18) and

$$\nabla_2^2 F = F_{,rr} + \frac{1}{r} F_{,r} + F_{,zz}, \quad \nabla_3^2 F = \left(\nabla_2^2 - \frac{1}{r^2} \right) F. \tag{5.20.5}$$

An overbar denotes the Fourier transform with respect to time, i.e.,

$$\bar{F}(r, z, \omega) = (2\pi)^{-1/2} \int_{-\infty}^{\infty} F(r, z, t) e^{i\omega t} \, dt. \tag{5.20.6}$$

General solutions of (5.20.4) are of the form

$$\bar{\sigma} = f_1(\xi r) \exp(\pm |\eta_1| z),$$
$$\bar{U} = f_2(\xi r) \exp(\pm |\eta_2| z), \quad \bar{\phi} = f_4(\xi r) \exp(\pm |\eta_4| z), \tag{5.20.7}$$

where η_α are given by

$$\eta_\alpha^2 = \xi^2 - k_\alpha^2, \quad \alpha = 1, 2, 4. \tag{5.20.8}$$

Here $f_1(\xi r)$ is a linear combination of Bessel's functions $\{ J_0(\xi r), Y_0(\xi r) \}$ or $\{ H_0^{(1)}(\xi r), H_0^{(2)}(\xi r) \}$. Similarly $f_2(\xi r)$ and $f_4(\xi r)$ are linear combinations of $\{ J_1(\xi r), Y_1(\xi r) \}$ or $\{ H_1^{(1)}(\xi r), H_1^{(2)}(\xi r) \}$. For solutions bounded at $r = 0$, $f_1 = J_0(\xi r)$. For outgoing waves $f_1 = H_0^{(2)}(\xi r)$. For displacements vanishing at $z = \infty$, we keep $\exp(-|\eta_\alpha| z)$ and drop $\exp(+|\eta_\alpha| z)$.

Half-Space Subject to Surface Loads

In this case, the general solution is of the form

$$\bar{\sigma} = \int_0^\infty A\xi J_0(\xi r) \exp(-\eta_1 z) \, d\xi,$$
$$\bar{U} = \int_0^\infty [B \exp(-\eta_2 z) + C \exp(-\eta_4 z)] \xi J_1(\xi r) \, d\xi,$$
$$\bar{\phi} = \int_0^\infty [B\nu_2 \exp(-\eta_2 z) + C\nu_4 \exp(-\eta_4 z)] \xi J_1(\xi r) \, d\xi, \tag{5.20.9}$$

where A, B, and C are functions of ξ and ω and

$$\nu_\beta = \frac{c_2^2 k_\beta^2 - \omega^2}{j\omega_0^2/2}, \quad \beta = 2, 4. \tag{5.20.10}$$

The displacement field $(\bar{u}, \bar{w})$ follows from (5.20.3)

$$\bar{u} = \int_0^\infty \left(-A\xi e^{-\eta_1 z} + B\eta_2 e^{-\eta_2 z} + C\eta_4 e^{-\eta_4 z} \right) \xi J_1(\xi r) \, d\xi,$$
$$\bar{w} = \int_0^\infty \left(-A\eta_1 e^{-\eta_1 z} + B\xi e^{-\eta_2 z} + C\xi e^{-\eta_4 z} \right) \xi J_0(\xi r) \, d\xi. \tag{5.20.11}$$

The stress components $(\bar{t}_{zz}, \bar{t}_{zr})$ and couple $\bar{m}_{z\theta}$ are obtained from (5.20.2)

$$\bar{t}_{zz} = \int_0^\infty \left(AP_1 e^{-\eta_1 z} + BP_2 e^{-\eta_2 z} + CP_3 e^{-\eta_4 z}\right) \xi J_0(\xi r)\, d\xi,$$

$$\bar{t}_{zr} = \int_0^\infty \left(AQ_1 e^{-\eta_1 z} + BQ_2 e^{-\eta_2 z} + CQ_3 e^{-\eta_4 z}\right) \xi J_1(\xi r)\, d\xi,$$

$$\bar{m}_{z\theta} = \int_0^\infty \left(BR_2 e^{-\eta_2 z} + CR_3 e^{-\eta_4 z}\right) \xi J_1(\xi r)\, d\xi, \qquad (5.20.12)$$

where

$$P_1 = (\lambda + 2\mu + \kappa)\eta_1^2 - \lambda\xi^2, \quad P_2 = -(2\mu + \kappa)\xi\eta_2, \quad P_3 = -(2\mu + \kappa)\xi\eta_4,$$
$$Q_1 = (2\mu + \kappa)\xi\eta_1, \quad Q_2 = -(\mu + \kappa)\eta_2^2 - \mu\xi^2 - \kappa\nu_2,$$
$$Q_3 = -(\mu + \kappa)\eta_4^2 - \mu\xi^2 - \kappa\nu_4, \quad R_2 = -\gamma\nu_2\eta_2, \quad R_3 = -\gamma\nu_4\eta_4.$$

$$(5.20.13)$$

Given the Fourier transforms of surface tractions on the surface $z = 0$, we have, for the boundary conditions,

$$T^0(\xi,\omega) = \int_0^\infty \bar{t}_{zz}^0(r,\omega)\xi J_0(\xi r)\, dr,$$

$$\{S^0(\xi,\omega), M^0(\xi,\omega)\} = \int_0^\infty \{\bar{t}_{zr}^0, \bar{m}_{z\theta}^0\}\xi J_1(\xi r)\, d\xi. \qquad (5.20.14)$$

Constants A, B, and C are now determined by solving three simultaneous equations

$$P_1 A + P_2 B + P_3 C = T^0(\xi,\omega),$$
$$Q_1 A + Q_2 B + Q_3 C = S^0(\xi,\omega),$$
$$R_2 B + R_3 C = M^0(\xi,\omega). \qquad (5.20.15)$$

The solution of this system is given by

$$DA = (Q_2 R_3 - Q_3 R_2)T^0 - (P_2 R_3 - P_3 R_2)S^0 + (P_2 Q_3 - P_3 Q_2)M^0,$$
$$DB = -Q_1 R_3 T^0 + P_1 R_3 S^0 - (P_1 Q_3 - Q_1 P_3)M^0,$$
$$DC = Q_1 R_2 T^0 - P_1 R_2 S^0 + (P_1 Q_2 - Q_1 P_2)M^0,$$
$$D \equiv P_1(Q_2 R_3 - Q_3 R_2) - Q_1(P_2 R_3 - P_3 R_2). \qquad (5.20.16)$$

The displacement field is determined by (5.20.11) and microrotation by $(5.20.9)_3$. The formal solution is complete.

However, a major task remains in calculating the integrals with respect to ξ and ω. Remembering the difficulties involved in Lamb's problem of classical elasticity (Eringen and Şuhubi [1975, p. 752]), exact calculations of these integrals can only be achieved for very special types of surface loads, e.g., harmonic loads that are distributed radially according to some Bessel functions. Even impulsive concentrated loads defy solutions in closed forms. Clearly computer calculations are called for.

A. Impulsive Concentrated Load

$$t_{zz}^0 = -T_0 \frac{\delta(r)}{2\pi r}\delta(t), \quad t_{zr}^0 = m_{z\theta}^0 = 0. \qquad (5.20.17)$$

In this case, (5.20.16) gives $T^0 = -T_0, S^0 = M^0 = 0$ and, from (5.20.16), we have

$$DA = (Q_3 R_2 - Q_2 R_3)T_0, \quad DB = Q_1 R_3 T_0,$$
$$DC = -Q_1 R_2 T_0, \qquad (5.20.18)$$

and D is given by $(5.20.16)_4$.

B. Concentrated Impulsive Couple

$$t_{zz}^0 = t_{zr}^0 = 0, \quad m_{z\theta}^0 = -M_0 \frac{\delta(r)}{2\pi r}\delta(t), \qquad (5.20.19)$$

with these, $T^0 = S^0 = 0$ and $M^0(\xi, \omega) = -M_0 = \text{const}$. From (5.20.16), we have

$$DA = (P_3 Q_2 - P_2 Q_3)M_0, \quad DB = (P_1 Q_3 - Q_1 P_3)M_0,$$
$$DC = (Q_1 P_2 - Q_2 P_1)M_0. \qquad (5.20.20)$$

There is no corresponding case to this one in classical elasticity.

Even though, in both Problems A and B, applied loads contribute only constant terms T_0 and M_0 to the solutions u, w, and ϕ, the dependence of P_i, Q_i, and R_i on ξ and ω makes the evaluation of the double integrals with respect to ξ and ω extremely complicated.

C. Static Axisymmetric Problems for Half-Space

Axisymmetric problems for an isotropic micropolar half-space have been considered by several investigators, Nowacki [1969], Puri [1971], Khan and Dhaliwal [1977].

The solutions of the micropolar field equations do not follow from the dynamical case by setting $\omega = 0$, since a part of the solution is lost in this process. While there exists a method to obtain the correct solution by means of differentiation with respect to a parameter, it is simpler and more pedalogical to follow a direct solution of the field equations (5.6.16) and (5.6.17).

In cylindrical coordinates, the field equations read

$$(\lambda + \mu)\frac{\partial e}{\partial r} + (\mu + \kappa)(\nabla^2 - r^{-2})u - \kappa\frac{\partial\phi}{\partial z} = 0,$$
$$(\lambda + \mu)\frac{\partial e}{\partial z} + (\mu + \kappa)\nabla^2 w + \frac{\kappa}{r}\frac{\partial}{\partial r}(r\phi) = 0,$$
$$\gamma(\nabla^2 - r^{-2})\phi + \kappa\left(\frac{\partial u}{\partial z} - \frac{\partial w}{\partial r}\right) - 2\kappa\phi = 0, \qquad (5.20.21)$$

where $\phi = \phi_\theta$ and

$$e = \frac{1}{r}\frac{\partial}{\partial r}(ru) + \frac{\partial w}{\partial z}, \quad \nabla^2 = \frac{\partial^2}{\partial r^2} + \frac{1}{r}\frac{\partial}{\partial r} + \frac{\partial^2}{\partial z^2}. \tag{5.20.22}$$

The physical components of the stress and couple stress tensors are given by (5.20.2). An appropriate solution for this problem is obtained by means of Hankel transforms $\bar{u}(\xi, z), \bar{w}(\xi, z)$, and $\bar{\phi}(\xi, z)$, i.e., we take

$$u(r, z) = \int_0^\infty \xi \bar{u}(\xi, z) J_1(\xi r)\, d\xi,$$

$$w(r, z) = \int_0^\infty \xi \bar{w}(\xi, z) J_0(\xi r)\, d\xi,$$

$$\phi(r, z) = \int_0^\infty \xi \bar{\phi}(\xi, z) J_1(\xi r)\, d\xi. \tag{5.20.23}$$

Substituting (5.20.23) into (5.20.21), we obtain differential equations whose solution, appropriate to half-space $z \geq 0$, is

$$\bar{u}(\xi, z) = \left[A - \frac{\lambda + 3(\mu + \kappa/2)}{\lambda + 2\mu + \kappa} \xi^{-1} B + \frac{\lambda + \mu + \kappa/2}{\lambda + 2\mu + \kappa} z B \right] e^{-\xi z}$$
$$- \frac{\gamma}{2\mu + \kappa} \eta C e^{-\eta z},$$

$$\bar{w}(\xi, z) = \left[A + \frac{\lambda + \mu + \kappa/2}{\lambda + 2\mu + \kappa} z B \right] e^{-\xi z} - \frac{\gamma}{2\mu + \kappa} \xi C e^{-\eta z},$$

$$\bar{\phi}(\xi, z) = B e^{-\xi z} + C e^{-\eta z}, \tag{5.20.24}$$

where A, B, and C are arbitrary constants and

$$\eta^2 = \xi^2 + \frac{(2\mu + \kappa)\kappa}{\gamma(\mu + \kappa)}. \tag{5.20.25}$$

The surface tractions are given by

$$\bar{t}_{zz}(\xi, z) = (2\mu + \kappa)\left[\left(-\xi A + \frac{\mu + \kappa/2}{\lambda + 2\mu + \kappa} B - \frac{\lambda + \mu + \kappa/2}{\lambda + 2\mu + \kappa} B \xi z \right) e^{-\xi z} \right.$$
$$\left. + \frac{\gamma}{2\mu + \kappa} C \xi \eta e^{-\eta z} \right],$$

$$\bar{t}_{zr}(\xi, z) = (2\mu + \kappa)\left[\left(-\xi A + B - \frac{\lambda + \mu + \kappa/2}{\lambda + 2\mu + \kappa} B \xi z \right) e^{-\xi z} \right.$$
$$\left. + \frac{\gamma}{2\mu + \kappa} C \xi^2 e^{-\eta z} \right],$$

$$\bar{m}_{z\theta}(\xi, z) = -\gamma[B\xi e^{-\xi z} + C\eta e^{-\eta z}]. \tag{5.20.26}$$

Substituting these into boundary conditions

$$\bar{t}_{zz}(\xi, 0) = T^0(\xi), \quad \bar{t}_{zr}(\xi, 0) = S^0(\xi), \quad \bar{m}_{z\theta}(\xi, 0) = M^0(\xi), \tag{5.20.27}$$

We obtain three simultaneous equations to determine $A, B,$ and C in terms of T^0, S^0 and M^0:

$$A = \frac{\Theta}{2\mu'} \left\{ 2(1-\nu)\left(-\frac{\eta}{\xi} + l^2\xi^2\right) T^0 + \left[(1-2\nu)\frac{\eta}{\xi} - 2(1-\nu)l^2\xi\eta\right] S^0 \right.$$

$$\left. + \left[(1-2\nu)\xi - 2(1-\nu)\eta\right]M^0 \right\},$$

$$B = -\frac{2(1-\nu)\Theta}{2\mu'}[\eta(T^0 + S^0) + \xi(\eta - \xi)M^0],$$

$$C = \frac{\Theta}{2\mu'}[2(1-\nu)(T^0 - S^0)\xi - l^{-2}M^0], \tag{5.20.28}$$

where

$$\Theta \equiv [\eta + 2(1-\nu)l^2\xi^2(\eta - \xi)]^{-1}, \quad \mu' \equiv \mu + \kappa/2,$$

$$\nu \equiv \frac{\lambda}{2(\lambda + \mu')}, \quad l^2 = \gamma/2\mu'. \tag{5.20.29}$$

This completes the formal solution, since displacement and stress fields follow from (5.20.24) and (5.20.2). The integrations over ξ, however, present major difficulties. Computer calculations are required.

Concentrated Load

A uniform pressure of total magnitude P, applied over a circular region of radius a is given by

$$t_{zz}(r, 0) = -\frac{P}{\pi a^2} H(a - r). \tag{5.20.30}$$

The Hankel transform of this is

$$\bar{t}_{zz}(\xi, 0) = -\frac{P}{\pi a} \frac{J_1(a\xi)}{\xi}. \tag{5.20.31}$$

If now we let $a \to 0$, this gives

$$\bar{t}_{zz}(\xi, 0) = -\frac{P}{2\pi}. \tag{5.20.32}$$

Substituting this into (5.20.29) with $S^0 = M^0 = 0$, we obtain $A, B,$ and C. The displacement and stress fields are obtained through (5.20.24) and (5.20.2). Here we give the displacement fields

$$u(r, z) = \frac{P}{4\pi\mu'} \int_0^\infty [(2\nu - 1 + \xi z)\xi e^{-\xi z} - 2(1-\nu)l^2\xi^2(\xi e^{-\xi z} - \eta e^{-\eta z})]$$

$$\Theta(\xi)J_1(\xi r)\,d\xi,$$

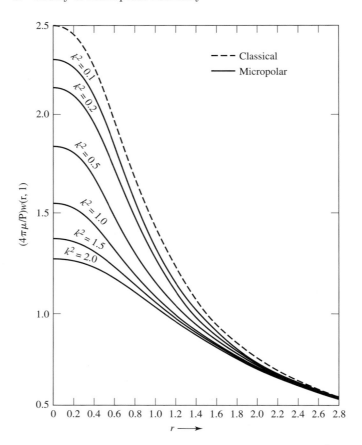

FIGURE 5.20.1. Normal displacement $4\pi\mu/P)w(r,1)$ vs. r, with k^2 varying, on the plane $z=1$, of the half-space under the action of concentrated force at the origin. (From Khan & Dhaliwal [1977])

$$w(r,z) = \frac{P}{4\pi\mu'}\int_0^\infty [(2-2\nu+\xi z)\eta e^{-\xi z} - 2(1-\nu)l^2\xi^3(e^{-\xi z}-e^{-\eta z})]$$

$$\Theta(\xi)J_0(\xi r)\,d\xi,$$

$$\phi(r,z) = \frac{P(1-\nu)}{2\pi\mu'}\int_0^\infty (\eta e^{-\xi z} - \xi e^{-\eta z})\xi\Theta(\xi)J_1(\xi r)\,d\xi. \qquad (5.20.33)$$

Khan and Dhaliwal [1977] have computed the normal displacement w, the stress component t_{zz}, and the couple stress component $m_{z\phi}$ for various κ at the plane $z=1$. Numerical computations were carried out for $\nu = 0.25$, $l^2 = 1$, $0 < r < 3$. Comparison is made for κ in the interval $0 \le \kappa \le \infty$, where $\kappa = 0$ corresponds to the case of classical elasticity. The interval of κ corresponds to $0 \le k^2 \le 2$ where $k^2 = 2\kappa/(\mu+\kappa)$. Figures 5.20.1 to 5.20.4 represent $w(k,1), \phi(r,1), t_{zz}(r,1)$, and $m_{z\phi}(r,1)$ for various values k^2.

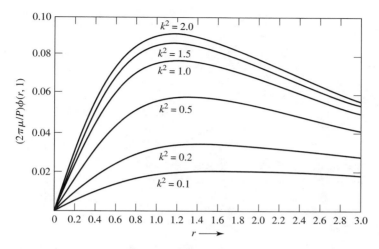

FIGURE 5.20.2. Rotation $(2\pi\mu/P)\phi(r,1)$ vs. r, with k^2 varying, on the plane $z = 1$, of the half-space under the action of concentrated force at the origin. (From Khan & Dhaliwal [1977])

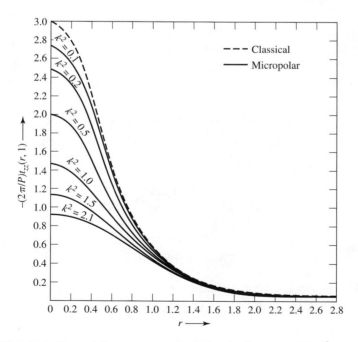

FIGURE 5.20.3. Normal force-stress $-(2\pi/P)t_{zz}(r,1)$ vs. r, with k^2 varying, on the plane $z = 1$, of the half-space under the action of concentrated force at the origin. (From Khan & Dhaliwal [1977])

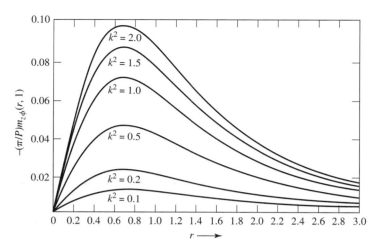

FIGURE 5.20.4. Couple-stress $(-\pi/P)m_{z\phi}(r,1)$ vs. r, with k^2 varying, on the plane $z = 1$, of the half-space under the action of concentrated force a the origin. (From Khan & Dhaliwal [1977])

5.21   Penny-Shaped Crack

A homogeneous isotropic micropolar solid containing a penny-shaped crack[12] at $z = 0\pm$, $0 \le r \le c$ is subject to uniform constant tension $t_{zz} = p_0$ at infinity. We wish to determine the stress and couple stress fields in the body. As in classical elasticity, the solution of this problem requires the superposition of two problems:

(i) The crack-free solid subject to uniaxial tension $t_{zz} = p_0$; and

(ii) the problem of a crack opened by constant normal pressure p_0.

The solution of problem (i) is

$$\mathbf{u} = \frac{p_0}{2\mu'}\left(-\frac{\lambda}{3\lambda + 2\mu'}\mathbf{i} + \frac{2\lambda + 2\mu'}{3\lambda + 2\mu'}z\mathbf{k}\right), \quad \boldsymbol{\phi} = \mathbf{0}, \tag{5.21.1}$$

$$\mu' \equiv \mu + \kappa/2, \quad \lambda \equiv 2\mu'\nu/(1 - 2\nu),$$

where μ' and ν are, respectively, the shear modulus and Poisson's ratio for micropolar solids.

For the second problem, we have the boundary conditions

$$t_{zz}(r,0) = -p_0, \quad m_{z\theta}(r,0) = 0, \quad 0 \le r \le c, \tag{5.21.2}$$

$$w(r,0) = 0, \quad \phi(r,0) = 0, \quad c < r < \infty, \tag{5.21.3}$$

$$t_{zr}(r,0) = 0, \quad 0 \le r < \infty. \tag{5.21.4}$$

---

[12]We follow Paul and Sridharan [1980].

Employing the general solution (5.20.23) of the static axisymmetric problem, and (5.21.4), we have

$$\xi A = B + \frac{\gamma}{2\mu'}\xi^2 C. \tag{5.21.5}$$

Using this and the nondimensional variables ρ, x, y,

$$r = c\rho, \quad \xi = x/c, \quad \eta = y/c = c^{-1}(x^2 + M^2), \quad M^2 = 2\mu'\kappa/\gamma(\mu' + \kappa/2), \tag{5.21.6}$$

we have for the boundary values of $t_{zz}, m_{z\theta}, w$, and ϕ:[13]

$$t_{zz}(\rho, 0) = -\mu'c^{-2} \int_0^\infty \left[(1-\nu)^{-1} B(x) + \frac{\gamma}{c^2\mu'}x(x-y)C(x)\right] xJ_0(x\rho)\, dx, \tag{5.21.7}$$

$$m_{z\theta}(\rho, 0) = -\gamma c^{-3} \int_0^\infty [xB(x) + yC(x)]\, xJ_1(x\rho)\, dx, \tag{5.21.8}$$

$$w(\rho, 0) = c^{-1} \int_0^\infty B(x)J_0(x\rho)\, dx, \tag{5.21.9}$$

$$\phi(\rho, 0) = c^{-2} \int_0^\infty [B(x) + C(x)]xJ_1(x\rho)\, dx. \tag{5.21.10}$$

The boundary conditions (5.21.2) and (5.21.3) read

$$\int_0^\infty \left[(1-\nu)^{-1} B(x) + \frac{\gamma}{c^2\mu'}x(x-y)C(x)\right] xJ_0(x\rho)\, dx = \frac{c^2 p_0}{\mu'}, \quad 0 \leq \rho \leq 1, \tag{5.21.11}$$

$$\int_0^\infty [xB(x) + yC(x)]\, xJ_1(x\rho)\, dx = 0, \quad 0 \leq \rho \leq 1, \tag{5.21.12}$$

$$\int_0^\infty B(x)J_0(x\rho)\, dx = 0, \quad \rho > 1, \tag{5.21.13}$$

$$\int_0^\infty [B(x) + C(x)]xJ_1(x\rho)\, dx = 0, \quad \rho > 1. \tag{5.21.14}$$

The problem is then reduced to two pairs of dual integral equations for the determination of $B(x)$ and $C(x)$. Paul and Sridharan [1980] have succeeded in reducing this system to a single pair of dual integral equations as follows:

---

[13]A misprint appears in the expressions $t_{zz}(r, z)$ and $t_{zr}(r, z)$ given by Paul and Sridharan [1980], namely the terms zB should be replaced by $z\xi B$. This does not affect the calculations made at $z = 0$.

(1) Multiply (5.21.11) and (5.21.9) by ρ and integrate with respect to ρ leading to

$$\int_0^\infty \left[(1-\nu)^{-1}B(x) + \frac{\gamma}{c^2\mu'}x(x-y)C(x) \right] J_1(\rho x)\,dx = \tfrac{1}{2}\rho c^2 \tfrac{p_0}{\mu'},\ \ 0 \le \rho \le 1,$$

(5.21.15)

$$\int_0^\infty x^{-1}B(x)J_1(x\rho)\,dx = C_0\rho^{-1},\quad \rho > 1,$$

(5.21.16)

$$C_0 \equiv c \int_0^1 \rho w(\rho,0)\,d\rho.$$

(5.21.17)

(2) Add $\gamma/(c^2\mu')$ times (5.21.12)–(5.21.15) and; $\gamma^2/(c^2\mu')$ times (5.21.14) to $(1-\nu)^{-1}$ times (5.21.16) leading to

$$\int_0^\infty xV(x)J_1(x\rho)\,dx = \tfrac{1}{2}\rho c^2 \tfrac{p_0}{\mu'},\quad 0 \le \rho \le 1,$$

(5.21.18)

$$\int_0^\infty V(x)J_1(x\rho)\,dx = \frac{C_0}{1-\nu}\frac{1}{\rho},\quad \rho > 1,$$

(5.21.19)

where

$$V(x) \equiv [(1-\nu)^{-1}x^{-1} + x\gamma/(c^2\mu')]B(x) + (\gamma/c^2\mu')xC(x).$$

(5.21.20)

(3) The solution of (5.21.18) and (5.21.19) is

$$V(x) = \frac{1}{3}\frac{c^2 p_0}{\mu'}\left(\frac{2}{\pi x}\right)^{1/2} J_{5/2}(x) + \frac{C_0}{1-\nu}\frac{\sin x}{x},$$

(5.21.21-a)

$$= \frac{2}{\pi}\frac{c^2 p_0}{\mu'}\left[\frac{\sin x - x\cos x}{x^3} - \frac{\sin x}{3x}\right] + \frac{C_0}{1-\nu}\frac{\sin x}{x}.$$

(5.21.21-b)

(4) With the use of (5.21.20), reduce the systems (5.21.11) and (5.21.13) to

$$\int_0^\infty x[1+G(x)]B_1(x)J_0(x\rho)\,dx = F(\rho),\quad 0 \le \rho \le 1,$$

(5.21.22)

$$\int_0^\infty B_1(x)J_0(x\rho)\,dx = 0,\quad \rho > 1,$$

(5.21.23)

where

$$B_1(x) = \frac{\pi\mu'}{2p_0(1-\nu)c^2 a},$$

$$G(x) = a[yx^{-1} + (\gamma/c^2\mu')(1-\nu)x(y-x)] - 1,$$

$$F(\rho) = \frac{\pi}{2} + \frac{\mu'}{c^2 p_0}\frac{\pi}{2}\int_0^\infty x(y-x)V(x)J_0(x\rho)\,dx,$$

$$a = \left[1 + (1-\nu)N^2\right]^{-1},\quad N^2 = \frac{\kappa}{\mu + \kappa}.$$

(5.21.24)

Equations (5.21.22) and (5.21.23) reduce to their classical counterparts in elasticity when the coupling parameter $N \to 0$.

(5) To obtain an integral equation from (5.21.22) and (5.21.23), introduce auxiliary function $\psi(t)$ (subject to $\psi(0) = 0$ to be verified later)

$$B_1(x) = \int_0^1 \psi(t) \sin xt \, dt. \qquad (5.21.25)$$

With this, (5.21.23) is satisfied identically. Using this, and $B_1(x)$ given by (5.21.25), in (5.21.9), we obtain

$$w(\rho, 0) = \frac{2}{\pi} \frac{cp_0}{\mu'}(1 - \nu)a \int_\rho^1 \frac{\psi(t)}{(t^2 - \rho^2)^{1/2}} \, dt. \qquad (5.21.26)$$

Integrating by parts, it can be seen that $w(\rho, 0)$ satisfies the edge condition

$$w(r, 0) = (0\sqrt{c - r}) \quad \text{as } r \to c^-. \qquad (5.21.27)$$

Substituting (5.21.26) into (5.21.17) leads to

$$C_0 = \frac{2}{\pi} \frac{c^2 p_0}{\mu'}(1 - \nu)a \int_0^1 t\psi(t) \, dt. \qquad (5.21.28)$$

(6) Using (5.21.10) and B_1, given by (5.21.24), we express $\phi(\rho, 0)$ in the form

$$(\gamma/c^2\mu')\phi(\rho, 0) = -\frac{2p_0a}{\pi\mu'} \int_0^\infty B_1(x)x^{-1}J_1(x\rho) \, dx + c^{-2} \int_0^\infty V(x)J_1(x\rho) \, d\rho. \qquad (5.21.29)$$

Upon substituting for $B_1(x)$ and $V(x)$ from (5.21.25) and (5.21.21), and integrating, we obtain

$$(\gamma/c^2\mu')\phi(\rho, 0) = -\frac{2p_0a}{\pi\mu'} \left[\frac{1}{\rho} \int_0^\rho \psi(t)t \, dt + \rho \int_\rho^1 \frac{\psi(t) \, dt}{t + \sqrt{t^2 - \rho^2}} \right]$$
$$+ \frac{2p_0}{3\pi\mu'}\rho(1 - \rho^2)^{1/2} + \frac{C_0}{c^2(1 - \nu)}\frac{1}{\rho}, \quad \rho < 1$$
$$\phi(\rho, 0) = 0, \quad \rho > 1 \qquad (5.21.30)$$

where we used Weber–Schafheitlin integrals given in Bateman [1953, Vol. II, p. 92].

From (5.21.30), it is clear that the boundary condition (5.21.14) is satisfied, and that, for the edge condition, we have

$$\lim_{\rho \to 1^-} \phi(\rho, 0) = 0. \qquad (5.21.31)$$

(7) Multiplying (5.21.22) by $\rho/(t^2 - \rho^2)^{1/2}$ and integrating with respect to ρ from 0 to t, we obtain

$$\int_0^\infty [1 + G(x)]B_1(x) \sin xt \, dx = \int_0^t \frac{\rho F(\rho)}{(t^2 - \rho^2)^{1/2}} \, d\rho, \quad 0 \le t \le 1. \quad (5.21.32)$$

From this, a regular Fredholm integral equation of the second kind is obtained by using (5.21.25)

$$\psi(t) + \int_0^1 K(v,t)\psi(v)\,dv = f_1(t) + C_0' f_2(t), \quad 0 \le t \le 1, \qquad (5.21.33)$$

where

$$K(v,t) \equiv \frac{2}{\pi} \int_0^\infty G(x) \sin(xv) \sin(xt)\,dx,$$

$$f_1(t) = t + \frac{2}{\pi} \int_0^\infty [(x^2 + M^2)^{1/2} - x] \left[\frac{\sin x - x \cos x}{x^3} - \frac{\sin x}{3x} \right] \sin xt\,dx,$$

$$f_2(t) = \frac{2}{\pi} \int_0^\infty x^{-1} \left[(x^2 + M^2)^{1/2} - x \right] \sin x \sin xt\,dx,$$

$$C_0' = \frac{\pi}{2} \frac{\mu'}{c^2 p_0} \frac{C_0}{1 - \nu}. \qquad (5.21.34)$$

In the limit $t \to 0$, (5.21.33) gives $\psi(0) = 0$, satisfying the earlier assumption.

(8) Finally, irrespective of the constant C_0', the function $\psi(t)$ given by

$$\psi(t) = \psi_1(t) + C_0' \psi_2(t) \qquad (5.21.35)$$

is a solution of (5.21.33) provided

$$\psi_i(t) + \int_0^1 K(v,t)\psi_i(v)\,dv = f_i(t), \quad 0 \le t \le 1, \quad i = 1, 2. \qquad (5.21.36)$$

These two integral equations have the same kernel which is symmetric and continuous. These equations are solved by numerical computations to obtain $\psi_i(t)$. Then C_0' is calculated from (5.21.34) and (5.21.28), i.e.,

$$C_0' = \left[a \int_0^1 t\psi_1(t)\,dt \right] / \left[1 - a \int_0^1 t\psi_2(t)\,dt \right]. \qquad (5.21.37)$$

Once $\psi(t)$ is known, the problem can be regarded as solved.

The stress and couple stress, $t_{zz}(\rho, 0)$ and $m_{z\theta}(\rho, 0)$, are calculated by using (5.21.7) and (5.21.8). With the use of (5.21.25), (5.21.20), and (5.21.24)$_1$, they can be put into the forms

$$t_{zz}(\rho, 0) = \frac{2}{\pi} p_0 \psi(1) \frac{H(\rho - 1)}{(\rho^2 - 1)^{1/2}} - \frac{2}{\pi} p_0 \int_0^{\min(\rho,1)} \frac{\psi'(t)}{\sqrt{\rho^2 - 1}}\,dt$$

$$- \frac{2}{\pi} p_0 \int_0^\infty x G(x) B_1(x) J_0(x\rho)\,dx$$

$$- \frac{\mu'}{c^2} \int_0^\infty (x - y) x V(x) J_0(x\rho)\,dx, \qquad (5.21.38)$$

$$m_{z\theta}(\rho,0) = \frac{2p_0 c}{\pi \rho} \int_0^1 \frac{t\psi(t)}{\sqrt{\rho^2-1}} dt + \frac{2}{\pi}p_0 c \int_0^1 \psi(t)dt \int_0^\infty G(x)\sin xt J_1(x\rho)dx$$

$$- \frac{\mu'}{c}\int_0^\infty y V(x)J_1(x\rho)\, dx, \quad \rho > 1. \tag{5.21.39}$$

In (5.21.38), the second and third terms are regular at $\rho = 1$. Thus, in the neighborhood of $\rho = 1$, we have

$$t_{zz}(\rho,0) = (2/\pi)p_0\psi(1)(\rho^2-1)^{-1/2} + \text{terms regular at } \rho = 1, \quad \rho > 1. \tag{5.21.40}$$

An examination (5.21.39) indicates that the singular behavior of $m_{z\theta}(\rho,0)$ at the crack tip is governed by the last term. With the use of (5.21.21-a) this term can be put into the form

$$- \tfrac{1}{3}(2/\pi)^{1/2}cp_0 \int_0^\infty [1 + O(x^{-2})]x^{1/2}J_{5/2}(x)J_1(x\rho)\, dx$$

$$- \frac{\mu'}{c}\frac{C_0}{1-\nu}\int_0^\infty [1 + O(x^{-2})]J_1(x\rho)\sin x\, dx. \tag{5.21.41}$$

Evaluating these two integrals, we have

$$m_{z\theta}(\rho,0) = \frac{2cp_0}{3\pi\rho}\frac{1}{\sqrt{\rho^2-1}} - \frac{\mu'}{c}\frac{C_0}{1-\nu}\frac{1}{\rho\sqrt{\rho^2-1}}$$

$$+ \text{terms regular at } \rho = 1+, \quad \rho > 1. \tag{5.21.42}$$

Consequently, the stress and couple stress have the same order singularity, viz. $(\rho-1)^{-1/2}$. Defining

$$\left\{ \begin{matrix} K_1 \\ K_2 \end{matrix} \right\} = \sqrt{c} \lim_{\rho \to 1+} \sqrt{\rho-1} \left\{ \begin{matrix} t_{zz}(\rho,0) \\ m_{z\theta}(\rho,0) \end{matrix} \right\}, \tag{5.21.43}$$

we have for the *stress intensity factors*

$$K_1 = \frac{\sqrt{2}}{\pi}p_0\psi(1)\sqrt{c}, \quad K_2 = \frac{\sqrt{2}}{\pi}p_0 c^{3/2}\left(\tfrac{1}{3} - C_0'\right). \tag{5.21.44}$$

We observe that K_1 and K_2 depend on ν, N, and $M = N/(l/c)$, i.e., Poisson's ratio, the material parameter, and the internal characteristic length.

The work done in opening the crack is given by

$$W = p_0 \int_0^c 2\pi r w(r,0)\, dr = 4\frac{c^3 p_0^2}{\mu'}(1-\nu)C_0'. \tag{5.21.45}$$

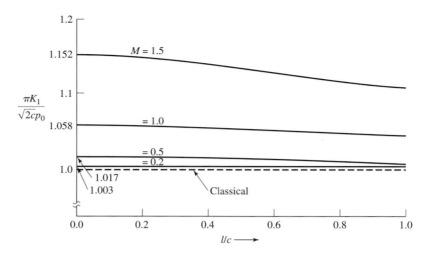

FIGURE 5.21.1. Force stress intensity factor vs. micropolar length-parameter. Poisson's ratio=0.25. (From Paul & Sridharan [1980])

As $\kappa \to 0$, K_1 goes to the classical value in elasticity and $K_2 \to 0$. Numerical calculations were given in the reference cited for K_1, K_2, and W as functions of M and l/c for $\nu = 0.25$. These are shown in Figures 5.21.1–5.21.3. Also shown in Figure 5.21.4 is the crack opening displacement w as a function of l/c and M for fixed Poisson's ratio $\nu = 0.25$.

5.22   Stress Distribution Around an Elliptic Hole

The static balance laws relevant to this problem[14] are

$$t_{kl,k} = 0, \quad m_{kl,k} + \epsilon_{lmn}t_{mn} = 0. \tag{5.22.1}$$

For the plane-strain problems, the general solution of these equations is given by stress and couple-stress functions F and G:

$$t_{kl} = \epsilon_{3ki}\epsilon_{3lj}F_{,ij} - \epsilon_{3ki}G_{,il}, \quad m_{k3} = G_{,k}, \tag{5.22.2}$$

where ϵ_{3ki} is the permutation symbol.

From (5.5.24), we solve for ϵ_{kl} and γ_{kl}

$$\epsilon_{kl} = \frac{1}{2\mu + \kappa} \left(-\frac{\lambda}{\lambda + 2\mu + \kappa}t_{rr} + \frac{\mu + \kappa}{\kappa}t_{kl} - \frac{\mu}{\kappa}t_{lk} \right),$$

$$\phi_{3,k} = \frac{1}{\gamma}m_{k3}. \tag{5.22.3}$$

---

[14]Kim and Eringen [1973].

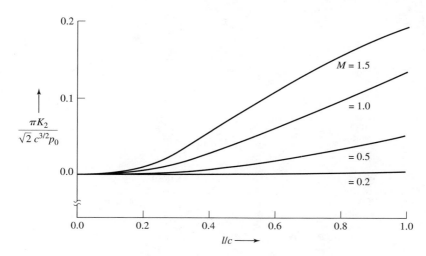

FIGURE 5.21.2. Couple stress intensity factor vs. micropolar length-parameter. Poisson's ratio = 0.25. (From Paul & Sridharan [1980])

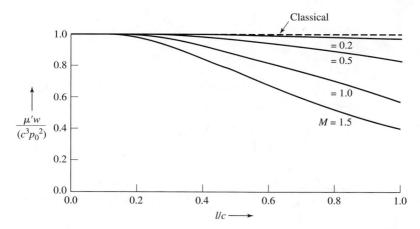

FIGURE 5.21.3. Work done in opening the crack vs. micropolar length-parameter. Poisson's ratio = 0.25. (From Paul & Sridharan [1980])

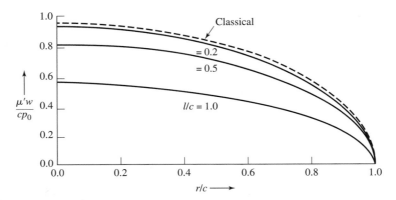

FIGURE 5.21.4. Shape of the crack. $M = 1$, Poisson's ratio = 0.25. (From Paul & Sridharan[1980])

Compatibility conditions given by (1.7.10), are also expressed in terms of linear strain measures as

$$\epsilon_{ik,j} - \epsilon_{jk,i} + \Gamma_{ikj} - \Gamma_{jki} = 0,$$
$$\epsilon_{klr}(\Gamma_{klm,n} - \Gamma_{kln,m}) = 0, \qquad (5.22.4)$$

where

$$\Gamma_{ijk} = \epsilon_{ijm}\phi_{m,k}. \qquad (5.22.5)$$

Combination of (5.22.2)–(5.22.5) leads to

$$(G - c^2\nabla^2 G)_{,1} = -2(1 - \nu)b^2(\nabla^2 F)_{,2},$$
$$(G - c^2\nabla^2 B)_{,2} = 2(1 - \nu)b^2(\nabla^2 F)_{,1}, \qquad (5.22.6)$$

where

$$c^2 = \frac{\gamma(\mu + \kappa)}{\kappa(2\mu + \kappa)}, \quad b^2 = \frac{\gamma}{2(2\mu + \kappa)}, \quad \nu = \frac{\lambda}{2\lambda + 2\mu + \kappa}. \qquad (5.22.7)$$

By cross differentiation, (5.22.6) gives

$$\nabla^4 F = 0, \qquad (5.22.8)$$
$$\nabla^2(c^2\nabla^2 G - G) = 0. \qquad (5.22.9)$$

Thus, the solution of these equations is required for the solution of the static plane-strain problem. The appropriate coordinate system is the elliptic coordinates system, α, β, and z which is related to the rectangular coordinates by

$$x = C\cosh\alpha\cos\beta,$$
$$y = C\sinh\alpha\sin\beta,$$
$$z = z, \qquad (5.22.10)$$

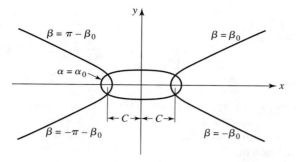

FIGURE 5.22.1. Elliptic Coordinate System

where C is one-half of the distance between the foci of the elliptic hole. In elliptic coordinates $\alpha =$ constant, curves represent the confocal ellipses, and $\beta =$ constant curves are hyperboles orthogonal to these ellipses (see Figure 5.22.1).

The boundary conditions read

$$t_{\alpha\alpha} = t_{\alpha\beta} = m_{\alpha z} = 0 \quad \text{at } \alpha = \alpha_0$$

$$\left. \begin{array}{l} t_{\alpha\alpha} = \dfrac{T}{2}(1 - \cos 2\beta) \\[2mm] t_{\alpha\beta} = \dfrac{T}{2}\sin 2\beta \\[2mm] m_{\alpha z} = 0 \end{array} \right\} \quad \text{at } \alpha = \infty, \qquad (5.22.11)$$

where T is the uniform tension applied at infinity. From the symmetry of the loading and geometry, it also follows that normal stresses are even functions of β and shear stresses are odd functions of β. Also both stresses are periodic in β with a period π. The general solution of (5.22.8) is given by

$$F(\alpha, \beta) = \sum_{n=2,4,\ldots}^{\infty} \left(c_n e^{n\alpha} + d_n e^{-n\alpha}\right) \cos n\beta + g_1 + g_2 \alpha$$

$$+ \frac{f_1}{8} C^2 \left(\cosh 2\alpha + \cos 2\beta\right)$$

$$+ \sum_{n=2,4,\ldots}^{\infty} \left\{ C^2 \frac{a_n}{16(1+n)} e^{(n+2)\alpha} \cos n\beta + \frac{a_n}{16(1-n)} e^{(n-2)\alpha} \cos n\beta \right.$$

$$+ \frac{b_n}{16(1+n)} e^{(-n+2)\alpha} \cos n\beta + \frac{b_n}{16(1-n)} e^{(-n-2)\alpha} \cos n\beta$$

$$+ \frac{a_n}{16(1+n)}e^{n\alpha}\cos(n+2)\beta + \frac{a_n}{16(1-n)}e^{n\alpha}\cos(n-2)\beta$$

$$+ \frac{b_n}{16(1+n)}e^{-n\alpha}\cos(n+2)\beta + \frac{b_n}{16(1-n)}e^{-n\alpha}\cos(n-2)\beta \bigg\},$$

$$(5.22.12)$$

where $a_n, b_n, c_n, d_n, g_1, g_2$, and f_1 are integration constants.

Next we employ the method of separation of variables to solve (5.22.9) for G. In the homogenous equation for G, the β-dependent part gives the *Mathieu Equation* and the α-dependent part reduces to the *Radial Mathieu Equation*. Since G is an odd function of β with period $\beta = \pi$, we choose the even integer order *Sine Mathieu Functions* such as $SO_2, SO_4, SO_6, \ldots$ for the β-dependent solution, i.e.,

$$So_m = \sum_{n=2,4,\ldots}^{\infty} D(m,n)\sin n\beta, \quad m = 2,4,6,\ldots. \quad (5.22.13)$$

These are the only possible solutions for the β-dependent part. There are two independent solutions for the radial Mathieu equation. These are expressible in terms of *Bessel Functions*, for each separation constant corresponding to (5.22.13). The general solution is therefore the superposition of these solutions, i.e.,

$$G(\alpha,\beta) = -A \sum_{n=2,4,\ldots}^{\infty} \left(a_n e^{n\alpha} - b_n e^{-n\alpha}\right)\sin n\beta + f_2$$

$$+ \sum_{m=2,4,\ldots}^{\infty} \left[\sum_{n=2,4,\ldots}^{\infty} D(m,n)\sin n\beta\right]\tanh\alpha$$

$$\left[k_{m1}\sum_{p=2,4,\ldots}^{\infty} pD(m,p)I_p\left(\frac{C}{c}\cosh\alpha\right)\right.$$

$$\left. + k_{m2}\sum_{p=2,4,\ldots}^{\infty} pD(m,p)K_p\left(\frac{C}{c}\cosh\alpha\right)\right], \quad (5.22.14)$$

where I_p and K_p are, respectively, *Modified Bessel Functions* of order p; f_2, k_{m1}, and k_{m2} are integration constants, and

$$A = (1-\nu)\frac{\gamma}{2\mu+\kappa}.$$

The application of the boundary condition at $\alpha = \infty$ drops the terms containing α such as a_n, c_n, and k_{m1}. To satisfy the boundary conditions at the elliptic hole $\alpha = \alpha_0$, we write the expressions of F and G in the form

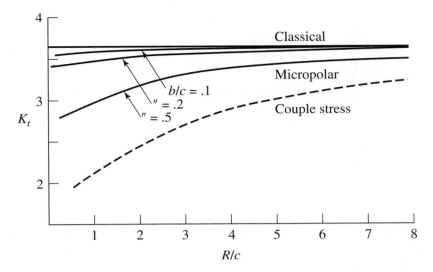

FIGURE 5.22.2. Stress Concentration Factors: $K_t = t_{\beta\beta}/T(\alpha_0 = 1, \nu = .00)$ (From Kim and Eringen)

of a *Fourier Series*. Since the contributions of the terms containing b_n, d_n, and k_{m2} are exponentially decreasing, only the first few terms suffice for the calculations. As the elliptic hole becomes flatter, the number of terms needed for accuracy becomes larger. The boundary conditions at $\alpha = \alpha_0$ are satisfied by a numerical process in which a given number of terms is retained and the convergence for a given ellipse is tested by increasing this number.

Calculations were performed by Kim and Eringen. The stress concentration K_t is presented as a function R/c for different values of b/c and Poisson's ratio ν in Figures 5.22.2 and 5.22.3. Here K_t is the ratio of $t_{\beta\beta}$ at the end of the major axis of the ellipse to the $t_{\alpha\alpha}$ applied at infinity. R/c is the ratio of the radius of the circle (from which the ellipse is generated by boundary perturbation) to the internal characteristic length and $b/c = \kappa/\mu$.

In Figures 5.22.4 and 5.22.5, are plotted K_t, and stress and couple stress distributions along the major axis direction. From these figures it can be deduced that, in the limit of a crack, the stress singularity is of the order $1/\sqrt{\rho}$ where ρ is the radius of curvature at the crack tip. Figure 5.22.6 displays an interesting phenomenon, namely a dip in $t_{\beta\beta}$. This dip is accompanied by the concentration of $t_{\alpha\alpha}$ and couple stress at that point. This indicates the release of the stress concentration in the direction along which the stress is highly concentrated. Measurement of this dip may provide determination of some of the micropolar moduli.

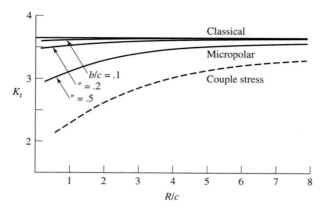

FIGURE 5.22.3. Stress Concentration Factors: $K_t = t_{\beta\beta}/T(\alpha_0 = 1, \nu = .25)$ (From Kim and Eringen)

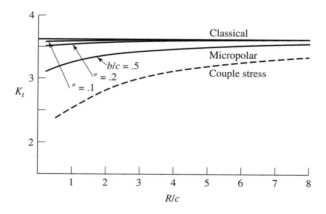

FIGURE 5.22.4. Stress Concentration Factors: $K_t = t_{\beta\beta}/T(\alpha_0 = 1, \nu = .50)$ (From Kim and Eringen)

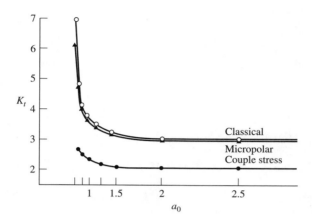

FIGURE 5.22.5. Stress Concentration Profile ($R/c = 1.$, $b/c = .2$, $\nu = .25$) (From Kim and Eringen)

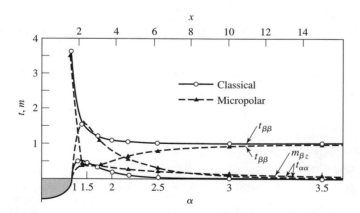

FIGURE 5.22.6. Stress Dostrobution along Major Axial Direction ($\alpha_0 = 1$, $R/c = 1.$, $b/c = .2$, $\nu = .25$) (From Kim and Eringen)

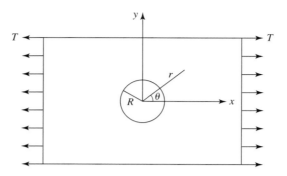

FIGURE 5.23.1. Plate with Circular Hole in Tension

5.23   Stress Concentration Around a Circular Hole

A plate with a circular hole of radius R is subjected to a uniform tension T at $x = \infty$ (Figure 5.23.1). Determine the stress and couple stress field in the plate.[15]

The equations of compatibility (5.22.6) in plane polar coordinates read

$$\frac{\partial}{\partial r}(G - c^2 \nabla^2 G) = -2(1 - \nu)b^2 \frac{1}{r}\frac{\partial}{\partial \theta}(\nabla^2 F),$$

$$\frac{1}{r}\frac{\partial}{\partial \theta}(G - c^2 \nabla^2 G) = (1 - \nu)b^2 \frac{\partial}{\partial r}(\nabla^2 F), \qquad (5.23.1)$$

where

$$c^2 = \frac{\gamma(\mu + \kappa)}{\kappa(2\mu + \kappa)}, \quad b^2 = \frac{\gamma}{2(2\mu + \kappa)}, \quad \nu = \frac{\lambda}{2\lambda + 2\mu + \kappa}. \qquad (5.23.2)$$

The stress potentials F and G satisfy (5.22.8) and (5.22.9), i.e.,

$$\nabla^4 F = 0, \quad \nabla^2(c^2 \nabla^2 G - G) = 0. \qquad (5.23.3)$$

Components of the stress and couple stress tensors are expressed in terms F and G by

$$t_{rr} = \frac{1}{r}\frac{\partial F}{\partial r} + \frac{1}{r^2}\frac{\partial^2 F}{\partial \theta^2} - \frac{1}{r}\frac{\partial^2 G}{\partial r \partial \theta} + \frac{1}{r^2}\frac{\partial G}{\partial \theta},$$

$$t_{\theta\theta} = \frac{\partial^2 F}{\partial r^2} + \frac{1}{r}\frac{\partial^2 G}{\partial r \partial \theta} - \frac{1}{r^2}\frac{\partial G}{\partial \theta},$$

$$t_{r\theta} = -\frac{1}{r}\frac{\partial^2 F}{\partial r \partial \theta} + \frac{1}{r^2}\frac{\partial F}{\partial \theta} - \frac{1}{r}\frac{\partial G}{\partial r} - \frac{1}{r^2}\frac{\partial^2 G}{\partial \theta^2},$$

$$t_{\theta r} = -\frac{1}{r}\frac{\partial^2 F}{\partial r \partial \theta} + \frac{1}{r^2}\frac{\partial F}{\partial \theta} + \frac{\partial^2 G}{\partial r^2},$$

$$m_{rz} = \frac{\partial G}{\partial r}, \quad m_{\theta z} = \frac{1}{r}\frac{\partial G}{\partial \theta}. \qquad (5.23.4)$$

---

[15]Kaloni and Ariman [1967].

The boundary conditions are

$$t_{rr} = t_{r\theta} = m_{rz} = 0, \quad r = R,$$

$$\left.\begin{array}{c} t_{rr} = \dfrac{T}{2}(1 + \cos 2\theta) \\[2mm] t_{r\theta} = -\dfrac{T}{2}\sin 2\theta \\[2mm] m_{rz} = 0 \end{array}\right\} \quad r = \infty. \qquad (5.23.5)$$

An appropriate solution of (5.23.3), for this problem is

$$F = \frac{T}{4}r^2(1 - \cos 2\theta) + A_1 \ln r + (A_2 r^{-2} + A_3)\cos 2\theta,$$

$$G = [A_4 r^{-2} + A_5 K_2(r/c)]\sin 2\theta. \qquad (5.23.6)$$

The equation of compatibility (5.23.1) is satisfied by

$$A_4 = 8(1 - \nu)c^2 A_3. \qquad (5.23.7)$$

The stress and couple stress fields are given by

$$t_{rr} = \frac{T}{2}(1 + \cos 2\theta) + \frac{A_1}{r^2} - \left(\frac{6A_2}{r^4} + \frac{4A_3}{r^2} - \frac{6A_4}{r^4}\right)\cos 2\theta$$
$$+ \frac{2A_5}{cr}\left[\frac{3c}{r}K_0(r/c) + \left(1 + \frac{6c^2}{r^2}\right)K_1(r/c)\right]\cos 2\theta,$$

$$t_{\theta\theta} = \frac{T}{2}(1 - \cos 2\theta) - \frac{A_1}{r^2} + \left(\frac{6A_2}{r^4} - \frac{6A_4}{r^4}\right)\cos 2\theta$$
$$- \frac{2A_5}{cr}\left[\frac{3c}{r}K_0(r/c) + \left(1 + \frac{6c^2}{r^2}\right)K_1(r/c)\right]\cos 2\theta,$$

$$t_{r\theta} = -\left(\frac{T}{2} + \frac{6A_2}{r^4} + \frac{2A_3}{r^2} - \frac{6A_4}{r^4}\right)\sin 2\theta$$
$$+ \frac{A_5}{cr}\left[\frac{6c}{r}K_0(r/c) + \left(1 + \frac{12c^2}{r^2}\right)K_1(r/c)\right]\sin 2\theta,$$

$$t_{\theta r} = -\left(\frac{T}{2} + \frac{6A_2}{r^4} + \frac{2A_3}{r^2} - \frac{6A_4}{r^4}\right)\sin 2\theta$$
$$+ \frac{A_5}{c^2}\left[\left(1 + \frac{6c^2}{r^2}\right)K_0(r/c) + \left(\frac{3c}{r} + \frac{12c^3}{r^3}\right)K_1(r/c)\right]\sin 2\theta,$$

$$m_{rz} = -\frac{2A_4}{r^3}\sin 2\theta - \frac{A_5}{c}\left[\frac{2c}{r}K_0(r/c) + \left(1 + \frac{4c^2}{r^2}\right)K_1(r/c)\right]\sin 2\theta,$$

$$m_{\theta z} = \left\{\frac{2A_4}{r^3} + \frac{2A_5}{r}\left[K_0(r/c) + \frac{2c}{r}K_1(r/c)\right]\right\}\cos 2\theta. \qquad (5.23.8)$$

The constants A_1 to A_5 follow from the boundary conditions (5.23.5)

$$A_1 = -\frac{T}{2}R^2, \quad A_2 = \frac{TR^4(1 - F_1)}{4(1 + F_1)},$$

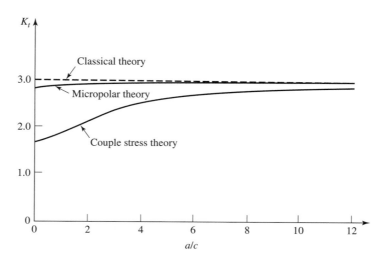

FIGURE 5.23.2. Stress-concentration factors for $b/c = 0.20, \nu = .00$ (From Kaloni and Ariman [1967])

$$A_3 = \frac{TR^2}{2(1+F_1)}, \quad A_4 = \frac{4(1-\nu)R^2 b^2 T}{1+F}, \quad A_5 = -\frac{TRcF_1}{(1+F_1)K_1(R/c)},$$

$$F_1 \equiv 8(1-\nu)\frac{b^2}{c^2}\left[4 + \frac{R^2}{c^2} + \frac{2R}{c}\frac{K_0(R/c)}{K_1(R/c)}\right]^{-1}. \tag{5.23.9}$$

The value of $t_{\theta\theta}$ at the periphery of the circular hole is of great interest

$$t_{\theta\theta} = T\left(1 + \frac{2\cos 2\theta}{1+F_1}\right). \tag{5.23.10}$$

The maximum value of this occurs at $\theta = \pm\pi/2$, leading to the *stress concentration factor*

$$K_t = t_{\theta\theta\,\max}/T = (3+F_1)/(1+F_1). \tag{5.23.11}$$

The stress concentration as a function of R/c for several values of b/c and ν is displayed in Figures 5.23.2–5.23.5. From these, it is clear that in a micropolar solid, the stress concentration is smaller than that of the classical elastic solid. $t_{\theta\theta}$ given by (5.23.10) reduces to the value obtained in classical elasticity when the coupling constant $\kappa/\mu \to 0$. As usual, micropolar theory introduces an internal characteristic length

$$c = [\gamma(\mu+\kappa)/\kappa(2\mu+\kappa)]^{1/2}, \tag{5.23.12}$$

and the stress concentration here depends on the ratio R/c of external and internal characteristic lengths.

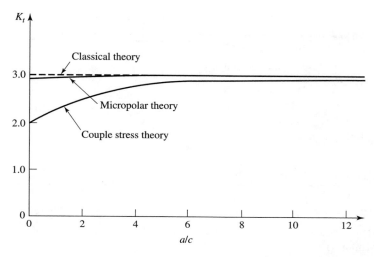

FIGURE 5.23.3. Stress-concentration factors for $b/c = 0.20, \nu = .50$ (From Kaloni and Ariman [1967])

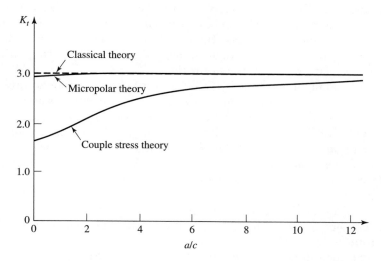

FIGURE 5.23.4. Stress-concentration factors for $b/c = 0.10, \nu = .00$ (From Kaloni and Ariman [1967])

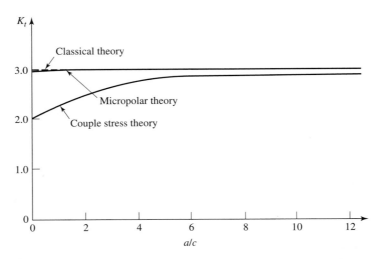

FIGURE 5.23.5. Stress-concentration factors for $b/c = 0.10, \nu = .50$ (From Kaloni and Ariman [1967])

5.24   Nonlinear Waves

Field equations of nonlinear micropolar elasticity are too complicated for exact solutions. Even in the simplest cases, such as one-dimensional motions, the nonlinear partial differential equations are not amenable for a rigorous mathematical analysis. However, under some restrictions, it has been possible to make some progress, displaying some interesting physical phenomena that are lost in the linear theory.

Recently, Maugin and Miled [1986][16] studied one-dimensional waves associated with the microrotation angles. They found solitary waves resembling the motions of ferromagnetic and ferroelectric domain walls. Erbay and Şuhubi [1989] considered finite amplitude weakly nonlinear acoustical waves in one dimension. They showed that these waves are modeled asymptotically by two coupled modified Kortweg–de Vries equations. Erbay et al. [1991] studied nonlinear wave modulations in longitudinal and transverse waves.

The full use of the nonlinear theory leads to long and tedious expressions. For brevity, we consider here only a one-dimensional longitudinal microrotation wave (LO-wave). The use of a material frame-of-reference provides some simplicity. we assume that the transverse components of the displacement vector vanish and the axis of rotation is the X-axis. Thus, we

---

[16]Errors found in the published work have been discussed with one of the authors, and have been, hereby, corrected.

take

$$x = X + U(\mathbf{X}, t), \quad y = Y, \quad z = Z, \tag{5.24.1}$$

$$\chi_{kK} = \begin{bmatrix} 1 & 0 & 0 \\ 0 & \cos\phi & -\sin\phi \\ 0 & \sin\phi & \cos\phi \end{bmatrix}, \tag{5.24.2}$$

where U is the axial displacement and $\phi(\mathbf{X}, t)$ is the rotation angle about the X-axis.

The deformation tensor $\mathfrak{C}_{KL}$ and wryness tensor Γ_{KL} are given by

$$\mathfrak{C}_{KL} = x_{k,K}\chi_{lL} = \begin{bmatrix} 1 + \frac{\partial U}{\partial X} & 0 & 0 \\ 0 & \cos\phi & -\sin\phi \\ 0 & \sin\phi & \cos\phi \end{bmatrix},$$

$$\Gamma_{KL} = \tfrac{1}{2}\epsilon_{KLM}\chi_{kM,L}\chi_{kN} = \begin{bmatrix} \frac{\partial\phi}{\partial X} & 0 & 0 \\ 0 & 0 & 0 \\ 0 & 0 & 0 \end{bmatrix}. \tag{5.24.3}$$

Constitutive equations are given by

$$T_{Kl} = \frac{\partial\Sigma}{\partial\mathfrak{C}_{KL}}\chi_{lL}, \quad M_{Kl} = \frac{\partial\Sigma}{\partial\Gamma_{LK}}\chi_{lL}, \tag{5.24.4}$$

where Σ is the free energy. For isotropic solids, Σ is a function of the invariants of $\mathfrak{C}_{KL}$ and Γ_{KL}. For a second-degree polynomial, it has the form

$$\Sigma = \tfrac{1}{2}\lambda(1 + U_{,X} + 2\cos\phi)^2 + \tfrac{1}{2}\mu[(1 + U_{,X})^2 + 2\cos^2\phi - \sin^2\phi]$$
$$+ \frac{\mu + \kappa}{2}[(1 + U_{,X})^2 + 2\cos^2\phi + \sin^2\phi] + \alpha(\phi_{,X})^2. \tag{5.24.5}$$

This will be positive semidefinite if

$$\lambda \geq 0, \quad \mu \geq 0, \quad \kappa \geq 0, \quad \alpha \geq 0. \tag{5.24.6}$$

We remark that while (5.24.5) is a quadratic in strain measures, unlike the linear theory, the strain measures are exact and not linear in the displacement and rotations. Consequently, λ, μ, κ, and α do not have direct relations to the material constants of the linear theory. Also in writing a quadratic polynomial for Σ, we have ignored the invariants that are of higher degree in the strain measures. Thus, the theory is valid for mildly finite deformations.

Using (5.24.2)–(5.24.5), we calculate

$$T_{Xx} = (\lambda + 2\mu + \kappa)(1 + U_{,X} + 2\cos\phi),$$
$$T_{Yz} = -T_{Zy} = \lambda(1 + U_{,X} + 2\cos\phi)\sin\phi + \mu\sin 2\phi,$$
$$T_{Yy} = T_{Zz} = \lambda(1 + U_{,X} + 2\cos\phi)\cos\phi + \mu\cos 2\phi + \mu + \kappa,$$
$$M_{Xx} = \alpha\phi_{,X}, \quad \text{all other } T_{Kl} \text{ and } M_{Kl} = 0. \tag{5.24.7}$$

Equations of motion are

$$T_{Kl,K} + \rho_0 f_l = \rho_0 \frac{\partial v_l}{\partial t},$$
$$M_{Kl,K} + \epsilon_{lmn} x_{m,K} T_{Kn} + \rho_0 l_l = \rho_0 \sigma_l. \tag{5.24.8}$$

With vanishing body forces and couples, the only surviving equations of (5.24.8) are

$$\frac{\partial T_{Xx}}{\partial X} = \rho_0 \frac{\partial^2 U}{\partial t^2}, \quad \frac{\partial M_{Xx}}{\partial X} + T_{Yz} - T_{Zy} = \rho_0 J \frac{\partial^2 \phi}{\partial t^2}, \tag{5.24.9}$$

where we considered only a single component of the constant microinertia J. Substituting from (5.24.7), we obtain two equations for the determination of $U(\mathbf{X}, t)$ and $\phi(X, t)$,

$$(\lambda + 2\mu + \kappa)\left(\frac{\partial^2 U}{\partial X^2} - 2\sin\phi \frac{\partial \phi}{\partial X}\right) = \rho_0 \frac{\partial^2 U}{\partial t^2},$$
$$\alpha\frac{\partial^2 \phi}{\partial X^2} + 2\lambda(1 + U_{,X} + 2\cos\phi)\sin\phi + \mu\sin 2\phi = \rho_0 J \frac{\partial^2 \phi}{\partial t^2}. \tag{5.24.10}$$

These equations may be written in nondimensional forms

$$\frac{\partial^2 \bar{U}}{\partial t^2} - V_L^2 \frac{\partial^2 \bar{U}}{\partial \bar{X}^2} = a\frac{\partial}{\partial \bar{X}}\left(\cos\frac{\bar{\phi}}{2}\right),$$
$$\frac{\partial^2 \bar{\phi}}{\partial \tau^2} - \frac{\partial^2 \bar{\phi}}{\partial \bar{X}^2} - \sin\bar{\phi} = e\sin\frac{\bar{\phi}}{2} + b\frac{\partial \bar{U}}{\partial X}\sin\frac{\bar{\phi}}{2}, \tag{5.24.11}$$

where

$$\bar{\phi} = 2\phi, \quad \tau = \omega_M t, \quad \bar{X} = X/\delta, \quad \bar{U} = U/\sqrt{J},$$
$$\omega_M^2 = 2(2\lambda + \mu)/\rho_0 J, \quad \delta^2 = \frac{\alpha}{2(2\lambda + \mu)},$$
$$e = \frac{2\lambda}{2\lambda + \mu}, \quad V_L^2 = \frac{J}{\alpha}(\lambda + 2\mu + \kappa), \quad a = \frac{(\lambda + 2\mu + \kappa)\sqrt{2J}}{\sqrt{\alpha(2\lambda + \mu)}},$$
$$b = 2\lambda\sqrt{2J}/\sqrt{\alpha(2\lambda + \mu)}. \tag{5.24.12}$$

The coupled nonlinear equations (5.24.11) consist of a d'Alembert equation and a double sine–Gordon equation for $\bar{\phi}$, as shown by Maugin and Miled. These equations possess a solitary-wave solution for special values of

$$\frac{\partial \bar{U}}{\partial \bar{X}} \to 0, \quad \bar{\phi} \to \bar{\phi}_0 = \pm 2\pi \quad \text{as } \xi \to \mp\infty. \tag{5.24.13}$$

Substituting

$$\xi = q\bar{X} - \omega\tau + \xi_0, \quad \xi_0 = \text{const.} \tag{5.24.14}$$

into (5.22.11), we have

$$(\omega^2 - \omega_L^2)\frac{\partial^2 \bar{U}}{\partial \xi^2} = a\frac{\partial}{\partial \xi}\left(\cos\frac{\bar\phi}{2}\right),$$

$$\phi_{,\xi\xi}(\omega^2 - q^2) - \sin\bar\phi - e\sin\frac{\bar\phi}{2} = b\frac{\partial \bar{U}}{\partial \xi}\sin\frac{\bar\phi}{2}. \qquad (5.24.15)$$

Under the conditions (5.24.13), we have the solutions

$$\phi = 4\tan^{-1}(\sqrt{-p}\sinh\xi), \qquad (5.24.16)$$

$$U = \frac{2a}{\omega^2 - \omega_L^2}\left[\frac{1}{\sqrt{-1-p}}\tan^{-1}\left(\sqrt{-p-1}\tanh\xi\right)\right] + U_0, \quad (5.24.17)$$

where

$$p = \frac{1}{4(\omega^2 - q^2)}\left(e + \frac{ab}{\omega^2 - \omega_L^2}\right), \qquad (5.24.18)$$

and the pseudo-dispersion relation

$$(\omega^2 - q^2)(e+4) - 2 = 0. \qquad (5.24.19)$$

The axial component of the stress-tensor is given by

$$T_{Xx} = (\lambda + 2\mu + \kappa)\left[1 + \frac{1}{1-p\sinh\xi}\left(\frac{2a}{\omega^2 - \omega_L^2} + 8p\sinh^2\xi\right)\right]. \quad (5.24.20)$$

5.25 Fundamental Solutions in Micropolar Elastostatics

Fundamental solution in the static case cannot be obtained from the dynamic case presented in Section 5.18. There exist several other methods of approach based on the Galerkin-type of representation and stress function approach cf. Sandru [1966], Ieşan [1971], Eringen [1968a]. Here we follow Dragoş [1984] who employed the Fourier transform technique. The fundamental solutions are the solutions of the field equations in an infinite domain under singular loads

$$(\lambda + 2\mu + \kappa)\nabla\nabla\cdot\mathbf{u} - (\mu+\kappa)\nabla\times\nabla\times\mathbf{u} + \kappa\nabla\times\boldsymbol{\phi} = -\mathbf{F}\delta(\mathbf{x}),$$
$$(\alpha+\beta+\gamma)\nabla\nabla\cdot\boldsymbol{\phi} - \gamma\nabla\times\nabla\times\boldsymbol{\phi} + \kappa\nabla\times\mathbf{u} - 2\kappa\boldsymbol{\phi} = -\mathbf{L}\delta(\mathbf{x}),$$
$$\lim_{r\to\infty}(\mathbf{u},\boldsymbol{\phi})\to\mathbf{0}; \qquad r = (x_1^2 + x_2^2 + x_3^2)^{1/2}, \qquad (5.25.1)$$

where $\mathbf{F} = \rho\mathbf{f}$, $\mathbf{L} = \rho j\mathbf{l}$, and thermal effects are ignored.

Applying the three-dimensional Fourier transforms to (5.25.1), we have

$$(\mu+\kappa)k^2\bar{\mathbf{u}} + (\lambda+\mu)\mathbf{k}(\mathbf{k}\cdot\bar{\mathbf{u}}) + \kappa i\mathbf{k}\times\bar{\boldsymbol{\phi}} = \mathbf{F},$$
$$(\gamma k^2 + 2\kappa)\bar{\boldsymbol{\phi}} + (\alpha+\beta)\mathbf{k}(\mathbf{k}\cdot\bar{\boldsymbol{\phi}}) + \kappa i\mathbf{k}\times\bar{\mathbf{u}} = \mathbf{L}. \qquad (5.25.2)$$

The inner and cross products of these equations by $\mathbf{k}$ give

$$\mathbf{k} \cdot \bar{\mathbf{u}} = \frac{\mathbf{k} \cdot \mathbf{F}}{(\lambda + 2\mu + \kappa)k^2}, \qquad \mathbf{k} \cdot \bar{\phi} = \frac{\mathbf{k} \cdot \mathbf{L}}{(\alpha + \beta + \gamma)k^2 + 2\kappa},$$
$$(\mu + \kappa)k^2 \mathbf{k} \times \bar{\mathbf{u}} + \kappa i[\mathbf{k}(\mathbf{k} \cdot \bar{\phi}) - k^2 \bar{\phi}] = \mathbf{k} \times \mathbf{F},$$
$$(\gamma k^2 + 2\kappa)\mathbf{k} \times \bar{\phi} + \kappa i[\mathbf{k}(\mathbf{k} \cdot \bar{\mathbf{u}}) - k^2 \bar{\mathbf{u}}] = \mathbf{k} \times \mathbf{L}. \qquad (5.25.3)$$

Replacing $\mathbf{k} \cdot \bar{\mathbf{u}}$ and $\mathbf{k} \times \bar{\phi}$ in (5.25.2), we determine

$$\bar{\mathbf{u}} = \left(\frac{2}{(2\mu + \kappa)k^2} + \frac{A}{k^2 + a^2} \right) \mathbf{F} - \frac{i}{2\mu + \kappa} \left(\frac{1}{k^2} - \frac{1}{k^2 + a^2} \right) \mathbf{k} \times \mathbf{L}$$
$$- \left(\frac{B}{k^4} - \frac{C}{k^2} + \frac{C}{k^2 + a^2} \right) \mathbf{k}(\mathbf{k} \cdot \mathbf{F}),$$

$$\bar{\phi} = \frac{1}{\gamma} \frac{1}{k^2 + a^2} - \frac{i}{2\mu + \kappa} \left(\frac{1}{k^2} - \frac{1}{k^2 + a^2} \right) \mathbf{k} \times \mathbf{F}$$
$$+ \left[\frac{D}{k^2 + a^2} - \frac{E}{k^2 + b^2} - \frac{1}{(4\mu + 2\kappa)k^2} \right] \mathbf{k}(\mathbf{k} \cdot \mathbf{L}), \qquad (5.25.4)$$

where

$$a^2 = \frac{(2\mu + \kappa)\kappa}{(\mu + \kappa)\gamma}, \qquad b^2 = \frac{2\kappa}{\alpha + \beta + \gamma}, \qquad A = \frac{\kappa}{(\mu + \kappa)(2\mu + \kappa)},$$
$$B = \frac{2\lambda + 2\mu + \kappa}{(\lambda + 2\mu + \kappa)(2\mu + \kappa)}, \qquad C = \frac{\gamma}{(2\mu + \kappa)^2}, \qquad D = \frac{\mu + \kappa}{(2\mu + \kappa)\kappa},$$
$$E = \frac{1}{2\kappa}. \qquad (5.25.5)$$

The inverse Fourier transforms of (5.25.4) give

$$u_i = G_{ij}^1 F_j + G_{ij}^2 L_j, \qquad \phi_i = H_{ij}^1 F_j + H_{ij}^2 L_j, \qquad (5.25.6)$$

where, in the three-dimensional case,

$$G_{ij}^1 = \frac{\delta_{ij}}{4\pi r} \left(\frac{2}{2\mu + \kappa} - Ae^{-ar} \right) + \frac{1}{4\pi} \frac{\partial^2}{\partial x_i \partial x_j} \left[\frac{1}{r} \left(-\frac{Br^2}{2} - C + Ce^{-ar} \right) \right],$$

$$G_{ij}^2 = H_{ij}^1 = \frac{1}{4\pi(2\mu + \kappa)} \epsilon_{ijk} \frac{\partial}{\partial x_k} \left(\frac{e^{-ar} - 1}{r} \right),$$

$$H_{ij}^2 = \frac{\delta_{ij}}{\gamma} \frac{e^{-ar}}{4\pi r} + \frac{1}{4\pi} \frac{\partial^2}{\partial x_i \partial x_j} \left[\frac{1}{r} \left(-De^{-ar} + Ea^{-br} + \frac{1}{2(2\mu + \kappa)} \right) \right]$$
$$(i, j, k = 1, 2, 3), \quad (5.25.7)$$

and, in the two-dimensional case,

$$G_{ij}^1 = -\frac{\delta_{ij}}{\pi(2\mu + \kappa)} \left[\ln r_0 + I + A\mu K_0(ar_0) \right]$$

$$+ \frac{1}{2\pi} \frac{\partial^2}{\partial x_i \partial x_j} \left[\left(\frac{1}{4} B r_0^2 + C \right) \ln r_0 + C K_0(a r_0) \right],$$

$$G_{i3}^2 = \frac{1}{2\pi(2\mu + \kappa)} \epsilon_{i3k} \frac{\partial}{\partial x_k} \left[\ln r_0 + K_0(a r_0) \right],$$

$$H_{i3}^1 = \frac{1}{2\pi(2\mu + \kappa)} \epsilon_{3ij} \frac{\partial}{\partial x_j} \left[\ln r_0 + K_0(a r_0) \right],$$

$$H_{33}^2 = \frac{K_0(a r_0)}{2\pi\gamma}; \quad \frac{\partial}{\partial x_3} = 0, \quad i,j = 1,2. \tag{5.25.8}$$

Significance of elements G_{ij}^α and H_{ij}^α can be seen from (5.25.6). For example, for $\mathbf{L} = \mathbf{0}$, G_{ij}^1 gives the displacement field along the x_i-axis produced by a unit concentrated force directed along the x_j-axis.

Using the reciprocal theorem

$$\int_{\mathcal{V}} \left(F_i^0 u_i + L_i^0 \phi_i \right) dv - \int_{\mathcal{V}} \left(F_i u_i^0 + L_i w_i^0 \right) dv$$

$$= \int_{\partial \mathcal{V}} \left(t_{(\mathbf{n})i} u_i^0 + m_{(\mathbf{n})i} \phi_i^0 \right) da - \int_{\partial \mathcal{V}} \left(t_{(\mathbf{n})i}^0 u_i + m_{(\mathbf{n})i}^0 \phi_i \right) da, \quad (5.25.9)$$

and taking $F_i^0 = \delta_{ij}\delta(\mathbf{x} - \mathbf{y})$, $L_i^0 = 0$, we have

$$u_i^0 = G_{ij}^1(\mathbf{x} - \mathbf{y}), \quad \phi_i = H_{ij}^1(\mathbf{x} - \mathbf{y}), \tag{5.25.10}$$

and we obtain

$$u_j(\mathbf{y}) = \int_{\mathcal{V}} \left[F_i(\mathbf{x}) G_{ij}^1(\mathbf{x} - \mathbf{y}) + L_i(\mathbf{x}) H_{ij}^1(\mathbf{x} - \mathbf{y}) \right] dv(\mathbf{x})$$

$$+ \int_{\partial \mathcal{V}} \left[t_{(\mathbf{n})i}(\mathbf{x}) G_{ij}^1(\mathbf{x} - \mathbf{y}) + m_{(\mathbf{n})i}(\mathbf{x}) H_{ij}^1(\mathbf{x} - \mathbf{y}) \right] da(\mathbf{x})$$

$$- \int_{\partial \mathcal{V}} \left[u_i(\mathbf{x}) t_{(\mathbf{n})i}^0(\mathbf{x} - \mathbf{y}) + \phi_i(\mathbf{x}) m_{(\mathbf{n})i}^0(\mathbf{x} - \mathbf{y}) \right] da(\mathbf{x}), \ (5.25.11)$$

where $t_{(\mathbf{n})i}$ and $m_{(\mathbf{n})i}$ are given at the boundary $\partial \mathcal{V}$ of $\mathcal{V}$, and $t_{(\mathbf{n})i}^0$ and $m_{(\mathbf{n})i}^0$ are calculated by means of the fundamental solution, i.e.,

$$t_{(\mathbf{n}i}^0 = \lambda G_{kj,j}^1 n_i + (2\mu + \kappa) G_{ij,k}^1 n_k + \mu(G_{kj,i}^1 - G_{ij,k}^1) n_k + \kappa \epsilon_{ikl} n_k H_{lj}^1,$$

$$m_{(\mathbf{n})i}^0 = \beta H_{kj,k}^1 n_i + (\beta + \gamma) H_{ij,k}^1 n_k + \beta n_k \left(H_{kj,i}^1 - H_{ij,k}^1 \right). \tag{5.25.12}$$

Similarly, for $\mathbf{F}^0 = \mathbf{0}$, $L_i^0 = \delta_{ij}\delta(\mathbf{x} - \mathbf{y})$, we have

$$u_i^0 = G_{ij}^2(\mathbf{x} - \mathbf{y}), \quad \phi_i = H_{ij}^2(\mathbf{x} - \mathbf{y}), \tag{5.25.13}$$

and (5.25.9) gives

$$\phi_j(\mathbf{y}) = \int_{\mathcal{V}} \left[F_i(\mathbf{x}) G_{ij}^2(\mathbf{x} - \mathbf{y}) + L_i(\mathbf{x}) H_{ij}^2(\mathbf{x} - \mathbf{y}) \right] dv(\mathbf{x})$$

$$+ \int_{\partial V} \left[t_{(\mathbf{n})i}(\mathbf{x}) G_{ij}^2(\mathbf{x} - \mathbf{y}) + m_{(\mathbf{n})i}(\mathbf{x}) H_{ij}^2(\mathbf{x} - \mathbf{y}) \right] da(\mathbf{x})$$

$$- \int_{\partial V} \left[u_i(\mathbf{x}) t_{(\mathbf{n})i}^0(\mathbf{x} - \mathbf{y}) + \phi_i(\mathbf{x}) m_{(\mathbf{n})i}^0(\mathbf{x} - \mathbf{y}) \right] da(\mathbf{x}). \quad (5.25.14)$$

Equations (5.25.11) and (5.25.14) are the integral representations of the solution of equations of micropolar elastostatics for arbitrary loads $(\mathbf{F}, \mathbf{L})$ and for the bounded domain $\mathcal{V}$.

5.26 Dislocations and Disclinations

A. Dynamic Case

All materials contain certain defects. Generally, these defects consist of voids, impurity atoms, dislocations, and disclinations. Dislocations are the results of incompatibility of the strain measures ϵ_{kl} and disclinations are those of the wryness tensor γ_{kl}. These are the cause of plastic distorsions in a body.

Here we discuss the elastic fields produced in an anisotropic micropolar elastic solid due to distributions of dislocations and disclinations.[17]

The linear strain measures of the micropolar elasticity may be decomposed into elastic and plastic parts

$$\epsilon_{ij} = \epsilon_{ij}^E + \epsilon_{ij}^P, \quad \gamma_{ji} = \kappa_{ij} = \kappa_{ij}^E + \kappa_{ij}^P. \quad (5.26.1)$$

The plastic parts $\boldsymbol{\epsilon}^P, \boldsymbol{\kappa}^P$ are the cause of dislocations and disclinations, with densities defined by (see (1.7.10))

$$\alpha_{ij} \equiv -\epsilon_{ipq}(\epsilon_{qj,p}^P + \epsilon_{qjs}\kappa_{ps}^P), \quad \theta_{ij} \equiv -\epsilon_{ipq}\kappa_{qj,p}^P, \quad (5.26.2)$$

α_{ij} and θ_{ij} are called dislocation and disclination density tensors.

For simplicity, we consider a linear anisotropic solid with center of symmetry, so that material moduli $C_{ijkl} = 0$ and the constitutive equations are given by

$$t_{ij} = A_{ijkp}\epsilon_{kp}^E, \quad m_{ji} = B_{ijpk}\kappa_{kp}^E, \quad (5.26.3)$$

where A_{ijkp} and B_{ijkp} possess symmetry regulations

$$A_{ijkp} = A_{kpij}, \quad B_{ijpk} = B_{pkij}. \quad (5.26.4)$$

----

[17] The mathematical method for the dislocation and disclination problems in micropolar elasticity was established by Eringen and Claus Jr. [1970], Claus Jr. and Eringen [1971]. This method was used by various investigators in the field, e.g., Nowacki [1977], Minagawa [1977]. Here, we follow the work of Minagawa [1985]. Other related work includes Schaefer [1967], Kessel [1970], Nowacki [1977], Minagawa [1977]. For a discussion of the topic in classical elasticity, see DeWitt [1973] and Kossecka and DeWitt [1977].

Equations of motion

$$t_{ji,j} - \rho\ddot{u}_i = 0, \quad m_{ji,j} + \epsilon_{ijk}t_{jk} - \rho j_{ij}\ddot{\phi}_j = 0, \tag{5.26.5}$$

with the use of (5.26.1) and (5.26.4), lead to the field equations

$$L_i^1(\mathbf{u}) + L_i^2(\boldsymbol{\phi}) = f_i(\boldsymbol{\epsilon}^P),$$
$$M_i^1(\mathbf{u}) + M_i^2(\boldsymbol{\phi}) = g_i(\boldsymbol{\epsilon}^P, \boldsymbol{\gamma}^P), \tag{5.26.6}$$

where the linear operators L_i^α and M_i^α and fictitious body loads f_i and g_i are defined by

$$L_i^1(\mathbf{u}) \equiv A_{jikp}u_{p,kj} - \rho\ddot{u}_i, \quad L_i^2(\boldsymbol{\phi}) \equiv -A_{jikp}\epsilon_{kpq}\phi_{q,j},$$
$$M_i^1(\mathbf{u}) \equiv B_{ijpk}\phi_{p,kj} - \epsilon_{ijk}A_{jkpq}\epsilon_{pqr}\phi_r - \rho j_{ij}\ddot{\phi}_j, \quad M_j^2(\mathbf{u}) \equiv \epsilon_{ijk}A_{jkpq}u_{q,p},$$
$$f_i(\boldsymbol{\epsilon}^P) \equiv A_{jikp}\epsilon_{kp,j}^P, \quad g_i(\boldsymbol{\epsilon}^P, \boldsymbol{\kappa}^P) \equiv B_{ijpk}\kappa_{kp,j}^P + \epsilon_{ijk}A_{jkpq}\epsilon_{pq}^P. \tag{5.26.7}$$

The formal solution of (5.26.6) is found in terms of Green's tensors G_{ps}^α and H_{ps}^α, $\alpha = 1, 2$,

$$u_i(\mathbf{u}, t) = -\int G_{is}^1(\mathbf{x} - \mathbf{x}', t - t')f_s(\boldsymbol{\epsilon}'^P)$$
$$+ G_{is}^2(\mathbf{x} - \mathbf{x}', t - t')g_s(\boldsymbol{\epsilon}'^P, \boldsymbol{\kappa}'^P)\,d^3\mathbf{x}'\,dt',$$
$$\phi_i(\mathbf{x}, t) = -\int H_{is}^1(\mathbf{x} - \mathbf{x}', t - t')f_s(\boldsymbol{\epsilon}'^P)$$
$$+ H_{is}^2(\mathbf{x} - \mathbf{x}', t - t')g_s(\boldsymbol{\epsilon}'^P, \boldsymbol{\kappa}'^P)\,d^3\mathbf{x}'\,dt'. \tag{5.26.8}$$

A prime placed on ϵ^P and γ^P denotes the replacement of $\mathbf{x}$ by $\mathbf{x}'$ and t by t' in the arguments of these functions (differentiations included).

Green's functions are the solutions of the equations

$$L_i^1(\mathbf{G}_s^1) + L_i^2(\mathbf{H}_s^1) = -\delta_{si}\delta(\mathbf{x})\delta(t),$$
$$M_i^1(\mathbf{G}_s^1) + M_i^2(\mathbf{H}_s^1) = 0, \tag{5.26.9}$$

$$L_i^1(\mathbf{G}_s^2) + L_i^2(\mathbf{H}_s^2) = 0,$$
$$M_i^1(\mathbf{G}_s^2) + M_i^2(\mathbf{H}_s^2) = -\delta_{si}\delta(\mathbf{x})\delta(t), \tag{5.26.10}$$

where δ_{ij} is the Kronecker delta and $\delta(\mathbf{x})$ is Dirac's delta measure.

Green's functions $\mathbf{G}^\alpha$ and $\mathbf{H}^\alpha$ are calculated by introducing four-dimensional Fourier transforms

$$\{\mathbf{G}^\alpha(\mathbf{x}, t), \mathbf{H}^\alpha(\mathbf{x}, t)\} = \frac{1}{(2\pi)^4}\int \{\mathbf{G}^{*\alpha}(\mathbf{k}, \omega), \mathbf{H}^{*\alpha}(\mathbf{k}, \omega)\}$$
$$\times \exp\left[i(\mathbf{k} \cdot \mathbf{x} - \omega t]\,d^3\mathbf{k}\,d\omega. \tag{5.26.11}$$

Substitution of (5.26.11) into (5.26.9) and (5.26.10) leads to two pairs of two linear equations which can be solved for $\mathbf{G}^{*\alpha}$ and $\mathbf{H}^{*\alpha}$ in the forms

$$G_{ij}^{*1} = \frac{D_{ij}}{D}, \quad H_{ij}^{*1} = \frac{D_{(3+i)j}}{D}, \quad G_{ij}^{*2} = \frac{D_{i(3+j)}}{D}, \quad H_{ij}^{*2} = \frac{D_{(3+i)(3+j)}}{D},$$
(5.26.12)

where D is the determinant of the 6×6 Hermitian matrix

$$\begin{bmatrix} A_{piqj}k_p k_q - \rho\omega^2 \delta_{ij} & \vdots & \sqrt{-1}A_{qikp}\epsilon_{kpj}k_q \\ \cdots & \cdots & \cdots \\ -\sqrt{-1}A_{qjkp}\epsilon_{kpi}k_q & \vdots & B_{ipjq}k_p k_q + \epsilon_{irs}A_{rspq}\epsilon_{pqj} - \rho j_{ij}\omega^2 \end{bmatrix},$$

and $D_{\alpha\beta}(\alpha, \beta = 1, 2, \ldots, 6)$ is its $\alpha\beta$-cofactor. From (5.26.12) it can be verified that

$$G_{ij}^{*1} = G_{ji}^{*1}, \quad G_{ij}^{*2} = -H_{ji}^{*1}, \quad H_{ij}^{*2} = H_{ji}^{*2}.$$
(5.26.13)

Consequently,

$$G_{ij}^{1} = G_{ji}^{1}, \quad G_{ij}^{2} = -H_{ji}^{1}, \quad H_{ij}^{2} = H_{ji}^{2}.$$
(5.26.14)

With these, (5.26.8) gives $u_i(\mathbf{x}, t)$ and $\phi_i(\mathbf{x}, t)$. Determinations of $\boldsymbol{\epsilon}^E$ and $\boldsymbol{\gamma}^E$ require differentiations. In this process, we use (5.26.9), (5.26.10), and introduce the following functions:

$$G_{nsj} = G_{ns,j}^{1} + \epsilon_{qjs}G_{nq}^{2}, \quad H_{nsj} = H_{ns,j}^{1} + \epsilon_{qjs}H_{nq}^{2},$$
$$\mathcal{G}_{nsj} = \mathcal{G}_{ns,j}^{1} + \epsilon_{qjs}\mathcal{G}_{nq}^{2}, \quad \mathcal{H}_{nsj} = \mathcal{H}_{ns,j}^{1} + \epsilon_{qjs}\mathcal{H}_{nq}^{2},$$
$$\{\mathcal{G}^{\alpha}(\mathbf{x}, t), \mathcal{H}^{\alpha}(\mathbf{x}, t)\} = \int \{G^{\alpha}(\mathbf{x}', t), H^{\alpha}(\mathbf{x}', t)\} (4\pi R)^{-1} d^3\mathbf{x}',$$
$$\Gamma(\mathbf{x}, t) = \int \delta(\mathbf{x}')\delta(t)(4\pi R)^{-1} d^3\mathbf{x}', \quad R = |\mathbf{x} - \mathbf{x}'|.$$
(5.26.15)

The resulting expressions for ϕ_r, ϵ_{mn}^{E}, and κ_{mn}^{E} are

$$\phi_r(\mathbf{x}, t) = \int [(\rho\dot{\mathcal{H}}_{rp,k}^{1}\dot{\epsilon}_{kp}^{P} + \epsilon_{cbp}A_{pqjs}\mathcal{H}_{rsj,c}\alpha_{bq})$$
$$+ \epsilon_{kpt}(\rho\dot{\mathcal{H}}_{rp,k}^{1}\dot{\kappa}_{kt}^{P} + \epsilon_{kqd}A_{dpjs}\mathcal{H}_{rsj}\theta_{qt})$$
$$+ (\rho j_{pj}\dot{\mathcal{H}}_{jr,k}^{2}\dot{\kappa}_{kp}^{P} + \epsilon_{cbd}B_{pdqj}\mathcal{H}_{rq,jc}^{2}\theta_{bp}) - \Gamma\kappa_{kr,k'}^{P}] d^3\mathbf{x}' dt',$$
(5.26.16)

$$\epsilon_{mn}^{E}(\mathbf{x}, t) = -\int \Big[(\rho\dot{G}_{pn}^{1}\dot{\epsilon}_{mp}^{P} + \epsilon_{mqk}A_{jskp}G_{nsj}\alpha_{qp})$$
$$+ \epsilon_{mnr}\left(\rho\dot{\mathcal{H}}_{rp,k}^{1}\dot{\epsilon}_{kp}^{P} + \epsilon_{cbp}A_{pqjs}\mathcal{H}_{rsj,c}\alpha_{bp}\right)$$
$$- \epsilon_{mpq}\left(\rho\dot{\mathcal{G}}_{pn,k}^{1}\dot{\kappa}_{kq}^{P} + \epsilon_{cbr}A_{rpjs}\mathcal{G}_{nsj,c}\theta_{bp}\right)$$

$$+ \, \epsilon_{mnr}\epsilon_{kpt}\left(\rho\dot{\mathcal{H}}^1_{rp}\dot{\kappa}^P_{kt} + \epsilon_{kqd}A_{dpjs}\mathcal{H}_{rsj}\theta_{qt}\right)$$

$$+ \left(\rho j_{pj}\dot{G}^2_{nj}\dot{\kappa}^P_{mp} + \epsilon_{mqk}B_{sjpk}G^2_{ns,j}\theta_{qp}\right)$$

$$+ \, \epsilon_{mnr}\left(\rho j_{pj}\dot{\mathcal{H}}^2_{jr,k}\dot{\kappa}^P_{kp} + \epsilon_{cbd}B_{pdqj}\mathcal{H}^2_{rq,jc}\theta_{bp}\right)\Big] \, d^3\mathbf{x}' \, dt',$$

$$(5.26.17)$$

$$\kappa^E_{mn}(\mathbf{x},t) = -\int\Big[\left(\rho\dot{H}^1_{pn}\dot{\epsilon}^P_{mp} + \epsilon_{mqk}A_{jskp}H_{nsj}\alpha_{qp}\right)$$

$$- \, \epsilon_{mpq}\left(\rho\dot{\mathcal{H}}^1_{np,k}\dot{\kappa}^P_{kq} + \epsilon_{cbr}A_{rpjs}\mathcal{H}_{nsj,c}\theta_{bq}\right)$$

$$+ \left(\rho j_{pj}\dot{H}^2_{nj}\dot{\kappa}^P_{mp} + \epsilon_{mqk}B_{sjpk}H^2_{ns,j}\theta_{qp}\right)\Big] \, d^3\mathbf{x}' \, dt'.$$

$$(5.26.18)$$

These give the elastic strains, since $\dot{\epsilon}^P_{ij}$ and $\dot{\gamma}^P_{ij}$ can be computed from the dislocation and disclination currents.

When the disclinations are absent ($\theta_{ij} = 0$), there exist ϕ^E_i and ϕ^P_i such that

$$\phi_i = \phi^E_i + \Phi^P_i, \quad \kappa^E_{ji} = \phi^E_{i,j}, \quad \kappa^P_{ij} = \phi^P_{j,i},$$

$$\epsilon^E_{ij} = \beta^E_{ij} - \epsilon_{ijk}\phi^E_k, \quad \epsilon^P_{ij} = \beta^P_{ij} - \epsilon_{ijk}\phi^P_k, \qquad (5.26.19)$$

and

$$\alpha_{ij} = -\epsilon_{ipq}\beta^P_{qj,i}, \qquad (5.26.20)$$

where β^E_{ij} and β^P_{ij} are, respectively, the elastic and plastic distortions. In this case, (5.26.16) and (5.26.17) reduce to

$$\phi^E_r(\mathbf{x},t) = \int\Big[\rho\dot{\mathcal{H}}_{rp,k}\dot{\beta}^P_{kp} + \epsilon_{ibp}A_{pq,js}\mathcal{H}_{rsj,c}\alpha_{bq} - \rho j_{pj}\dot{H}^2_{jr}\dot{\phi}^P_p\Big] \, d^3\mathbf{x}' \, dt',$$

$$(5.26.21)$$

$$\beta^E_{mn}(\mathbf{x},t) = -\int[\rho\dot{G}^1_{pn}\dot{\beta}^P_{mp} + \epsilon_{mqk}A_{jskp}G_{nsj}\alpha_{qp} + \rho j_{pj}\dot{G}^2_{nj,m}\dot{\phi}^P_p] \, d^3\mathbf{x}' \, dt'.$$

$$(5.26.22)$$

B. Static Case

In fact, for the static case, these Green's functions have already been given by (5.25.7) for the three-dimensional case and (5.25.8) for the two-dimensional case. The solution is given by

$$u_i(\mathbf{x}) = -\int_V\big[G^1_{ij}(\mathbf{x}-\mathbf{x}')f_j(\epsilon'^P) + G^2_{ij}(\mathbf{x}-\mathbf{x}')g_j(\epsilon'^P,\kappa'^P)\big] \, d^3\mathbf{x}',$$

$$\phi_i(\mathbf{x}) = -\int_V\big[H^1_{ij}(\mathbf{x}-\mathbf{x}')f_j(\epsilon'^P) + H^2_{ij}(\mathbf{x}-\mathbf{x}')g_j(\epsilon'^P,\kappa'^P)\big] \, d^3\mathbf{x}'.$$

Substituting for $f_i(\epsilon'^P)$ and $g_i(\epsilon'^P, \kappa'^P)$ from (5.26.7), after integrating by parts and assuming that the boundary terms vanish, we obtain

$$u_i(\mathbf{x}) = \int_V \left[(G^1_{im,l} - \epsilon_{jlm} G^2_{ij}) A_{lmkp} \epsilon^P_{kp} + B_{mlpk} G^2_{im,l} \kappa^P_{kp} \right] d^3\mathbf{x}',$$

$$\phi_i(\mathbf{x}) = \int_V \left[(H^1_{im,l} - \epsilon_{jlm} H^2_{ij}) A_{lmkp} \epsilon^P_{kp} + B_{mlpk} H^2_{im,l} \kappa^P_{kp} \right] d^3\mathbf{x}'.$$

$$(5.26.23)$$

(a) Dislocations

Dislocations can be expressed in terms of the slip (Burger's vector) b_i of a plane S^+ whose normal vector $\mathbf{n}$ is directed toward S^-:

$$\epsilon^P_{ij} = -b_j n_i \delta(\mathbf{s} - \mathbf{x}), \qquad \kappa^P_{ij} = 0, \qquad (5.26.24)$$

where $\delta(\mathbf{s} - \mathbf{x})$ is the one-dimensional Dirac delta function in the normal direction to $\mathbf{s}$.

(b.) Disclination

Disclination is an incompatibility in the rotation

$$\epsilon^P_{ij} = -\epsilon_{ijl} \omega_k (x_l - x^0_l) n_i \delta(\mathbf{S} - \mathbf{x}),$$
$$\kappa_{ij} = -\omega_i n_j \delta(\mathbf{S} - \mathbf{x}). \qquad (5.26.25)$$

If (5.26.24) and (5.26.25) are substituted into (5.26.23), explicit expressions result from (5.26.23).

The foregoing analysis, which was first introduced by Eringen and Claus Jr., has been pursued by various authors for discussions of static and dynamic dislocations and disclinations. Nowacki [1977], Minagawa [1977], Anthony [1970], Cheng and He [1995], among others, have contributed extensively to this field. The last authors also discussed a defect problem in the form of spherical inclusion.

5.27   Theory of Micropolar Plates[18]

Consider a thin plate of thickness $2h$, having $x_3 = 0$ as its median plane, Figure 5.27.1. Let $\mathcal{C}$ denote the boundary curve of the median plane.

The derivation of the equations of motion of thin plates is based on the following theory of integration of the linear field equations

$$L(\mathbf{u}, \phi) = 0, \qquad (5.27.1)$$

---

[18]The present section is based on the micropolar plate theory first given by Eringen [1967d].

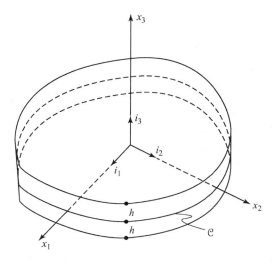

FIGURE 5.27.1. A plate of thickness $2h$

where $\mathbf{u}, \boldsymbol{\phi}$ are functions of x_1, x_2, x_3, and t. Assume that $\mathbf{u}$ and $\boldsymbol{\phi}$ are analytic functions of $z = x_3/h$, e.g.,[19]

$$\mathbf{u} = \sum_{n=0} \mathbf{u}_n(x_1, x_1, t) z^n, \tag{5.27.2}$$

then form, for all n,

$$\int_{-1}^{1} L(\mathbf{u}, \boldsymbol{\phi}) z^n \, dz = 0. \tag{5.27.3}$$

This gives a hierarchy of linear equations

$$L_{(n)}(\mathbf{u}_i, \boldsymbol{\phi}_i) = 0, \quad n = 0, 1, \dots. \tag{5.27.4}$$

The solution of this set (if it converges) gives the solution of (5.27.1). The solution of the set of equations (5.27.4) is not any simpler than the original equations (5.27.1). However, it suggests a systematic approximation. For thin plates, we may consider $n = 0$ and $n = 1$ to obtain approximate plate theory for the lowest and next order.

Based on this procedure, along with other additional assumptions on the order of magnitudes of some components of stress and couple stress tensors, plate and shell theories have been constructed in classical elasticity and continuously argued in favor of one or the other. Unfortunately, often these assumptions cannot be rigorously justified.

Here we present two lowest-order plate theories ($n = 0$ and $n = 1$). The case $n = 0$ will be obtained as a special case $n = 1$.

---

[19]Instead of the factor z^n, other orthogonal functions are also used, e.g., e^{inz}, etc.

A. A First-Order Micropolar Plate Theory

To construct a first-order plate theory, we perform the following integrations:

(i) We integrate equations of balance of momenta with respect to x_3 over the thickness of the plate; and

(ii) We take the cross product of the equations of the balance of linear momentum with $x_3\mathbf{i}_3$ and integrate over the thickness of the plate.

The results of (i) are

$$\bar{t}_{kl,k} + \rho(F_l - \ddot{\bar{u}}_l) = 0, \tag{5.27.5}$$
$$\bar{t}_{k3,k} + \rho(F_3 - \ddot{w}) = 0, \tag{5.27.6}$$
$$\bar{m}_{kl,k} + \epsilon_{kl3}(\bar{t}_{3k} - \bar{t}_{k3}) + \rho(L_l - j\ddot{\phi}_l) = 0, \tag{5.27.7}$$
$$\bar{m}_{k3,k} + \epsilon_{kl3}\bar{t}_{kl} + \rho(L_3 - j\ddot{\phi}_3) = 0. \tag{5.27.8}$$

These four equations constitute lowest-order (zeroth-order) theory. A first-order theory incorporates to these equations the result of (ii), i.e.,

$$M_{kl,k} - 2h\bar{t}_{3l} + (T_l - I\ddot{v}_l) = 0. \tag{5.27.9}$$

The mean stress, couple stress, and plate loads are defined by

$$(\bar{t}_{kl}, \bar{m}_{kl}) = \frac{1}{2h}\int_{-h}^{h} (t_{kl}, m_{kl})\,dx_3,$$
$$(M_{kl}, L_l^1, Iv_l) = \int_{-h}^{h} (t_{kl}, f_l, u_l)x_3\,dx_3, \quad I \equiv \tfrac{2}{3}, h^3,$$
$$\rho F_l = \frac{1}{2h}[t_{3l}]_{-h}^{h} + \rho\bar{f}_l,$$
$$\rho F_3 = \frac{p}{2h} + \rho f_3 = \frac{1}{2h}[t_{33}]_{-h}^{h} + \rho f_3,$$
$$\rho L_l = \frac{1}{2h}[m_{3l}]_{-h}^{h} + \rho\bar{l}_l,$$
$$\rho L_3 = \frac{1}{2h}[m_{33}]_{-h}^{h} + \rho\bar{l}_3,$$
$$\rho T_l = [x_3 t_{3l}]_{-h}^{h} + \rho L_l^1. \tag{5.27.10}$$

Equations (5.27.5)–(5.27.9) constitute the first-order balance equations of micropolar plates.

Constitutive equations of thin plate theory are obtained by taking

$$\mathbf{u} = [\bar{u}_k(x_1,x_2,t) + x_3 v_k(x_1,x_2,t)]\mathbf{i}_k + w(x_1,x_2,t)\mathbf{i}_3,$$
$$\boldsymbol{\phi} = \phi_k(x_1,x_2,t)\mathbf{i}_k + \phi(x_1,x_2,t)\mathbf{i}_3, \tag{5.27.11}$$

where $\mathbf{i}_k$ and $\mathbf{i}_3$ are the Cartesian base vectors.

In plate theory, it is assumed that $t_{33} = 0$. With this assumption, the strain component ϵ_{33} is given by[20]

$$\epsilon_{33} = -\lambda\epsilon_{kk}/(\lambda + 2\mu + \kappa).$$

Using this and the linear constitutive equations, we obtain

$$\bar{t}_{kl} = \frac{E}{1 - \nu^2}\left[\nu\bar{u}_{r,r}\delta_{kl} + \frac{1 - \nu}{2}\left(\bar{u}_{k,l} + \bar{u}_{l,k}\right)\right] - \frac{\kappa}{2}\left(\bar{u}_{k,l} - \bar{u}_{l,k} + 2\epsilon_{kl3}\phi\right),$$

$$M_{kl} = D\left[\nu v_{r,r}\delta_{kl} + \frac{1 - \nu}{2}\left(v_{k,l} + v_{l,k}\right)\right] - \frac{\kappa I}{2}\left(v_{k,l} - v_{l,k}\right), \qquad (5.27.12)$$

where

$$D = \frac{EI}{1 - \nu^2}, \quad \lambda = \frac{\nu E}{(1 + \nu)(1 - 2\nu)}, \quad 2\mu + \kappa \equiv 2G = \frac{E}{1 + \nu}. \quad (5.27.13)$$

The total stress and couple stress fields are given by

$$t_{kl} = \bar{t}_{kl} + \frac{x_3}{I}M_{kl}, \quad t_{k3} = \bar{t}_{k3} = (\mu + \kappa)(v_k + w_{,k}) - \kappa(v_k - \epsilon_{kl3}\phi_l),$$

$$t_{3k} = \bar{t}_{3k} = (\mu + \kappa)(v_k + w_{,k}) - \kappa(w_{,k} + \epsilon_{kl3}\phi_l), \quad t_{33} = 0,$$

$$m_{kl} = \alpha\phi_{r,r}\delta_{kl} + \beta\phi_{k,l} + \gamma\phi_{l,k}, \quad m_{k3} = \gamma\phi_{,k}, \quad m_{3k} = \beta\phi_{,k},$$

$$m_{33} = \alpha\phi_{k,k}, \quad k, l = 1, 2. \qquad (5.27.14)$$

In the zeroth-order theory, we set $\mathbf{v} = \mathbf{0}$ which gives $M_{kl} = 0$. This appears to eliminate bending moment. However, the presence of m_{kl} restores the effect of the couple. As we shall see, this limit suggests a special value for the internal characteristic length.

Substituting the expression of stress and couple stress given by (5.27.12) and (5.27.14) into the balance equation (5.27.5)–(5.27.8), we obtain the field equations:

(a) Extensional Motions:

$$\frac{1}{2}\left(\frac{E}{1 - \nu} - \kappa\right)\bar{u}_{k,lk} + \frac{1}{2}\left(\frac{E}{1 + \nu} + \kappa\right)\bar{u}_{l,kk} - \kappa\epsilon_{kl3}\phi_{,k} + \rho\left(F_l - \ddot{u}_l\right) = 0,$$

$$\hspace{11cm} (5.27.15)$$

$$\gamma\phi_{,kk} - 2\kappa\phi - \kappa\epsilon_{kl3}\bar{u}_{k,l} + \rho\left(L_3 - j\ddot{\phi}\right) = 0. \qquad (5.27.16)$$

---

[20]The case of plane strain $\epsilon_{33} = 0$ was discussed by Eringen [1967d].

(b) Flexural Motions:

$$\frac{I}{2}\left(\frac{E}{1-\nu}-\kappa\right)v_{k,lk}+\frac{I}{2}\left(\frac{E}{1+\nu}+\kappa\right)v_{l,kk}-2h\left(G-\frac{\kappa}{2}\right)w_{,l}$$
$$-2h\left(G+\frac{\kappa}{2}\right)v_l+2\kappa h\epsilon_{lk3}\phi_k+\rho\left(T_l+I\ddot{v}_l\right)=0,$$

$$(5.27.17)$$

$$\left(G-\frac{\kappa}{2}\right)v_{k,k}+\left(G+\frac{\kappa}{2}\right)w_{,kk}+\kappa\epsilon_{kl3}\phi_{l,k}+\rho\left(F_3-\ddot{w}_3\right)=0, (5.27.18)$$

$$(\alpha+\beta)\,\phi_{k,lk}+\gamma\phi_{l,kk}+\kappa\epsilon_{kl3}(v_k-w_{,k})-2\kappa\phi_l+\rho(L_l-j\ddot{\phi}_l)=0,$$
$$k,l=1,2.\ (5.27.19)$$

A set of boundary and initial conditions can be derived by the averaging process (i) and (ii):

(a) Extensional Motions:

$$\begin{aligned}
\bar{t}_{kl}n_k &= \bar{t}_l, & m_{k3}n_k &= \bar{m}_3 \quad \text{on } \mathcal{C}_L, \\
\bar{u}_k &= \bar{u}_{0k}, & \phi &= \phi_0 \quad \text{on } \mathcal{C}-\mathcal{C}_L, \\
\bar{u}_k(t=0) &= U_{0k}(x_1,x_2), & \phi(t=0) &= \Phi_0(x_1,x_2), \\
\dot{\bar{u}}(t=0) &= V_{0k}(x_1,x_2), & \dot{\phi}(t=0) &= \nu_0(x_1,x_2) \quad \text{on } \mathcal{S}.
\end{aligned}$$

$$(5.27.20)$$

(b) Flexural Motions:

$$\begin{aligned}
M_{kl}n_k &= \bar{M}_l, \quad \bar{t}_{k3}n_k = \bar{t}_3, \quad m_{kl}n_k = \bar{m}_l \quad \text{on } \mathcal{C}_L, \\
v_k &= \bar{v}_{0k}, \quad w = \bar{w}_0, \quad \phi_k = \bar{\phi}_{0k} \quad \text{on } \mathcal{C}-\mathcal{C}_L, \\
v_k(t=0) &= \tilde{v}_k(x_1,x_2), \qquad w(t=0) = \hat{w}(x_1,x_2), \\
\phi_k(t=0) &= \hat{\phi}_k(x_1,x_2), \qquad \dot{v}_k(t=0) = \tilde{v}_k(x_1,x_2) \quad \text{on } \mathcal{S}, \quad (5.27.21) \\
\dot{w}(t=0) &= \tilde{w}(x_1,x_2), \qquad \dot{\phi}_k(t=0) = \nu_k(x_1,x_2).
\end{aligned}$$

Here the quantities on the right-hand sides are prescribed, on the boundary $\mathcal{C}$ or on the surface $\mathcal{S}$. From these equations, it is clear that the stretching and flexures of plates can be treated independently of each other.

Similar to the decomposition of the displacement and microrotation fields in the three-dimensional case, we take

$$v_k = v_{,k} + \epsilon_{kl3}V_{,l}, \quad \phi_k = \phi_{,k} + \epsilon_{kl3}\Phi_{,l}, \quad (5.27.22)$$

where v, V, ϕ, and Φ are functions of x_1, x_2, and t. Upon substituting (5.24.22) into the field equations (5.27.15) and (5.27.19), we calculate the divergence and curl of these equations. Eliminating v and V among these equations, we arrive at the equations

$$-D\nabla^4 w + \frac{4\kappa Gh}{G+\kappa/2}\nabla^2 w + \rho\left(\frac{D}{G+\kappa/2}+I\right)\nabla^2\ddot{w} - \frac{\rho^2 I}{G+\kappa/2}\frac{\partial^4 w}{\partial t^4}$$

$$-2\rho h \ddot{w} + \frac{\kappa D}{G + \kappa/2} \nabla^4 \Phi - \frac{4\kappa G h}{G + \kappa/2} \nabla^2 \Phi - \frac{\kappa \rho I}{G + \kappa/2} \nabla^2 \ddot{\Phi}$$

$$-\frac{D}{2h(G + \kappa/2)} \nabla^2 P + \frac{\rho I}{2h(G + \kappa/2)} \ddot{P} + P + \rho \frac{G - \kappa/2}{G + \kappa/2} T_{k,k} = 0,$$

$$(5.27.23)$$

$$\gamma \nabla^4 \Phi - \frac{2\kappa G}{G - \kappa/2} \nabla^2 \Phi - \rho j \nabla^2 \ddot{\Phi} + \frac{2\kappa G}{G - \kappa/2} \nabla^2 w - \frac{\kappa \rho}{G - \kappa/2} \ddot{w}$$

$$+ \frac{\kappa}{2h(G - \kappa/2)} P + \rho \epsilon_{kl3} L_{k,l} = 0, \quad (5.27.24)$$

$$\kappa \nabla^2 v = \gamma \nabla^4 \Phi - 2\kappa \nabla^2 \Phi - \rho j \nabla^2 \ddot{\Phi} + \kappa \nabla^2 w + \rho \epsilon_{kl3} L_{k,l}, \quad (5.27.25)$$

$$\kappa \nabla^2 V = -(\alpha + \beta + \gamma)\nabla^4 \phi + 2\kappa \nabla^2 \phi + \rho j \nabla^2 \ddot{\phi} - \rho L_{k,k}, \quad (5.27.26)$$

$$-I(\alpha + \beta + \gamma)\nabla^6 \phi + 2[\kappa I + h(\alpha + \beta + \gamma)]\nabla^4 \phi - \frac{4\kappa G h}{G + \kappa/2} \nabla^2 \phi$$

$$+ \rho I \left(j + \frac{\alpha + \beta + \gamma}{G + \kappa/2} \right) \nabla^4 \ddot{\phi} - 2\rho \left(hj + \frac{\kappa I}{G + \kappa/2} \right) \nabla^2 \ddot{\phi} - \frac{\rho^2 j I}{G + \kappa/2} \nabla^2 \frac{\partial^4 \phi}{\partial t^4}$$

$$- \rho I \nabla^2 L_{k,k} + 2\rho h L_{k,k} + \frac{\rho^2 I}{G + \kappa/2} \ddot{L}_{k,k} + \frac{\kappa \rho}{G + \kappa/2} \epsilon_{kl3} T_{k,l} = 0,$$

$$P \equiv 2\rho h F_3. \quad (5.27.27)$$

In terms of w and potentials v, V, ϕ, and Φ, the stress, couple stress, and transverse shear take the forms

$$M_{kl} = D[\nu \nabla^2 v + (1 - \nu)v_{,kl}] + \left(G - \frac{\kappa}{2} \right) I \epsilon_{kr3} V_{,rl} + \left(G + \frac{\kappa}{2} \right) I \epsilon_{lr3} V_{,rk},$$

$$(5.27.28)$$

$$\bar{m}_{kl} = \alpha \nabla^2 \phi \delta_{kl} + (\beta + \gamma)\phi_{k,l} + \beta \epsilon_{kr3} \Phi_{,rl} + \gamma \epsilon_{rl3} \Phi_{,rk}, \quad (5.27.29)$$

$$\bar{t}_{k3} = (G - \kappa/2)v_{,k} + (G + \kappa/2)w_{,k} - \kappa \Phi_{,k} + \epsilon_{kr3}[(G - \kappa/2)V_{,r} + \kappa \phi_{,r}]. \quad (5.27.30)$$

The boundary and initial conditions are given by (5.27.20) and (5.27.21), in which we insert the expressions of $M_{kl}, m_{kl}, \bar{t}_{k3}$ given by (5.27.12) and (5.27.14).

The formulation is now complete. These equations can be used to derive various approximate theories. Here, we consider only the case of negligible rotatory inertia ($\rho I = 0$) and shear deformation ($G \rightarrow \infty$). We also set $L_k = T_k = 0$,

$$-D\nabla^4 w + 4\kappa h \nabla^2 w - 2\rho h \ddot{w} - 4\kappa h \nabla^2 \Phi + P = 0, (5.27.31)$$

$$\gamma \nabla^4 \Phi - 2\kappa \nabla^2 \Phi - \rho j \nabla^2 \ddot{\Phi} + 2\kappa \nabla^2 w = 0, (5.27.32)$$

$$\nabla^2(v + w) = 0, (5.27.33)$$

$$\kappa \nabla^2 V + (\alpha + \beta + \gamma)\nabla^4 \phi - 2\kappa \nabla^2 \phi - \rho j \nabla^2 \ddot{\phi} = 0, (5.27.34)$$

$$-(\alpha + \beta + \gamma)\nabla^6 \phi + [2\kappa + 3(\alpha + \beta + \gamma)h^{-2}]\nabla^4 \phi - 6\kappa h^{-2} \nabla^2 \phi$$

$$-3\rho j h^{-2} \nabla^2 \ddot{\phi} = 0. (5.27.35)$$

The internal energy density of the plate is obtained by integrating the three-dimensional equation

$$2\rho\epsilon = t_{\alpha\beta}[e_{\alpha\beta} + \epsilon_{\alpha\beta\gamma}(r_\gamma - \phi_\gamma)] + m_{\alpha\beta}\phi_{\beta,\alpha}, \quad \alpha, \beta, \gamma = 1, 2, 3, \quad (5.27.36)$$

across the thickness of the plate, where

$$e_{\alpha\beta} = \tfrac{1}{2}(u_{\alpha,\beta} + u_{\beta,\alpha}), \quad r_\alpha = \tfrac{1}{2}\epsilon_{\alpha\beta\gamma}u_{\gamma,\beta}. \quad (5.27.37)$$

In the case $t_{33} = 0$, this gives

$$\epsilon = 2h\bar{\epsilon}_p + \epsilon_b, \quad (5.27.38)$$

where

$$2\rho\epsilon_p = \frac{\nu E}{1 - \nu^2}\bar{e}_{kk}\bar{e}_{ll} + 2G\bar{e}_{kl}\bar{e}_{kl} + 2\kappa(\bar{r}_3 - \phi)^2 + \gamma\phi_{,k}\phi_{,k},$$

$$2\rho h\epsilon_b = I\left(\frac{\nu E}{1 - \nu^2}\tilde{e}_{kk}\tilde{e}_{ll} + 2G\tilde{e}_{kl}\tilde{e}_{kl} + 2\kappa\tilde{r}_3^2\right)$$
$$+ 2hG(v_k + w_{,k})(v_k + w_{,k}) + 4\kappa h(r_k - \phi_k)(r_k - \phi_k)$$
$$+ 2h(\alpha\phi_{k,k}\phi_{l,l} + \beta\phi_{k,l}\phi_{l,k} + \gamma\phi_{k,l}\phi_{k,l}), \quad (5.27.39)$$

$$2\bar{e}_{kl} \equiv \bar{u}_{k,l} + \bar{u}_{l,k}, \qquad 2\tilde{e}_{kl} \equiv v_{k,l} + v_{l,k},$$
$$2\bar{r}_3 \equiv \bar{u}_{2,1} - \bar{u}_{1,2}, \qquad 2\tilde{r}_3 \equiv v_{2,1} - v_{1,2}. \qquad (5.27.40)$$

Here $\bar{\epsilon}_p$ is the extensional energy and ϵ_b is the bending energy. Observing that these are quadratic forms, the following theorems hold:

Theorem 1 (Nonnegative Internal Energy $\bar{\epsilon}_p$). The necessary and sufficient conditions for $\bar{\epsilon}_p$ nonnegative (for $t_{33} = 0$) are

$$\frac{E}{1 - \nu} \geq 0, \quad 2G \geq 0, \quad \kappa \geq 0, \quad \gamma \geq 0. \quad (5.27.41)$$

Theorem 2 (Nonnegative Bending Energy). The necessary and sufficient conditions for ϵ_b to be nonnegative are:

$$\frac{\nu E}{1 - \nu^2} \geq 0, \quad G \geq 0, \quad \kappa \geq 0,$$
$$2\alpha + \beta + \gamma \geq 0, \quad \gamma + \beta \geq 0, \quad \gamma - \beta \geq 0. \quad (5.27.42)$$

B. The Lowest-Order Theory

The lowest-order plate equations are obtained from the foregoing equations by setting $v = 0$ and dropping the moment equation (5.27.9). In this case, then, we have:

(a) Extension of the Plate:

$$\bar{t}_{kl,k} + \rho(F_l - \ddot{u}_l) = 0, \qquad (5.27.43)$$

$$\bar{m}_{k3,k} + \epsilon_{kl3}\bar{t}_{kl} + \rho(L_3 - j\ddot{\phi}) = 0, \qquad (5.27.44)$$

are the balance laws and

$$\bar{t}_{kl} = \frac{E}{1-\nu^2}\left[\nu\bar{u}_{r,r}\delta_{kl} + \frac{1-\nu}{2}(\bar{u}_{k,l} + \bar{u}_{l,k})\right] - \frac{\kappa}{2}(\bar{u}_{k,l} - \bar{u}_{l,k}) + 2\epsilon_{kl3}\phi,$$

$$(5.27.45)$$

$$\bar{m}_{k3} = \gamma\phi_{,k}, \qquad (5.27.46)$$

are the constitutive equations.

The field equations are given by

$$\frac{1}{2}\left(\frac{E}{1-\nu} - \kappa\right)\bar{u}_{k,lk} + \frac{1}{2}\left(\frac{E}{1+\nu} + \kappa\right)\bar{u}_{l,kk} - \kappa\epsilon_{kl3}\phi_{,k} + \rho(F_l - \ddot{u}_l) = 0,$$

$$(5.27.47)$$

$$\gamma\phi_{,kk} - 2\kappa\phi - \kappa\epsilon_{kl3}\bar{u}_{k,l} + \rho(L_3 - j\ddot{\phi}) = 0, \qquad (5.27.48)$$

and the extensional energy by

$$\rho\bar{\epsilon}_p = \frac{1}{2}\left\{\frac{\nu E}{1-\nu^2}\bar{e}_{kk}\bar{e}_{ll} + 2G\bar{e}_{kl}\bar{e}_{kl} + 2\kappa(\bar{r}_3 - \phi)^2 + \gamma\phi_{,k}\phi_{,k}\right\}. \quad (5.27.49)$$

(b) Flexure of the Plate:

For flexure, the balance laws are given by

$$\bar{t}_{k3,k} + \rho(F_3 - \ddot{w}) = 0, \qquad (5.27.50)$$

$$\bar{m}_{kl,k} + \epsilon_{kl3}(\bar{t}_{3k} - \bar{t}_{k3}) + \rho(L_l - j\ddot{\phi}_l) = 0. \qquad (5.27.51)$$

Constitutive equations read:

$$\bar{t}_{k3} = (\mu + \kappa)w_{,k} + \kappa\epsilon_{kl3}\phi_l, \qquad \bar{t}_{3k} = \mu w_{,k} - \kappa\epsilon_{kl3}\phi_l,$$

$$m_{kl} = \alpha\phi_{r,r} + \beta\phi_{k,l} + \gamma\phi_{l,k}. \qquad (5.27.52)$$

The field equations take the forms

$$\left(G + \frac{\kappa}{2}\right)\nabla^2 w + \kappa\epsilon_{kl3}\phi_{l,k} + \rho(F_3 - \ddot{w}) = 0, \quad (5.27.53)$$

$$(\alpha + \beta)\phi_{k,lk} + \gamma\phi_{l,kk} - \kappa\epsilon_{kl3}w_{,k} - 2\kappa\phi_l + \rho(L_l - j\ddot{\phi}_l) = 0. \quad (5.27.54)$$

The energy equation (5.27.38) reduces to

$$\rho\epsilon_b = h(\alpha\phi_{k,k}\phi_{l,l} + \beta\phi_{k,l}\phi_{l,k} + \gamma\phi_{l,k}\phi_{l,k})$$
$$+ Ghw_{,k}w_{,k} + 2\kappa h(r_k - \phi_k)(r_k - \phi_k). \qquad (5.27.55)$$

The inequalities (5.27.41) and (5.27.42) remain valid.

C. Energy Theorems

Theorem. If a plate is in equilibrium under a given system of body and surface loads, then the total internal extensional energy ϵ_p, over the surface of the plate S, is equal to one-half of the work of the symmetrical part of the loads with respect to the median plane $x_3 = 0$, and the total flexure energy is equal to one-half of the work of the antisymmetrical part of the loads.

Mathematically

$$\int_S \rho(\mathbf{F} \cdot \mathbf{u} + L_3\phi)\, da + \oint_\mathcal{C} (\bar{t}_k \bar{u}_k + \bar{m}_3\phi)\, ds = 2 \int_S 2h\rho\bar{\epsilon}_p, \quad (5.27.56)$$

$$\int_S (\mathbf{T} \cdot \mathbf{v} + 2\rho h \mathbf{L} \cdot \boldsymbol{\phi} + F_3 w)\, da$$

$$+ \oint_\mathcal{C} (\bar{\mathbf{M}} \cdot \mathbf{v} + 2h\bar{\mathbf{m}} \cdot \boldsymbol{\phi} + 2h\bar{t}_3 w)\, ds = 2 \int \rho\epsilon_b\, da. \quad (5.27.57)$$

To prove (5.27.56), we form the first integral on the left by replacing $\mathbf{F}$ and L_3 from the expressions (5.27.5) and (5.27.8), i.e.,

$$V_p \equiv \int_S (\mathbf{F} \cdot \mathbf{u} + L_3\phi)\, da = - \int_S [\bar{t}_{kl,k}\bar{u}_l + (\bar{m}_{k3,k} + \epsilon_{kl3}\bar{t}_{kl})\phi]\, da.$$

Upon integration by parts and using Green–Gauss theorems, this gives

$$V_p = - \oint_\mathcal{C} (\bar{t}_k \bar{u}_k + \bar{m}_3\phi)\, ds + 2 \int_S 2\rho h \bar{\epsilon}_p\, da,$$

which proves the theorem.

The proof of (5.27.56) is similar. This theorem is the generalization of a theorem known as Clapeyron's theorem in classical plate theory. It finds important applications in obtaining approximate solutions and proving the uniqueness theorem for static solutions.

Theorem 3 (Uniqueness). If the conditions (5.27.41) and (5.27.42) are satisfied, $\bar{r}_3 \neq \phi_3$, $\tilde{r}_3 \neq (v_{2,1} - v_{1,2})$, $j \geq 0$, in a bounded domain S with boundary $\mathcal{C}$, there exists one $\bar{\mathbf{u}}$, one $\mathbf{v}$, one $\boldsymbol{\phi}$, one w, and one ϕ, all twice differentiable with respect to x_1, x_2 in $S + \mathcal{C}$ over $0 \leq t \leq \infty$, which satisfy (5.27.15)–(5.27.19), the initial and boundary conditions (5.27.20) and (5.27.21).

PROOF. Suppose that the contrary is valid and two solutions $\bar{\mathbf{u}}^{(\alpha)}$, $\mathbf{v}^{(\alpha)}$, $\boldsymbol{\phi}^{(\alpha)}$, $w^{(\alpha)}$, and ϕ^α, $\alpha = 1, 2$, exist satisfying (5.27.15)–(5.27.19) and (5.27.20) and (5.27.21).

Let

$$\bar{\mathbf{u}} = \bar{\mathbf{u}}^{(1)} - \bar{\mathbf{u}}^{(2)}, \mathbf{v} = \mathbf{v}^{(1)} - \mathbf{v}^{(2)}, \qquad \boldsymbol{\phi} = \boldsymbol{\phi}^{(1)} - \boldsymbol{\phi}^{(2)},$$
$$w = w^{(1)} - w^{(2)}, \qquad \phi = \phi^{(1)} - \phi^{(2)}. \quad (5.27.58)$$

Then clearly $\bar{\mathbf{u}}$, $\mathbf{v}$, $\boldsymbol{\phi}$, w, and ϕ satisfy (5.27.15)–(5.27.19) with $\mathbf{f} = \mathbf{L} = \mathbf{T} = 0$, $L_3 = F_3 = 0$, $\mathbf{U}_0 = \mathbf{V}_0 = \hat{\mathbf{v}} = \hat{\boldsymbol{\phi}} = \boldsymbol{\nu} = 0$, $\Phi_0 = \nu_0 = \hat{w} = \tilde{w} = 0$, i.e., vanishing applied loads and initial values. Now multiply (5.27.5) by $\dot{\bar{u}}_l$, (5.27.8) by $\dot{\phi}$, and add the two equations leading to

$$\frac{d}{dt} \int_S 2\rho h (\epsilon_p + \tfrac{1}{2}\dot{\bar{\mathbf{u}}} \cdot \dot{\bar{\mathbf{u}}} + \tfrac{1}{2}j\dot{\phi}^2)\, da = \oint_{\mathcal{C}} (\bar{t}_k \dot{\bar{u}}_k + \bar{m}_3 \dot{\phi})\, ds. \quad (5.27.59)$$

This is the equation of the total extensional energy of the plate.

Similarly, by multiplying (5.27.6) by $\dot{w}$, (5.27.9) by $\dot{v}_l$, and (5.27.7) by $\dot{\phi}_l$ and adding, we obtain the total bending energy of the plate

$$\frac{d}{dt} \int_S 2\rho h \left(\epsilon_b + \frac{I}{2}\dot{\mathbf{v}} \cdot \dot{\mathbf{v}} + \tfrac{1}{2}j\dot{\boldsymbol{\phi}} \cdot \dot{\boldsymbol{\phi}} + \tfrac{1}{2}\dot{w}^2 \right)\, da$$
$$= \oint_{\mathcal{C}} (\mathbf{M} \cdot \dot{\mathbf{v}} + 2h\bar{\mathbf{m}} \cdot \dot{\boldsymbol{\phi}} + 2h\bar{t}_3 \dot{w})\, ds. \quad (5.27.60)$$

These two expressions can also be obtained by integrating the energy equation

$$\rho \dot{\epsilon} = t_{\alpha\beta}(\dot{u}_{\beta,\alpha} - \epsilon_{\alpha\beta\gamma}\dot{\phi}_\gamma) + m_{\alpha\beta}\dot{\phi}_{\beta,\alpha} \quad (5.27.61)$$

across the thickness.

From the boundary conditions, the line integrals over $\mathcal{C}$ vanish. But at $t = 0$, from the initial conditions, the integrals over S must also vanish, i.e.,

$$\int_S 2\rho h (\epsilon_p + \tfrac{1}{2}\dot{\bar{\mathbf{u}}} \cdot \dot{\bar{\mathbf{u}}} + \tfrac{1}{2}j\dot{\phi}^2)\, da = 0, \quad (5.27.62)$$

$$\int_S 2\rho h \left(\epsilon_b + \frac{I}{2}\dot{\mathbf{v}} \cdot \dot{\mathbf{v}} + \tfrac{1}{2}j\dot{\boldsymbol{\phi}} \cdot \dot{\boldsymbol{\phi}} + \tfrac{1}{2}\dot{w}^2 \right)\, da = 0. \quad (5.27.63)$$

But both ϵ_p and ϵ_b are nonnegative definite. Consequently, we must have

$$\dot{\bar{\mathbf{u}}} = \dot{\mathbf{v}} = \dot{\boldsymbol{\phi}} = \mathbf{0}, \quad \dot{\phi} = \dot{w} = 0, \quad (5.27.64)$$

which shows that

$$\mathbf{u}^{(1)} = \mathbf{u}^{(2)}, \quad \mathbf{v}^{(1)} = \mathbf{v}^{(2)}, \quad \boldsymbol{\phi}^{(1)} = \boldsymbol{\phi}^{(2)}, \quad \phi^{(1)} = \phi^{(2)}, \quad w^{(1)} = w^{(2)},$$

and thus the proof of the theorem. Note that the extension problem and flexure problem independently possess unique solutions.

Clearly, if we set $\mathbf{v} = \mathbf{0}$ in the above equations, the theorem remains valid. Consequently, the uniqueness is valid for the lowest (zeroth)-order theory whose field equations are given by (5.27.47), (5.27.48), (5.27.53), and (5.27.54). $\square$

5.28   Flexural Waves

A. Phase Velocity of Flexural Waves[21]

Consider a harmonic flexural wave propagating in the $(x_1 = x)$-direction, i.e.,

$$w = w_0 \exp[i(\omega t + kx)], \quad \Phi = \Phi_0 \exp[i(\omega t + kx)], \qquad (5.28.1)$$

where w_0 and Φ_0 are constants. Substituting these into (5.27.23) and (5.27.24) with $P = 0, L_l = 0$, and negligible rotatory inertia $(\rho I = 0)$, we obtain the dispersion relations for the phase velocity $c = \omega/k$:

$$pc^4 - 2qc^2 + r = 0, \qquad (5.28.2)$$

where

$$p = 2h\rho^2 j \left[1 + \frac{D}{8h^3(G + \kappa/2)}\Lambda^{-2}\right]\Lambda^{-2},$$

$$q = \frac{\rho D}{8h^2}\left(j + \frac{\gamma}{G + \kappa/2}\right)\Lambda^{-4} + \rho\left(\frac{2\kappa Ghj + \kappa D}{G + \kappa/2} + h\gamma\right)\Lambda^{-2} + \frac{8\kappa G\rho h^3}{G + \kappa/2},$$

$$r = \frac{\gamma D}{4h^2}\Lambda^{-4} + \frac{2\kappa G}{G + \kappa/2}(D + 2h\gamma)\Lambda^{-2}, \quad \Lambda \equiv (2kh)^{-1}. \qquad (5.28.3)$$

Here Λ is the ratio of the wavelength to the thickness of the plate. The two roots of (5.28.2) are given by

$$c^2(\Lambda) = \frac{1}{p}(q \pm \sqrt{q^2 - pr}). \qquad (5.28.4)$$

When the microinertia j vanish, we will have $p = 0$ and (5.28.2) gives $c^2 = r/2q$. Assuming that j, or equivalently pr/q^2 is small as compared to unity, (5.28.4) gives approximately

$$c_1^2(\Lambda) = r/2q, \quad c_2^2(\Lambda) = \frac{2q}{p} - c_1^2(\Lambda). \qquad (5.28.5)$$

In Figure 5.28.1, the phase velocities $c_1(\Lambda)$ and $c_2(\Lambda)$ are plotted against Λ^{-1}. To get an idea about the trend, we take

$$c_1(0)/c_s = \frac{1}{\sqrt{1 + \epsilon}}, \quad c_1(\infty) = 0,$$

$$c_2(0)/c_s = 1 + \frac{1}{\sqrt{\epsilon(1 + \epsilon)}}, \quad c_2(\infty) = \infty, \qquad (5.28.6)$$

where

$$c_s \equiv \left(\frac{G + \kappa/2}{\rho}\right)^{1/2}, \quad \epsilon = \frac{j}{\gamma}(G + \kappa/2). \qquad (5.28.7)$$

---

[21] Eringen [1967d].

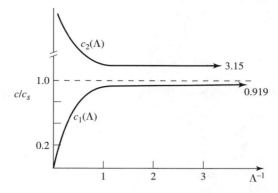

FIGURE 5.28.1. Flexural Wave Speeds in Micropolar Plates

Here c_s is the speed of equivolumnal elastic waves. Flexural wave speed $c_1(\Lambda)$ is similar to that of the Timoshenko beam theory. By adjusting ϵ so that $c_1(0)$ becomes equal to the surface wave velocity (with $\mu' = \mu + \kappa/2$, $\nu = 0.25$, consequently $c_1/c_3 = 0.919$), we obtain an excellent fit for the observed phase velocity $c_1(\Lambda)$. For this case, ϵ is found to be

$$\epsilon = \frac{1}{(0.919)^2[1 + 0.932\kappa/(2\mu + \kappa)]^2} - 1. \qquad (5.28.8)$$

In the classical elasticity limit $\kappa = 0$, we get $\epsilon \simeq 0.1840$. In this case then, the velocity of infinitely short flexural waves agrees exactly with that of the Rayleigh surface waves.

For a long time, it was fashionable to introduce a fudge factor to account for the shear deformability effect. Micropolar theory eliminates such arbitrary factors. Moreover, even with the shear deformability effect, the microrotational waves propogating with the phase velocity $c_2(\Lambda)$ are missing in all previous beam theories.

Micropolar plate theory also predicts a second class of microrotational waves governed by (5.27.27). The speed of these waves, for vanishing rotatory inertia, is obtained by setting $\phi = \phi_0 \exp(i\omega t + ikx)$:

$$c^2 = I\frac{\alpha + \beta + \gamma}{8\rho h^2 j}\Lambda^{-2} + \frac{\kappa I + (\alpha + \beta + \gamma)h}{\rho h j} + \frac{8\kappa G h^2}{\rho j(G + \kappa/2)}\Lambda^2. \qquad (5.28.9)$$

From this it is clear that both short and long waves are transmitted by large speeds. These waves are also dispersive.

B. Phase Velocity for the Lowest-Order Plate Theory

The phase velocity of flexural waves is obtained by substituting

$$w = w_0 \exp(ikx + i\omega t), \quad \phi_l = \phi_0 \exp(ikx + i\omega t)\delta_{l3}, \qquad (5.28.10)$$

into (5.27.53) and (5.27.54), with $F_3 = L_l = 0$, i.e.,

$$[-(G + \kappa/2)k^2 + \rho\omega^2]w_0 = 0,$$
$$(-\gamma k^2 - 2\kappa + \rho j\omega^2)\phi_0 = 0. \qquad (5.28.11)$$

From these there follows the dispersion relations

$$\omega_2^2 = \frac{G + \kappa/2}{\rho}k^2, \quad \omega_4^2 = \frac{2\kappa}{\rho j} + \frac{\gamma}{\rho j}k^2. \qquad (5.28.12)$$

These give, respectively, the phase velocities of the transverse acoustic and optic branches. It should be noted that $\omega_4(k = 0) = 2\kappa/\rho j$ is the cut-off frequency of the optical waves found for plane harmonic waves in the three-dimensional case.

C. Passage to the Classical Plate Theory

It is expected that, in the special case,

$$\kappa = 0, \quad \alpha = \beta = \gamma = 0, \quad L_k = 0, \quad j = 0, \qquad (5.28.13)$$

the present theory will go to the classical plate theory. In fact, this is the case, as can be seen from (5.27.31)–(5.27.36), in the case of negligible rotatory inertia and shear deformations. The only equation that survives is (5.27.31), which reduces to

$$-D\nabla^4 w - 2\rho h\ddot{w} + P = 0. \qquad (5.28.14)$$

For the case that accounts for the rotatory inertia and shear deformation, this limit can be read from (5.27.23).

Remark. Green and Naghdi [1967] employ a perturbation process using plate half-thickness as a perturbation parameter. This theory is equivalent to *The Lowest-Order Theory* discussed in Section 5.27B, except for an error in their equation $(4.20)_2$ corresponding to (5.27.51). However, it is important to note that, unlike the first-order theory, this theory does not go into the classical plate theory of flexure.

5.29   Other Contributions

Constanda [1974] derived a Galerkin representation for the solution of the bending of the micropolar plates of Eringen. By the method of potentials, he solved boundary value problems. Schiavone, in a series of papers, discussed existence and uniqueness problems. Existence theorems for extensional motions were presented in [1989a]; uniqueness in dynamical problems for bending in [1991]. A generalized Fourier approximation was presented in [1989b] and [1990]. Schiavone and Constanda [1989] gave a method by extending their previous work to prove existence and regular solutions for Dirichlet and Neumann-type problems in both interior and exterior domains. Dynamical problems were treated by Wang [1990] for both extensional and bending motions, also giving uncoupled equations for the treatment of other micropolar plate problems.

The micropolar plate theory brings special challenges to the structural engineering and mechanics community. It solves, in a natural fashion, the classical problem of shear deformation without any "fudge factor." It accounts for the rotational motions and, consequently, it provides optical branches and extends moderate wavelength regions. Much work remains for future researchers.

Chapter 5 Problems

5.1. Find the nonvanishing, independent components of the micropolar material moduli for the following cases:

   (a) cubic crystals;

   (b) hexagonal crystals; and

   (c) orthotropic solids.

5.2. Obtain material stability restrictions for the orthotropic solids.

5.3. For isotropic micropolar elastic solids, determine the independent invariants of the strain tensors $\mathfrak{C}_{KL}$ and Γ_{KL}.

5.4. Express field equations of micropolar elastic solids in spherical coordinates.

5.5. Express equations of motion and strain measures in spherical coordinates.

5.6. (Short-Term Paper) Find the symmetry group of the dynamical equations of micropolar elastic solids.

5.7. For experimental determination of the micropolar constants α, β and γ, some simple solutions of micropolar equations are required. Set up the necessary problems, obtain their solutions and discuss the possible experiments that may lead to the determination of α, β and γ.

5.8. Determine the dispersion relations of longitudinal micropolar waves reflected from the plane welded surfaces of two micropolar half-planes.

5.9. Determine the dispersion relations in a micropolar half-space overlayed by an ideal fluid.

5.10. By means of superposition, determine the displacement fields **u** and ϕ in a micropolar elastic solid of infinite extent, due to distributed constant pressure over a circular region in the interior of the solid.

5.11. Determine the stress field in the exterior of a spherical cavity in a micropolar elastic solid of infinite extent, due to uniform impulsive pressure at the surface of the cavity.

5.12. Obtain the dispersion relations in a micropolar plate due to:

(a) the longitudinal wave in the axial direction; and

(b) the flexural wave.

Compare with exact solutions.

5.13. (Short-Term Paper) Determine the singularity at the tip of a crack in a two-dimensional micropolar elastic solid.

5.14. (Short-Term Paper) Construct a micropolar shell theory following a similar step as in Section 5.27.

5.15. (Short-Term Paper) Obtain differential equations of a beam theory extending the Euler beam theory for buckling to micropolar beams.

5.16. (Short-Term Paper) Chopped fibers in a composite matrix are considered to change the microinertia tensor j_{ij}.

(a) Determine the micropolar elastic constants as functions of j_{ij}.

(b) If j_{ij} is considered to be a constitutive variable along with ϵ_{ij} and γ_{ij}, what would be the new forms of the constitutive equations that are linear in each of $\epsilon_{ij}, \gamma_{ij}$ and j_{ij}?

(c) The constitutive equations in (b) are restricted by the material stability regulations. Find these restrictions.

6
Microstretch Elasticity

6.0 Scope

This chapter is devoted to the discussion of linear microstretch elasticity.[1] The linear constitutive equations are obtained in Section 6.1, for both anisotropic and isotropic thermo-microstretch elastic solids. The field equations are given and boundary-initial value problems are formulated. The material stability condition (nonnegative strain energy) restricts the material moduli. In Section 6.2 we prove the uniqueness theorem.

Section 6.3 treats the problem of plane harmonic microstretch waves. Here are predicted two additional waves, over those predicted by micropolar elasticity: an acoustic mode and an optic mode with a cut-off frequency. With these two additional dispersion relations, microstretch elasticity models Bravais lattices with base. Thus, a linear lattice model discussed in Section 6.4 indicates the relevance of this model to the lattice structure. So do the microstructured continua which possess internal expansion and contraction (breathing) modes, in addition to micropolar modes.

In Section 6.5 we give the fundamental solutions of the field equation. Section 6.6 is devoted to the discussion of microstretch surface waves. At present, microstretch elasticity remains mostly dormant. A few other treatments exist, but unlike the micropolar theory, the number of studies in this field is rather limited. Some of the problems that have appeared in the

---

[1] Microstretch continuum theory was introduced by Eringen [1969a, 1971b, 1990].

literature are discussed briefly in Section 6.7. Here we have also indicated the potential areas of applications in this field.

6.1   Linear Constitutive Equations

Constitutive equations of microstretch elastic solids were obtained in Section 3.3. Here we present the linear theory.

Carrying out the linear approximations in the sense of (5.1.4), for microstretch, we write

$$
\begin{array}{ll}
x_{k,K} = (\delta_{kl} + u_{k.l})\delta_{lK} + O(\epsilon^2), & v_k = \dfrac{\partial u_k}{\partial t} + O(\epsilon^2), \\[2mm]
\overline{\chi}_{kK} = (\delta_{kl} - \epsilon_{klm}\phi_m)\delta_{lK} + O(\epsilon^2), & \nu_k = \dfrac{\partial \phi_k}{\partial t} + O(\epsilon^2), \\[2mm]
j = 1 + 3\phi + O(\epsilon^2), & \nu = \dfrac{\partial \phi}{\partial t} + O(\epsilon^2), \\[2mm]
\dfrac{\rho}{\rho_0} = 1 - u_{k,k} + O(\epsilon^2), & \theta = T_0 + T.
\end{array} \tag{6.1.1}
$$

With these, the deformation tensors (1.5.20) read

$$
\overline{\mathfrak{C}}_{KL} - \delta_{KL} \simeq \epsilon_{kl}\delta_{kl}\delta_{lL}, \quad \Gamma_{KL} \simeq \gamma_{kl}\delta_{kK}\delta_{lL}, \quad \Gamma_K \simeq 3\gamma_k\delta_{kK}, \tag{6.1.2}
$$

$$
\mathfrak{C}_{KL} - \delta_{KL} = 2e\delta_{KL},
$$

where

$$
\epsilon_{kl} = u_{l,k} + \epsilon_{lkm}\phi_m, \quad \gamma_{kl} = \phi_{k,l}, \quad \gamma_k = 3\phi_{,k},
$$

$$
e = 3\phi. \tag{6.1.3}
$$

We notice that ϵ_{kl} and γ_{kl} are also the linear strain measures of the micropolar elasticity.

A second-degree polynomial for $\Sigma \equiv \rho_0\psi$ in terms of the linear strain measures, and another for Φ, in terms of temperature gradient, leads to the linear constitutive equations

$$
\rho_0\psi = \Sigma = \Sigma_T + U, \qquad \Phi = \frac{1}{2T_0^2}K_{kl}T_{,k}T_{,l}, \tag{6.1.4}
$$

where

$$
\Sigma_T = \Sigma_0 - \rho_0\eta_0 T - \rho_0\frac{C_0}{T_0}T^2 - CT\phi - D_kT\phi_{,k} - A_{kl}T\epsilon_{kl} - B_{kl}T\gamma_{kl},
$$

$$
U \equiv \tfrac{1}{2}\rho_0\frac{C_0}{T_0}T^2 + \tfrac{1}{2}C^s\phi^2 + C_k^s\phi\phi_{,k} + A_{kl}^s\phi\epsilon_{kl} + B_{kl}^s\phi\gamma_{kl}
$$

$$
+ \tfrac{1}{2}C_{kl}^s\phi_{,k}\phi_{,l} + A_{klm}^s\phi_{,k}\epsilon_{lm} + B_{klm}^s\phi_{,k}\gamma_{lm} + \tfrac{1}{2}A_{klmn}\epsilon_{kl}\epsilon_{mn}
$$

$$
+ \tfrac{1}{2}B_{klmn}\gamma_{kl}\gamma_{mn} + C_{klmn}\epsilon_{kl}\gamma_{mn}. \tag{6.1.5}
$$

Here U is the strain energy density. The material moduli η_0, C_0, C, D_k, B_{kl}, C^s, C^s_k, A^s_{kl}, B^s_{kl}, C^s_{kl}, A^s_{klm}, B^s_{klm}, A_{klmn}, B_{klmn}, and C_{klmn} are functions of ρ_0 and T_0. The heat conduction coefficients are denoted by K_{kl}/T_0 and also depend on ρ_0 and T_0. The material moduli C, D_k and those carrying a superscript (s) in the expression of U are the extra elastic moduli, over the micropolar theory, for the anisotropic microstretch thermoelasticity.

From (6.1.4) and (6.1.5), the following symmetry regulations are observed:

$$K_{kl} = K_{lk}, \quad C^s_{kl} = C^s_{lk}, \quad A_{klmn} = A_{mnkl}, \quad B_{klmn} = B_{mnkl}. \tag{6.1.6}$$

From (3.3.15) and (3.3.17), there follow the linear constitutive equations

$$\eta = -\frac{1}{\rho_0}\frac{\partial \Sigma}{\partial T} = \eta_0 + \frac{C_0}{T_0}T + \frac{1}{\rho_0}(C\phi + D_k\phi_{,k} + A_{kl}\epsilon_{kl} + B_{kl}\gamma_{kl}),$$

$$t_{kl} = \frac{\partial \Sigma}{\partial \epsilon_{kl}} = -A_{kl}T + A^s_{kl}\phi + A^s_{mkl}\phi_{,m} + A_{klmn}\epsilon_{mn} + C_{klmn}\gamma_{mn},$$

$$m_{kl} = \frac{\partial \Sigma}{\partial \gamma_{lk}} = -B_{lk}T + B^s_{lk}\phi + B^s_{mlk}\phi_{,m} + C_{mnlk}\epsilon_{mn} + B_{lkmn}\gamma_{mn},$$

$$s - t = \frac{\partial \Sigma}{\partial \phi} = -CT + C^s\phi + C^s_k\phi_{,k} + A^s_{kl}\epsilon_{kl} + B^s_{kl}\gamma_{kl},$$

$$m_k = \frac{\partial \Sigma}{\partial \phi_{,k}} = -D_kT + C^s_k\phi + C^s_{kl}\phi_{,l} + A^s_{klm}\epsilon_{lm} + B^s_{klm}\gamma_{lm},$$

$$q_k = \frac{K_{kl}}{T_0}T_{,l}. \tag{6.1.7}$$

In the linear theory, equations of conservation of mass and microinertia are dropped and ρ, j_{kl}, and j_0 are considered constants, so that we only need to employ (2.2.40a)–(2.2.42a). This last equation is equivalent to (3.3.19). Thus, we need only the following four equations of balance laws:

$$t_{kl,k} + \rho(f_l - \ddot{u}_l) = 0 \quad \text{in } \mathcal{V} - \sigma, \quad \text{(6.1.8-a)}$$

$$[[t_{kl} - \rho\ v_l(v_k - u_k)]]\,n_k = 0 \quad \text{on } \sigma, \quad \text{(6.1.8-b)}$$

$$m_{kl,k} + \epsilon_{lmn}t_{mn} + \rho(l_l - j_{lm}\ddot{\phi}_m) \quad \text{in } \mathcal{V} - \sigma, \quad \text{(6.1.9-a)}$$

$$[[m_{kl} + \rho j_{pl}\nu_p(v_k - u_k)]]\,n_k = 0 \quad \text{on } \sigma, \quad \text{(6.1.9-b)}$$

$$m_{k,k} + t - s + \rho(l - \tfrac{1}{2}j_0\ddot{\phi}) = 0 \quad \text{in } \mathcal{V} - \sigma, \text{(6.1.10-a)}$$

$$\left[\left[m_k - \frac{1}{2}\rho j_0\nu(v_k - u_k)\right]\right]n_k = 0 \quad \text{on } \sigma, \quad \text{(6.1.10-b)}$$

$$-\rho\dot{\epsilon} + t_{kl}(v_{l,k} + \epsilon_{lkr}\nu_r)$$
$$+m_{kl}\nu_{l,k} + m_k\nu_{,k} + (s - t)\nu + q_{k,k} + \rho h = 0, \quad \text{in } \mathcal{V} - \sigma, \text{(6.1.11-a)}$$

$$\left\lVert t_{kl}v_l + m_{kl}v_l + m_k v + q_k \right.$$

$$-\rho\left(\epsilon + \frac{1}{2}\boldsymbol{v}\cdot\boldsymbol{v} + \frac{1}{2}j_{mn}v_m v_n + \frac{1}{4}j_0 v^2\right)$$

$$\left.\times(v_k - u_k)\right\rVert n_k = 0, \quad \text{on } \sigma. \qquad (6.1.11\text{-b})$$

The second set of equations (6.1.8-b), (6.1.9-b), (6.1.10-b), and (6.1.11-b) are the jump conditions on a discontinuity surface σ, which may be sweeping the body with its own velocity u_k.

Substituting from (6.1.7) into the balance equations (6.1.8-a)–(6.1.11-a), we obtain the field equation. Here, it is simpler to employ (3.3.19) which is equivalent to (6.1.11-a).

$$-A_{kl}T_{,k} + A^s_{kl}\phi_{,k} + A^s_{mkl}\phi_{,mk} + A_{klmn}(u_{n,mk} + \epsilon_{nmp}\phi_{p,k})$$
$$+ C_{klmn}\phi_{m,nk} + \rho(f_l - \ddot{u}_l) = 0, \qquad (6.1.12)$$

$$-B_{lk}T_{,k} + B^s_{lk}\phi_{,k} + B^s_{mlk}\phi_{,mk} + C_{mnlk}(u_{n,mk} + \epsilon_{nmp}\phi_{p,k})$$
$$+ B_{lkmn}\phi_{m,nk} + \epsilon_{lmn}[-A_{mn}T + A^s_{mn}\phi + A^s_{pmn}\phi_{,p}$$
$$+ A_{mnpq}(u_{q,p} + \epsilon_{qpr}\phi_r) + C_{mnpq}\phi_{p,q}] + \rho(l_l - j_{lm}\ddot{\phi}_m) = 0, \qquad (6.1.13)$$

$$-D_k T_{,k} + C^s_{kl}\phi_{,lk} + A^s_{klm}(u_{m,lk} + \epsilon_{mlr}\phi_{r,k}) + B^s_{klm}\phi_{l,mk}$$
$$+ CT - C^s\phi - A^s_{kl}(u_{l,k} + \epsilon_{lkm}\phi_m) - B^s_{kl}\phi_{k,l} + \rho(l - \frac{1}{2}j_0\ddot{\phi}) = 0, \qquad (6.1.14)$$

$$\rho_0 C_0 \dot{T} + T_0[C\dot{\phi} + D_k\dot{\phi}_{,k} + A_{kl}(\dot{u}_{l,k} + \epsilon_{lkm}\phi_m)$$
$$+ B_{kl}\dot{\phi}_{k.l}] - \frac{K_{kl}}{T_0}T_{,lk} - \rho h = 0. \qquad (6.1.15)$$

Equations (6.1.12)–(6.1.15) constitute eight field equations of anisotropic thermo-microstretch elastic solids needed to determine the eight unknown functions u_k, ϕ_k, ϕ, and T.

A set of boundary conditions accompanying these equations follows from the jump conditions (6.1.8-b)–(6.1.11-b)

$$
\begin{array}{llll}
u_k = \hat{u}_k & \text{on } \partial\mathcal{V}_u; & t_{kl}n_k = \hat{t}_{(n)l} & \text{on } \partial\mathcal{V}_t = \partial\mathcal{V} - \partial\mathcal{V}_u, \\
\phi_k = \hat{\phi}_k & \text{on } \partial\mathcal{V}_\phi; & m_{kl}n_k = \hat{m}_{(n)l} & \text{on } \partial\mathcal{V}_m = \partial\mathcal{V} - \partial\mathcal{V}_\phi, \\
\phi = \hat{\phi} & \text{on } \partial\mathcal{V}_\phi; & m_k n_k = \hat{\lambda}_{(n)} & \text{on } \partial\mathcal{V}_\lambda = \partial\mathcal{V} - \partial\mathcal{V}_\phi, \\
T = \hat{T} & \text{on } \partial\mathcal{V}_T; & q_k n_k = \hat{q}_{(n)} & \text{on } \partial\mathcal{V}_q = \partial\mathcal{V} - \partial\mathcal{V}_T.
\end{array}
$$
$$(6.1.16)$$

Initial conditions (Cauchy data) are

$$
\begin{aligned}
\boldsymbol{u}(\boldsymbol{x},0) &= \boldsymbol{u}^0(\boldsymbol{x}), & \dot{\boldsymbol{u}}(\boldsymbol{x},0) &= \boldsymbol{v}^0(\boldsymbol{x}), \\
\boldsymbol{\phi}(\boldsymbol{x},0) &= \boldsymbol{\phi}^0(\boldsymbol{x}), & \dot{\boldsymbol{\phi}}(\boldsymbol{x},0) &= \boldsymbol{\nu}^0(\boldsymbol{x}), \\
\phi(\boldsymbol{x},0) &= \phi^0(\boldsymbol{x}), & \dot{\phi}(\boldsymbol{x},0) &= \nu^0(\boldsymbol{x}), \\
T(\boldsymbol{x},0) &= T^0(\boldsymbol{x}), & \boldsymbol{x} &\in \mathcal{V},
\end{aligned}
$$
$$(6.1.17)$$

where quantities on the right-hand sides of (6.1.16) and (6.1.17) are pre-scribed.

A. Isotropic Solids

For materials possessing various symmetry groups, constitutive equations are simplified by subjecting material moduli to the appropriate symmetry regulations. Here we obtain constitutive equations of isotropic microstretch thermoelastic solids. In this case, odd-order material tensors vanish. Also γ_{kl} is an axial tensor, while ϵ_{kl}, ϕ_k, ϕ, and T are polar tensors; B_{kl}, B_{kl}^s, and C_{klmn} vanish. Consequently, we have

$$D_k = C_k^s = 0, \quad B_{kl} = B_{kl}^s = 0, \quad A_{klm}^s = 0, \qquad B_{klm}^s = b_0 \epsilon_{klm},$$
$$C_{klmn} = 0, \quad A_{kl} = \beta_0 \delta_{kl}, \quad C = \beta_1,$$
$$C^s = \lambda_1, \quad A_{kl}^s = \lambda_0 \delta_{kl}, \quad C_{kl}^s = a_0 \delta_{kl},$$
$$A_{klmn} = \lambda \delta_{kl} \delta_{mn} + (\mu + \kappa) \delta_{km} \delta_{ln} + \mu \delta_{kn} \delta_{lm},$$
$$B_{klmn} = \alpha \delta_{kl} \delta_{mn} + \beta \delta_{kn} \delta_{lm} + \gamma \delta_{km} \delta_{ln}. \tag{6.1.18}$$

The free energy Σ and the dissipation potential Φ are given by

$$\Sigma = \Sigma_T + U, \qquad \Phi = \frac{K}{2T_0^2} T_{,k} T_{,k}, \tag{6.1.19}$$

where

$$\Sigma_T = \Sigma_0 - \rho_0 \eta_0 T - \rho_0 \frac{C_0}{2T_0} T^2 - \beta_1 T \phi - \beta_0 T u_{k,k},$$
$$U = \frac{\rho_0 C_0}{2T_0} T^2 + U_\epsilon + U_\gamma,$$
$$U_\epsilon = \tfrac{1}{2} \lambda_1 \phi^2 + \lambda_0 \phi \epsilon_{kk} + \tfrac{1}{2} a_0 \phi_{,k} \phi_{,k} + \tfrac{1}{2} [\lambda \epsilon_{kk} \epsilon_{ll} + \mu \epsilon_{kl} \epsilon_{lk} + (\mu + \kappa) \epsilon_{kl} \epsilon_{kl}],$$
$$U_\gamma = \tfrac{1}{2} (\alpha \gamma_{kk} \gamma_{ll} + \beta \gamma_{kl} \gamma_{lk} + \gamma \gamma_{kl} \gamma_{kl}) + b_0 \epsilon_{klm} \phi_{,k} \gamma_{lm}. \tag{6.1.20}$$

Constitutive equations are given by

$$\eta = \eta_0 + \frac{C_0}{T_0} T + \frac{\beta_1}{\rho_0} \phi + \frac{\beta_0}{\rho_0} u_{k,k},$$
$$t_{kl} = (-\beta_0 T + \lambda_0 \phi + \lambda u_{r,r}) \delta_{kl} + \mu(u_{k,l} + u_{l,k}) + \kappa(u_{l,k} - \epsilon_{klr} \phi_r),$$
$$m_{kl} = \alpha \phi_{r,r} + \beta \phi_{k,l} + \gamma \phi_{l,k} + b_0 \epsilon_{mlk} \phi_{,m},$$
$$m_k = a_0 \phi_{,k} + b_0 \epsilon_{klm} \phi_{l,m},$$
$$s - t = -\beta_1 T + \lambda_1 \phi + \lambda_0 u_{k,k},$$
$$q_k = \frac{K}{T_0} T_{,k}. \tag{6.1.21}$$

The field equations of isotropic solids follow from (6.1.12)–(6.1.15)

$$-\beta_0 T_{,l} + \lambda_0 \phi_{,l} + (\lambda + \mu) u_{k,lk} + (\mu + \kappa) u_{l,kk} + \kappa \epsilon_{lkm} \phi_{m,k} + \rho(f_l - \ddot{u}_l) = 0,$$

$$(\alpha + \beta)\phi_{k,lk} + \gamma\phi_{l,kk} + \kappa\epsilon_{lmn}u_{n,m} - 2\kappa\phi_l + \rho(l_l - j\ddot{\phi}_l) = 0,$$

$$a_0\nabla^2\phi + \beta_1 T - \lambda_1\phi - \lambda_0 u_{k,k} + \rho(l - \tfrac{1}{2}j_0\ddot{\phi}) = 0,$$

$$\rho_0 C_0\dot{T} + \beta_1 T_0\dot{\phi} + \beta_0\dot{u}_{k,k} - \frac{K}{T_0}\nabla^2 T - \rho h = 0,$$

$$(6.1.22)$$

where we also wrote $j_{kl} = j\delta_{kl}$, assuming microisotropy. Boundary and initial conditions (6.1.16) and (6.1.17) remain valid.

B. Nonnegative Strain Energy

Material stability requires that the strain energy U given by $(6.1.5)_2$ for anisotropic solids and $(6.1.20)_2$ for isotropic solids must be positive semidefinite for all ϕ, ϕ_k, ϵ_{kl}, and γ_{kl}. U_ϵ and U_γ are uncoupled, so that each must be nonnegative independently. For $\boldsymbol{\epsilon} = \mathbf{0}$ and $\phi_{,k} = 0$, from $U_\epsilon \geq 0$, it follows that $\lambda_1 \geq 0$. Now write U_ϵ in the form

$$U_\epsilon = \tfrac{1}{2}a_0\phi_{,k}\phi_{,k} + \tfrac{1}{2}\{[(\lambda_0/\lambda_1^{1/2})\epsilon_{kk} + \lambda_1^{1/2}\phi]^2 + (\lambda - \lambda_0^2/\lambda_1)\epsilon_{kk}\epsilon_{ll}$$
$$+\mu\epsilon_{kl}\epsilon_{lk} + (\mu + \kappa)\epsilon_{kl}\epsilon_{kl}\}. \qquad (6.1.23)$$

The first term is independent of the terms enclosed in braces. Hence $a_0 \geq 0$. Moreover, the first term inside the braces is nonnegative and it is also independent of the remaining terms. The remaining terms in the braces are identical to the corresponding strain energy of the micropolar elasticity with λ replaced by $\lambda - \lambda_0^2/\lambda_1$. Hence, we have

$$3\lambda + 2\mu + \kappa \geq 3\lambda_0^2/\lambda_1, \qquad 2\mu + \kappa \geq 0, \qquad \kappa \geq 0.$$

For the condition $U_\gamma \geq 0$, we have the inequalities (5.3.9) obtained for the micropolar elasticity. Hence we have

Theorem (Nonnegative Strain Energy). The necessary and sufficient conditions for the strain energy to be nonnegative are

$$
\begin{array}{ccc}
3\lambda + 2\mu + \kappa \geq 3\lambda_0^2/\lambda_1, & 2\mu + \kappa \geq 0, & \kappa \geq 0, \\
3\alpha + \beta + \gamma \geq 0, & \gamma + \beta \geq 0, & \gamma - \beta \geq 0, \quad (6.1.24) \\
a_0 \geq 0, & \lambda_1 \geq 0, & C_0 \geq 0.
\end{array}
$$

C. Boundary-Initial Value Problems

The field equations (6.1.22) may be expressed in vectorial forms:

$$-\beta_0\nabla T + \lambda_0\nabla\phi + (\lambda + 2\mu + \kappa)\nabla\nabla \cdot \boldsymbol{u} - (\mu + \kappa)\nabla \times \nabla \times \boldsymbol{u}$$
$$+\kappa\nabla \times \boldsymbol{\phi} + \rho(\boldsymbol{f} - \ddot{\boldsymbol{u}}) = \mathbf{0}, \quad (6.1.25)$$

$$(\alpha + \beta + \gamma)\nabla\nabla \cdot \boldsymbol{\phi} - \gamma\nabla \times \nabla \times \boldsymbol{\phi} + \kappa\nabla \times \mathbf{u} - 2\kappa\boldsymbol{\phi}$$
$$+ \rho(\boldsymbol{l} - j\ddot{\boldsymbol{\phi}}) = \mathbf{0}, \quad (6.1.26)$$
$$a_0\nabla^2\phi + \beta_1 T - \lambda_1\phi - \lambda_0\nabla \cdot \boldsymbol{u} + \rho(l - \tfrac{1}{2}\mathrm{J}_0\ddot{\phi}) = 0, \quad (6.1.27)$$
$$\rho_0 C_0\dot{T} + \beta_1 T_0\dot{\phi} + \beta_0\nabla \cdot \dot{\boldsymbol{u}} - \frac{K}{T_0}\nabla^2 T - \rho h = 0. \quad (6.1.28)$$

These coupled partial differential equations under the boundary-initial conditions listed in (6.6.16) and (6.1.17) constitute the *boundary-initial value problems* of isotropic thermo-microstretch elasticity.

6.2   Uniqueness Theorem

The uniqueness theorem may be proven following the method presented in Section 5.7. However, it is simpler and pedagogically interesting to use a more direct approach.[2] To this end, we begin with the energy equation.

The local energy equation (3.3.9), with $_D\eta = 0$, reads

$$-\rho\theta\dot{\eta} + \nabla \cdot \boldsymbol{q} + \rho h = 0. \quad (6.2.1)$$

The global energy equation is given by (2.1.9), i.e.,

$$\frac{d}{dt}\int_V \rho(\epsilon + K)\,dv = \int_{\partial V}(t_{kl}v_l + m_{kl}\nu_l + m_k\nu + q_k)\,da_k$$
$$+ \int_V \rho(f_k v_k + l_k\nu_k + l\nu + h)\,dv. \quad (6.2.2)$$

From (6.1.4) and $\psi = \epsilon - \theta\eta$, it may be deduced that

$$\epsilon = \frac{U}{\rho_0} + \frac{\Sigma_0}{\rho_0} + \frac{C_0 T^2}{2T_0} + T_0\eta, \quad (6.2.3)$$

where we used $(6.1.5)_1$ and $(6.1.7)_1$. Substituting (6.2.3) into (6.2.2) and using (6.2.1), we have

$$\frac{d}{dt}\int_V \rho\left(\frac{U}{\rho_0} + \frac{C_0 T^2}{2T_0} + K\right)dv + \int_V \frac{q_k T_{,k}}{T_0}\,dv$$
$$= \int_{\partial V}\left(t_{kl}v_l + m_{kl}\nu_l + m_k\nu + \frac{q_k T}{T_0}\right)da_k$$
$$+ \int_V \rho\left(f_k v_k + l_k\nu_k + l\nu + \frac{hT}{T_0}\right)dv. \quad (6.2.4)$$

The uniqueness theorem may now be stated as:

---

[2]Eringen [1971, 1990].

Theorem (Uniqueness).   Let the boundary and initial conditions (6.1.16), (6.1.17), $\rho > 0$, $j \geq 0$, $j_{kl}p_k p_l \geq 0$, and $U \geq 0$ be satisfied in a bounded, regular domain $\mathcal{V}$ of space with boundary $\partial \mathcal{V}$, occupied by the body. There exist at most one $\boldsymbol{u}(\boldsymbol{x}, t)$, $\phi_k(\boldsymbol{x}, t)$, $\phi(\boldsymbol{x}, t)$, and $T(\boldsymbol{x}, t)$, each twice differentiable with respect to $\boldsymbol{x}$, in $\mathcal{V} + \partial \mathcal{V}$ and $0 \leq t \leq \infty$, which satisfy (6.1.12)–(6.1.15).

PROOF.   Suppose that the contrary is valid and two solutions $\boldsymbol{u}^{(\alpha)}$, $\phi_k^{(\alpha)}$, $\phi^{(\alpha)}$, and T^α ($\alpha = 1, 2$), exist satisfying the equations mentioned in the theorem. Let

$$\boldsymbol{u} = \boldsymbol{u}^{(1)} - \boldsymbol{u}^{(2)}, \qquad \boldsymbol{\phi} = \boldsymbol{\phi}^{(1)} - \boldsymbol{\phi}^{(2)},$$
$$\phi = \phi^{(1)} - \phi^{(2)}, \qquad T = T^{(1)} - T^{(2)}, \tag{6.2.5}$$

then $\boldsymbol{u}$, $\boldsymbol{\phi}$, ϕ, and T satisfy (6.1.12)–(6.1.15) with $\boldsymbol{f} = \boldsymbol{0}, \boldsymbol{l} = 0, l = 0, h = 0$, and the homogenous boundary conditions. This means that the right-hand side of (6.2.4) vanishes

$$\frac{d}{dt} \int_{\mathcal{V}} \rho \left(\frac{1}{\rho_0} U + \frac{C_0 T^2}{T_0} + K \right) dv = - \int_{\mathcal{V}} \frac{q_k T_{,k}}{T_0}. \tag{6.2.6}$$

In view of the entropy inequality $q_k T_{,k}/T_0 \geq 0$ and so

$$\frac{d}{dt} \int_{\mathcal{V}} \rho \left(\frac{1}{\rho_0} U + \frac{C_0 T^2}{T_0} + K \right) dv \leq 0. \tag{6.2.7}$$

Using the initial conditions, this integrates into

$$\int_{\mathcal{V}} \rho \left(\frac{1}{\rho_0} U + \frac{C_0 T^2}{T_0} + K \right) dv \leq 0. \tag{6.2.8}$$

Now each term in the integral is nonnegative. Consequently, the only possible solution is

$$\boldsymbol{u} = \boldsymbol{0}, \qquad \boldsymbol{\phi} = \boldsymbol{0}, \qquad \phi = 0, \qquad T = 0. \tag{6.2.9}$$

Hence the proof of the theorem.   □

Remark. In the static case:

(i) The first boundary-value problem (where $\boldsymbol{u}$, $\boldsymbol{\phi}$, ϕ, and T are specified on $\partial \mathcal{V}$) has at most one solution; and

(ii) any two solutions of the second boundary-value problem (where tractions and $q_{(n)}$ are specified) differ only by a rigid displacement characterized by

$$u_k^0 = \epsilon_{ijk} a_j x_k + b_i, \qquad \phi_k^0 = a_k, \quad \phi^0 = 0 \quad \text{on } \partial \mathcal{V}, \tag{6.2.10}$$

where a_k and b_k are arbitrary constants.

Existence theorems were given by Ieşan and Quintanilla [1994] and Bofill and Quintanilla [1995].

6.3   Plane Harmonic Microstretch Waves

The decompositions (5.11.1) of the displacement and microrotation vectors are valid for the linear microstretch continuum. The resulting field equation, for the nonthermal case, follows from (6.1.25)–(6.1.28). The field equations for τ, U, and Φ are identical to those given by (5.11.3). For potentials σ and ϕ, we have the following equations:

$$\lambda_0\phi + (\lambda + 2\mu + \kappa)\nabla^2\sigma - \rho\ddot{\sigma} = 0,$$
$$a_0\nabla^2\phi - \lambda_1\phi - \lambda_0\nabla^2\sigma - \tfrac{1}{2}\rho j\ddot{\phi} = 0. \qquad (6.3.1)$$

Applied loads and temperature are assumed to vanish.

For plane harmonic waves, (6.3.1) leads to the dispersion relations

$$(\omega^2 - k^2 c_1^2)(\omega^2 - mk^2 - n) - pk^2 = 0, \qquad (6.3.2)$$

where

$$m = \frac{2a_0}{\rho j}, \quad n \equiv \frac{2\lambda_1}{\rho j}, \quad p = \frac{2\lambda_0^2}{\rho^2 j}, \quad c_1^2 = \frac{\lambda + 2\mu + \kappa}{\rho}. \qquad (6.3.3)$$

According to (6.1.24), m, n, p, and c_1^2 are all nonnegative. Consequently, both roots (6.3.2) are real and one root is nonnegative. The sign of the second root is found by comparing the radical with the first term. This shows that the second root will also be nonnegative if

$$k^2 \geq \frac{\lambda_0^2}{a_0(\lambda + 2\mu + \kappa)} - \frac{\lambda_1}{a_0}. \qquad (6.3.4)$$

The two roots for ω are given by

$$\left.\begin{array}{c}\omega_1 \\ \omega_2\end{array}\right\} = \frac{1}{\sqrt{2}}\{(m + c_1^2)k^2 + n \mp [(mk^2 - c_1^2k^2 + n)^2 + 4pk^2]^{1/2}\}^{1/2}. \quad (6.3.5)$$

For $k = 0$, $\omega_1 = 0$, and $\omega_2 = \sqrt{n}$. Hence we have one acoustic mode represented by ω_1 and one optic mode with ω_2 having a cut-off frequency

$$\omega_{cr} = \sqrt{n} = (2\lambda_1/\rho j)^{1/2}. \qquad (6.3.6)$$

By measuring ω_{cr}, we can determine the material constant λ_1. The optic mode is coupled with the acoustic mode, so that vanishing ϕ implies vanishing σ.

6.4   A Lattice Model for Microstretch Continuum[3]

Dilatational motions of a microstretch continuum can be modeled by a one-dimensional linear chain consisting of two different masses M and m,

---

[3]This section is new, not published before.

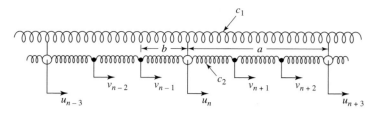

FIGURE 6.4.1. Lattice Model for Micro-stretch Solids

attached to each other by two different linear springs. The large masses M are attached to each other by two different linear springs (spring constant c_1), at equal intervals a. The interval between two large adjacent masses is shared equally by two small masses m, attached to each other and to large masses by linear springs of spring constant c_2, (Figure 6.4.1).

Equation of motions of M and the two adjacent masses m are:

$$M\ddot{u}_n = c_1(u_{n+3} + u_{n-3} - 2u_n) + c_2(v_{n+1} + v_{n-1} - 2u_n),$$
$$m\ddot{v}_{n+1} = c_2(u_n + v_{n+2} - 2v_{n+1}),$$
$$m\ddot{v}_{n-1} = c_2(u_n + v_{n-2} - 2v_{n-1}). \tag{6.4.1}$$

Here we are interested in microdilatational motions so that the last two equations are added, to obtain the equation of microdilatation (the centroidal motions of small masses at two sides of the large mass)

$$m(\ddot{v}_{n+1} + \ddot{v}_{n-1}) = c_2(2u_n + v_{n+2} + v_{n-2} - 2v_{n+1} - 2v_{n-1}). \tag{6.4.2}$$

The passage to continuum is made by using expansions of the type

$$u_{n\pm m}(t) = u(x \pm mb, t) = u(x,t) \pm mbu_x + \tfrac{1}{2}(mb)^2 u_{xx} + \dots.$$

Substituting this into $(6.4.1)_1$ and $(6.4.2)$, we obtain

$$M\ddot{u} = c_1 a^2 u_{xx} + 2c_2(v - u),$$
$$m\ddot{v} = c_2(u - v + b^2 v_{xx}), \tag{6.4.3}$$

where we set $u_n = u$, $v_n = v$, and $3b = a$.

We set

$$u = \sigma_x, \qquad v - u = \alpha\phi_x, \tag{6.4.4}$$

where α is to be determined. With these, $(6.4.3)$ may be transformed into the form

$$M\ddot{\sigma} = c_1 a^2 \sigma_{xx} + 2c_2\alpha\phi,$$
$$m\ddot{\phi} = c_2 b^2 \phi_{xx} - \frac{mc_1 a^2}{M\alpha}\left(1 - \frac{c_2 b^2 M}{c_1 a^2 m}\right)\sigma_{xx} - c_2\left(1 + \frac{2m}{M}\right)\phi. \tag{6.4.5}$$

Comparing these equations with (6.3.1), we identify

$$M/a^3 = \rho, \qquad c_1/a = \lambda + 2\mu + \kappa,$$

$$2c_2\alpha/a^3 = \lambda_0, \qquad m = \frac{1}{2}\rho j\beta,$$

$$\left(1 + \frac{2m}{M}\right)c_2 = \lambda_1\beta, \qquad c_2 b^2 = a_0\beta, \qquad \frac{mc_1 a^2}{Ma}\left(1 - \frac{Mc_2 b^2}{mc_1 a^2}\right) = \lambda_0\beta, \tag{6.4.6}$$

where β is to be determined. From these, it follows that

$$\alpha = \lambda_0 a^3/2c_2, \qquad \beta = 2m/\rho j,$$

$$\lambda_1 = \frac{\rho j c_2}{2m}\left(1 + \frac{2m}{M}\right), \qquad \lambda_0 = \pm\frac{1}{a^2}\left[c_1 c_2 j\left(1 - \frac{Mb^2 c_2}{ma^2 c_1}\right)\right]^{1/2},$$

$$a_0 = \frac{1}{2}\frac{c_2 b^2 Mj}{ma^3}. \tag{6.4.7}$$

These results show that $\lambda_1 \geq 0$, $a_0 \geq 0$, $\lambda + 2\mu + \kappa \geq 0$; in accordance with (6.1.24) and the existence of the acoustical mode. Although here $a = 3b$, in general, a can be any multiple of b by considering a larger number of small masses in the interval a.

We already know (see Section 5.12) that the micropolar modes of the microstretch continua can be modeled with linear chain. Consequently, the microstretch model is the continuum limit of a lattice model when the lattice possesses internal degrees of freedom represented by small masses, e.g., lattice with base. This situation is valid in more general forms, for the micromorphic continua where the internal degrees of freedom contain microshears as well.

The dispersion relations for the lattice model are obtained by setting

$$u_{n\pm m}(t) = u_0 \exp[i(\pm kmb - \omega t)]. \tag{6.4.8}$$

Equations $(6.4.1)_1$ and $(6.4.2)$ lead to

$$[M\omega^2 + 2c_1(\cos 3kb - 1) - 2c_2]u_0 + 2c_2 v_0 \cos kb = 0,$$

$$c_2 u_0 + [m\omega^2 \cos kb + c_2 \cos 2kb - 2c_2 \cos kb]v_0 = 0. \tag{6.4.9}$$

The coefficient determinant gives the dispersion relations

$$\omega^4 - (p + q)\omega^2 + pq - \frac{c_2^2}{mM} = 0, \tag{6.4.10}$$

where

$$p \equiv \frac{2c_2}{M} + \frac{2c_1}{M}(1 - \cos 3kb),$$

$$q \equiv \frac{2c_2}{m} - \frac{c_2 \cos 2kb}{m \cos kb}. \tag{6.4.11}$$

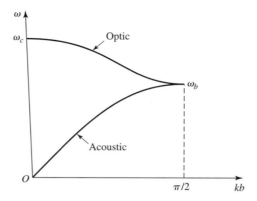

FIGURE 6.4.2. Dispersion Relations for the Couples Scalar Acoustic and Optic Modes in Micro-stretch Solids

The roots of (6.4.10) are

$$2\omega^2 = p + q \mp \left[(p+q)^2 - 4pq + \frac{4c_2^2}{mM}\right]^{1/2}. \qquad (6.4.12)$$

It can be seen that these two roots for ω^2 are real, by examining the discriminate Δ:

$$\Delta^2 = (p-q)^2 + \frac{4c_2^2}{mM} \geq 0. \qquad (6.4.13)$$

Moreover, an examination of

$$(p+q)^2 - \Delta^2 = 4\left(pq - \frac{c_2^2}{Mm}\right)$$

will show that it is nonnegative in the interval $|kb| < \pi/2$. Consequently, both ω_1^2 and ω_2^2 are nonnegative. For $k = 0$, we find that

$$\omega_1 = 0, \qquad \omega_2 = \omega_c = c_2\left(\frac{2}{M} + \frac{1}{m}\right). \qquad (6.4.14)$$

Hence ω_1 belongs to an acoustic mode and ω_2 to an optic mode. For $kb = \pi/2$, (6.4.9) gives $u_0 = v_0$, so that, at the end of the Brillouin zone, we have

$$\omega_B = \left(\frac{2c_1 + 2c_2}{M}\right)^{1/2}, \qquad |kb| = \frac{\pi}{2}. \qquad (6.4.15)$$

The sketch of the dispersion curves is displayed in Figure 6.4.2

6.5   Fundamental Solutions

Here we consider the general solution of the field equations of a static microstretch elastic solid in infinite extent. The decomposition (5.11.1) of the displacement and microrotation vectors uncouples the vector potentials. The fundamental solutions of the field equation for the vector potentials are the same as discussed in Section 5.25. Therefore we proceed to give solutions of

$$\lambda_0 \phi + (\lambda + 2\mu + \kappa)\nabla^2 \sigma + F = 0,$$
$$a_0 \nabla^2 \phi - \lambda_1 \phi - \lambda_0 \nabla^2 \sigma + L = 0, \tag{6.5.1}$$

where we set

$$\boldsymbol{u} = \nabla \sigma, \qquad \rho \boldsymbol{f} = \nabla F, \qquad \rho \boldsymbol{l} = \nabla L. \tag{6.5.2}$$

We employ the three-dimensional Fourier transform

$$\overline{\psi}(\boldsymbol{\xi}) \equiv (2\pi)^{-3/2} \int_{-\infty}^{\infty} \psi(\boldsymbol{x}) e^{i\boldsymbol{\xi}\cdot\boldsymbol{x}} d^3\boldsymbol{x} \tag{6.5.3}$$

in (6.5.1) and solve for $\overline{\phi}$ and $\overline{\sigma}$

$$\overline{\phi} = [a_0(\lambda + 2\mu + \kappa)]^{-1}[\lambda_0 \overline{F} + (\lambda + 2\mu + \kappa)\overline{L}](\xi^2 + p^2)^{-1},$$
$$\overline{\sigma} = [a_0(\lambda + 2\mu + \kappa)]^{-1}[-(\xi^2 a_0 + \lambda_1)\overline{F} + \lambda_0 \overline{L}][\xi^2(\xi^2 + p^2)]^{-1}, \tag{6.5.4}$$

where

$$p^2 \equiv \frac{(\lambda + 2\mu + \kappa)\lambda_1 - \lambda_0^2}{a_0(\lambda + 2\mu + \kappa)}. \tag{6.5.5}$$

The following inverse transform is known (cf. Sneddon [1972, p. 83]):

$$\mathcal{F}^{-1}\left(\frac{1}{\xi^2 + p^2}\right) = (\pi/2)^{1/2} \frac{e^{-pr}}{r}, \qquad r \equiv (\boldsymbol{x}\cdot\boldsymbol{x})^{1/2}.$$

Using the convolution theorem, we obtain

$$\phi(\boldsymbol{x}) = \left(\frac{\pi}{2}\right)^{1/2} [a_0(\lambda + 2\mu + \kappa)]^{-1} \frac{e^{-pr}}{r} * [\lambda_0 F + (\lambda + 2\mu + \kappa)L],$$
$$\sigma(\boldsymbol{x}) = -\left(\frac{\pi}{2}\right)^{1/2} (\lambda + 2\mu + \kappa)^{-1} \frac{e^{-pr}}{r} * F \tag{6.5.6}$$
$$+ \left(\frac{\pi}{2}\right)^{1/2} \frac{\lambda_1 F + \lambda_0 L}{a_0(\lambda + 2\mu + \kappa)p^2} * \frac{1 - e^{-pr}}{r},$$

where the asterisk (*) denotes the three-dimensional convolution, e.g.,

$$A(\boldsymbol{x}) * F(\boldsymbol{x}) = (2\pi)^{-3/2} \int_{-\infty}^{\infty} A(\boldsymbol{x} - \boldsymbol{\xi})F(\boldsymbol{\xi}) d^3\boldsymbol{\xi}. \tag{6.5.7}$$

The following two cases are of practical importance:

(i) *Concentrated force:*

$$F(\boldsymbol{x}) = F_0\delta(\boldsymbol{x}), \qquad L = 0,$$

$$\phi = \frac{1}{4\pi}\frac{\lambda_0}{a_0}\frac{F_0}{(\lambda + 2\mu + \kappa)}\frac{e^{-pr}}{r},$$

$$\sigma = -\frac{1}{4\pi}\frac{\lambda_1}{a_0(\lambda + 2\mu + \kappa)p^2}\frac{F_0}{r}\left[\left(\frac{p^2 a_0}{\lambda_1} - 1\right)e^{-pr} - 1\right]. \tag{6.5.8}$$

(ii) *Concentrated couple:*

$$F = 0, \qquad L = L_0\delta(\boldsymbol{x}),$$

$$\phi = \frac{1}{4\pi}\frac{L_0}{a_0}\frac{1}{r}e^{-pr},$$

$$\sigma = \frac{1}{4\pi}\frac{\lambda_0 L_0}{a_0(\lambda + 2\mu + \kappa)p^2}\frac{1 - e^{-pr}}{r}. \tag{6.5.9}$$

The displacement $\boldsymbol{u}$ is then given by (6.5.2).

In classical elasticity, to produce a point defect as a rigid inclusion, we need a body force that is proportional to the gradient of the Dirac delta function-type of singularity. This is known as the spherical dilatation. It is produced by three mutually orthogonal force-dipoles without moments. For microstretch solids, couple singularity also produces singular displacement fields and microstretch which fall outside the domain of classical elasticity.

6.6 Microstretch Surface Waves[4]

With decomposition (5.11.1), the field equations for τ, $\mathbf{U}$, and $\boldsymbol{\Phi}$ are decoupled from the microstretch ϕ, having the same form as (5.11.3), valid for the micropolar theory. The coupling occurs between the microstretch ϕ and displacement components σ through

$$\lambda_0\phi + (\lambda + 2\mu + \kappa)\nabla^2\sigma - \rho\ddot{\sigma} = 0,$$

$$a_0\nabla^2\phi - \lambda_1\phi - \lambda_0\nabla^2\sigma - \frac{1}{2}\rho j_0\ddot{\phi} = 0. \tag{6.6.1}$$

For the surface waves, σ and ϕ must be of the form

$$\{\sigma, \phi\} = \{\sigma_0, \phi_0\}\exp(-\delta y + ikx + i\omega t), \tag{6.6.2}$$

where σ_0 and ϕ_0 are constants. Substituting (6.6.2) into (6.6.1), we obtain

$$\frac{\lambda_0}{\rho c_1^2}\phi_0 + \left(\xi + \frac{\omega^2}{c_1^2}\right)\sigma_0 = 0,$$

$$\left(\xi - \frac{\lambda_1}{a_0} + \frac{\rho j_0}{2a_0}\omega^2\right)\phi_0 - \frac{\lambda_0}{a_0}\xi\sigma_0 = 0. \tag{6.6.3}$$

---

[4]Not published before.

where

$$\xi = \delta^2 - k^2. \tag{6.6.4}$$

Setting the determinant of the coefficients of ϕ_0 and σ_0 we obtain a quadratic equation for ξ whose solution is

$$\left.\begin{array}{l} \xi_1 \\ \xi_2 \end{array}\right\} = \frac{1}{2a_0}[-\beta \mp (\beta^2 - 4a_0\gamma)^{1/2}], \tag{6.6.5}$$

where

$$\beta = \left(\frac{a_0}{c_1^2} + \frac{\rho j_0}{2}\right)\omega^2 - \lambda_1 + \frac{\lambda_0^2}{\rho c_1^2},$$

$$\gamma = \frac{\omega^2}{c_1^2}\left(-\lambda_1 + \frac{\rho j_0}{2}\omega^2\right). \tag{6.6.6}$$

Hence we have two roots for δ_i^2, i.e.,

$$\delta_i^2 - k^2 = \xi_i, \quad i = 1, 2. \tag{6.6.7}$$

We note that

$$(\beta^2 - 4a_0\gamma)^{1/2} \le \beta \quad \text{for} \quad \omega^2 \ge \omega_{cr}, \tag{6.6.8}$$

where ω_{cr} is the cut-off frequency of the LO-branch, determined by

$$\omega_{cr}^2 = 2\lambda_1/\rho j. \tag{6.6.9}$$

Therefore, when $\omega > \omega_{cr}$, we have

$$\phi_{0i} = -Z_i\sigma_{0i}, \quad i = 1, 2, \tag{6.6.10}$$

where

$$Z_i = \frac{\rho c_1^2}{\lambda_0}\left(\xi_i + \frac{\omega^2}{c_1^2}\right), \quad i = 1, 2. \tag{6.6.11}$$

Expressions (6.6.2) give the solutions for σ and ϕ

$$\bar{\sigma} = \sigma_{01}\exp(-\delta_1 y + ikx) + \sigma_{02}\exp(-\delta_2 y + ikx),$$
$$\bar{\phi} = \phi_{01}\exp(-\delta_1 y + ikx) + \phi_{02}\exp(-\delta_2 y + ikx). \tag{6.6.12}$$

The vector waves are given by (5.16.3), namely

$$\bar{U} = [A\exp(-\lambda_2 y) + B\exp(-\lambda_4 y)]e^{ikx},$$
$$\bar{\phi}_3 = [\nu_2 A\exp(-\lambda_2 y) + \nu_4 B\exp(-\lambda_4 y)]e^{ikx}, \tag{6.6.13}$$

where λ_2, λ_4, ν_2, and ν_4 are given by (5.16.4).

The boundary conditions at the surface $y = 0$ of the half-space are

$$\bar{t}_{yy} = \lambda_0 \bar{\phi} + \lambda \nabla^2 \bar{\sigma} + (2\mu + \kappa)(\bar{\sigma}_{,yy} - \bar{U}_{,xy}) = 0,$$
$$\bar{t}_{yx} = (2\mu + \kappa)\bar{\sigma}_{,xy} - \mu\bar{U}_{,xx} + (\mu + \kappa)\bar{U}_{,yy} + \kappa\bar{\phi}_3 = 0,$$
$$\bar{m}_{yz} = \gamma\bar{\phi}_{3,y} = 0,$$
$$m_y = \phi_{,y} = 0 \quad \text{at } y = 0. \tag{6.6.14}$$

Of these, the last equation gives

$$\phi_{02} = -(\delta_1/\delta_2)\phi_{01}. \tag{6.6.15}$$

Using this in (6.6.10), we have

$$\sigma_{01} = -\phi_{01}/Z_1, \quad \sigma_{02} = (\delta_1/\delta_2 Z_2)\phi_{01}, \tag{6.6.16}$$

and (6.6.12) reduce to

$$\bar{\sigma} = \phi_{01}[-Z_1^{-1}\exp(-\delta_1 y) + (\delta_1/\delta_2 Z_2)\exp(-\delta_2 y)]e^{ikx},$$
$$\bar{\phi} = \phi_{01}[\exp(-\delta_1 y) - (\delta_1/\delta_2)\exp(-\delta_2 y)]e^{ikx}. \tag{6.6.17}$$

Substituting these and (6.6.13) into the remaining three boundary conditions in (6.6.14), we obtain three linear equations for A, B, and ϕ_{01}:

$$\bar{t}_{yy} = \alpha_1 A + \alpha_2 B + \alpha_3 \phi_{01} = 0,$$
$$\bar{t}_{yx} = \beta_1 A + \beta_2 B + \beta_3 \phi_{01} = 0,$$
$$\bar{m}_{yz} = \gamma_1 A + \gamma_2 B = 0, \tag{6.6.18}$$

where

$$\alpha_1 = (2\mu + \kappa)ik\lambda_2, \quad \alpha_2 = (2\mu + \kappa)ik\lambda_4,$$
$$\alpha_3 = \lambda_0\left(1 - \frac{\delta_1}{\delta_2}\right) - (\lambda + 2\mu + \kappa)\frac{\delta_1^2}{Z_1}\left(1 - \frac{\delta_2 Z_1}{\delta_1 Z_2}\right) + \frac{k^2\lambda}{Z_1}\left(1 - \frac{\delta_2 Z_1}{\delta_1 Z_2}\right),$$
$$\beta_1 = \mu k^2 + (\mu + \kappa)\lambda_2^2 + \kappa\nu_2, \quad \beta_2 = \mu k^2 + (\mu + \kappa)\lambda_4^2 + \kappa\nu_4,$$
$$\beta_3 = (2\mu + \kappa)ik\left(\frac{\delta_1}{Z_1} - \frac{\delta_1}{Z_2}\right),$$
$$\gamma_1 = -\gamma\nu_2\lambda_2, \quad \gamma_2 = -\gamma\nu_4\lambda_4. \tag{6.6.19}$$

Setting the determinant of the coefficients of A, B, and ϕ_{01} in (6.6.18) to zero, we obtain the dispersion relations of microstretch surface waves

$$\gamma_1(\alpha_2\beta_3 - \alpha_3\beta_2) - \gamma_2(\alpha_1\beta_3 - \alpha_3\beta_1) = 0. \tag{6.6.20}$$

The roots of this equation for ω^2/c_2^2 give the surface wave velocity. In the case $\lambda_0 = 0$, equation (6.6.1) is uncoupled and (6.6.5) gives

$$\xi_1 = -\omega^2/c_1^2, \quad \xi_2 = \frac{\lambda_1}{a_0} - \frac{\rho j_0}{a_0}\omega^2. \tag{6.6.21}$$

The first of these corresponds to the micropolar surface waves, discussed in Section 5.16. Consequently, for $\lambda_0 = 0$ the dispersion relations (6.6.20) must reduce to that of surface waves.

Calculations of the surface wave velocities could not be performed because of unavailability of the microstretch material moduli. However, from (6.6.20), it can be deduced that for $\omega > \omega_{cr}$, there will be an additional surface wave velocity corresponding to the LO-branch, over the micropolar surface wave.

6.7   Other Solutions

A cylinder whose lateral surfaces are free of tractions but whose end cross sections are subject to tractions is called *Saint Venant's Problems* in classical elasticity. This problem involves the treatment of *extension, torsion, and bending of a cylinder* subject to applied loads at the two end surfaces of the cylinder.

The Saint Venant's problem was formulated and solved for the micropolar theory of elasticity by Ieşan [1971]. Ieşan and Nappa [1994] tackled the same problem for the microstretch elastic solids. They also gave the necessary and sufficient conditions for the existence of a solution of the plane strain problem of microstretch elastostatics. The plane-strain problem consists of the determination of u_α ($\alpha = 1, 2$), ϕ_3, and ϕ which satisfy the equations of equilibrium

$$t_{\beta\alpha,\beta} + F_\alpha = 0, \qquad m_{\beta 3,\beta} + \epsilon_{\alpha\beta 3}t_{\alpha\beta} + G_3 = 0,$$
$$m_{\alpha,\alpha} + t - s + H = 0, \qquad \alpha, \beta = 1, 2 \text{ on } \Sigma, \qquad (6.7.1)$$

where all quantities are a function of x_1 and x_2 only. The constitutive equations for the plane-strain are of the form

$$t_{\alpha\beta} = \lambda e_{\rho\rho}\delta_{\alpha\beta} + (\mu + \kappa)\epsilon_{\alpha\beta} + \mu\epsilon_{\beta\alpha} + \lambda_0\phi\delta_{\alpha\beta},$$
$$m_{\alpha 3} = \gamma\phi_{3,\alpha}, \qquad m_\alpha = a_0\phi_{,\alpha},$$
$$t \equiv t_{\alpha\alpha}, \qquad s - t = \lambda_1\phi + \lambda_0 u_{\alpha,\alpha}, \qquad (6.7.2)$$

with strain measure

$$\epsilon_{\alpha\beta} = u_{\beta,\alpha} + \epsilon_{\beta\alpha 3}\phi_3. \qquad (6.7.3)$$

The boundary conditions are

$$t_{\alpha\beta}n_\alpha = \hat{t}_\beta, \qquad m_{\alpha 3}n_\alpha = \hat{m}, \qquad m_\alpha n_\alpha = \hat{\lambda} \quad \text{on } L,$$

where L is a closed curve (the boundary) of a generic cross section σ of the cylinder. It is assumed that F_α, G_3, $H \in C^\infty(\Sigma)$ and the domain Σ is C^∞ smooth.

Theorem. The necessary and sufficient conditions for the existence of a solution of the plane strain problem are given by

$$\int_\Sigma F_\alpha da + \int_L \hat{t}_\alpha \, ds = 0,$$

$$\int_\Sigma (\epsilon_{3\alpha\beta} x_\alpha F_\beta + G_3) da + \int_L (\epsilon_{3\alpha\beta} x_\alpha \hat{t}_\beta + \hat{m}) \, ds = 0. \qquad (6.7.4)$$

Ieşan and Nappa also gave solutions of the extension and bending problems and studied torsion and flexure in the above-mentioned reference and in their work [1995] on the plane-strain extension and bending for a cylinder having a circular cross section. Further treatment of the torsion and flexure problem for the microstretch elastic circular cylinder was given by De Cicco and Nappa [1997]. An existence theorem for the solution of microstretch elasticity (no thermal effects) was also studied. They also studied the asymptotic behavior for the solution of a homogeneous problem. The axisymmetric problem for the microstretch elastic half-space was dealt with by Kumar and Chadha [1985]. They showed that considerable changes occur in normal stress t_{zz}, couple stress, and microrotation with stretch. Chadha and Kumar [1988] also considered the steady-state thermoelastic problem on a microstretch half-space. Thermal stresses in microstretch elastic cylinders were considered by Nappa [1995] who assumed a steady distribution of temperature, reducing the problem to a plane-strain problem.

Tomar et al. [1998] discussed the reflections of longitudinal displacement waves, transverse displacement waves coupled with transverse microrotation waves, and longitudinal microstretch waves from the microstretch elastic half-space.

Ciarletta [1998] studied the linear theory of the bending of microstretch elastic plates. Spatial decay estimate and energetic measure, important to the dynamcial process, was discussed. A uniqueness theorem was given with no definiteness assumption on the elastic moduli.

Ieşan and Scalia [1997] established a decay estimate of the Saint-Venant type for static microstretch elasticity. They also studied the decay estimate of elastic energy for the dynamical case.

Wave propagation in a generalized thermo-microstretch elastic solid was studied by Singh and Kumar [1998]. Here, the term "generalized" denotes the inclusion of a second-order time derivative of the temperature field into the equation of heat conduction.

The power and potential of the microstretch theory remain dormant presently. The theory has promise in composite materials with chopped fibers. Porous solids with spherical inclusions and/or pressurized voids are candidates for microstretch elasticity. Clearly, spherical inclusions and defects fall under the domain of microstretch elasticity. Spot heat sources in a micropolar solid, hydrodynamic pressure exerted in porous elastic solids are also the type of problems that may be modeled by the microstretch theory. Animal bone carrying bone marrow and many other biological tissues, subject to micropressures arising from spherical microexpansions or contractions, are problems that fall in the domain of microstretch elasticity. Validity of these forecasts must await future developments.

Chapter 6 Problems

6.1. Prove the uniqueness theorem with the method used in Section 5.7.

6.2. Construct a variational principle for the microstretch elastic continua.

6.3. Determine the reflected microstretch waves from the free boundary of a half-space.

6.4. Display the surface wave velocities of the microstretch surface waves.

6.5. Determine the temperature field due to a heat spot in an infinite microstretch solid.

6.6. Determine the displacement and stress fields in a microstretch elastic solid with a spherical cavity. The surface of the cavity is subject to an impulsive pressure load.

6.7. Extend Euler buckling to a microstretch beam.

6.8. Formulate and obtain the solution of the point defect in a microstretch elastic solid.

7

Micromorphic Elasticity

7.0 Scope

This chapter is concerned with the linear theory of micromorphic thermoelastic solids.[1] Linear constitutive equations are given in Section 7.1. The boundary-initial value problems are formulated for anisotropic and isotropic solids. The passage made in Section 7.2 to the microstretch and micropolar cases, displays the relations of micromophic moduli to those of microstretch and micropolar. The discussion of the restrictions placed on the material constants by the nonnegative strain energy is rather complicated in micromorphic elasticity, since here a third-order strain tensor is to be considered. In Section 7.3 we obtain these restrictions by the requirement that the strain energy must be nonnegative for arbitrary strain measures. In Section 7.4. We discuss plane harmonic waves in micromorphic elasticity. Here we find that the dispersion curves possess seven branches:

  (a) longitudinal acoustic (LA) branch;

  (b) transverse acoustic (TA) branch; and

  (c) five optic branches.

---

[1] The origin of micromorphic elasticity goes back to two papers by Eringen and Şuhubi [1964]. The theory was completed with the introduction of the microinertia conservation law by Eringen [1964a]. Later contributions, reformulations, and generalizations include Eringen [1964b], [1965], [1968b], [1972a], [1992], and Twiss and Eringen [1971], [1972], [1992].

The presence of these branches resembles the branches observed in lattice dynamics. The extra branches over those of the micropolar elasticity indicate that the internal motions of ions in a lattice are represented more faithfully.

Section 7.5 provides some references to other published work in this field. Here we also discuss potential areas of application of the theory.

7.1   Linear Constitutive Equations

Constitutive equations of micromorphic elastic solids were obtained in Section 3.3. Here we present the linear theory.

The linear approximation may be obtained for the micromorphic solids in the sense of a norm similar to (5.1.4). We employ linear approximations

$$
\begin{aligned}
&x_{k,K} \simeq (\delta_{kl} + u_{k,l})\delta_{lK}, &\qquad &\chi_{kK} \simeq (\delta_{kl} + \phi_{kl})\delta_{lK}, \\
&\mathfrak{X}_{Kk} \simeq (\delta_{lk} - \phi_{lk})\delta_{Kl}, &\qquad &v_k \simeq \frac{\partial u_k}{\partial t}, \\
&\nu_{kl} \simeq \frac{\partial \phi_{kl}}{\partial t}, &\qquad &\sigma_{kl} \simeq i_{lm}\frac{\partial^2 \phi_{km}}{\partial t^2}, \\
&\rho/\rho_0 \simeq 1 - u_{k,k}, &\qquad &\theta = T_0 + T.
\end{aligned}
\tag{7.1.1}
$$

With these, the deformation tensors (1.5.11) read

$$
\mathfrak{C}_{KL} - \delta_{KL} \simeq \epsilon_{kl}\delta_{kK}\delta_{lL}, \quad C_{KL} - \delta_{KL} \simeq 2e_{kl}\delta_{kK}\delta_{lL},
$$
$$
\Gamma_{KLM} = \gamma_{klm}\delta_{kK}\delta_{lL}\delta_{mM},
\tag{7.1.2}
$$

where ϵ_{kl}, e_{kl}, and γ_{klm} are defined by

$$
\epsilon_{kl} = u_{l,k} - \phi_{lk}, \quad 2e_{kl} = \phi_{kl} + \phi_{lk},
$$
$$
\gamma_{klm} = \phi_{kl,m}, \quad \theta = T_0 + T, \quad T_0 > 0, \quad |T| \ll T_0.
\tag{7.1.3}
$$

Using these, the free energy ψ and the dissipation potential Φ (with $_D\eta \equiv 0$) are approximated by

$$
\begin{aligned}
\Sigma = \rho_0\psi = {}& \rho_0[\psi_0 - \eta_0 T - (C_0/2T_0)T^2] - TA_{kl}\epsilon_{kl} - TB_{kl}e_{kl} \\
& - TC_{klm}\gamma_{klm} + \tfrac{1}{2}A_{klmn}\epsilon_{kl}\epsilon_{mn} + \tfrac{1}{2}B_{klmn}e_{kl}e_{mn} + \tfrac{1}{2}C_{klmnpq}\gamma_{klm}\gamma_{npq} \\
& + E_{klmn}\epsilon_{kl}e_{mn} + F_{klmnp}\epsilon_{kl}\gamma_{mnp} + G_{klmnp}e_{kl}\gamma_{mnp},
\end{aligned}
$$
$$
\Phi = \frac{1}{T_0^2}K_{kl}T_{,k}T_{,l}.
\tag{7.1.4}
$$

From these, the following symmetry regulations are observed:

$$
B_{kl} = B_{lk}, \quad A_{klmn} = A_{mnkl}, \quad B_{klmn} = B_{mnkl} = B_{lkmn} = B_{nmkl},
$$
$$
C_{klmnpq} = C_{npqklm}, \quad E_{klmn} = E_{klnm}, \quad G_{klmnp} = G_{lkmnp}, \quad K_{kl} = K_{lk}.
\tag{7.1.5}
$$

Using (3.3.6) and (3.3.10), we obtain the linear constitutive equations of micromorphic thermoelastic solids:

$$t_{kl} \simeq \frac{\partial \Sigma}{\partial \epsilon_{kl}} = -A_{kl}T + A_{klmn}\epsilon_{mn} + E_{klmn}e_{mn} + F_{klmnp}\gamma_{mnp},$$

$$s_{kl} \simeq \frac{\partial \Sigma}{\partial e_{kl}} = -B_{kl}T + E_{mnkl}\epsilon_{mn} + B_{klmn}e_{mn} + G_{klmnp}\gamma_{mnp},$$

$$m_{klm} \simeq \frac{\partial \Sigma}{\partial \gamma_{lmk}} = -C_{lmk}T + F_{nplmk}\epsilon_{np} + G_{nplmk}e_{np} + C_{lmknpq}\gamma_{npq},$$

$$q_k = \frac{1}{T_0}K_{kl}T_{,l},$$

$$\eta \simeq -\frac{1}{\rho}\frac{\partial \Sigma}{\partial T} = \eta_0 + \frac{C_0}{T_0}T + \frac{1}{\rho_0}(A_{kl}\epsilon_{kl} + B_{kl}e_{kl} + C_{klm}\gamma_{klm}). \quad (7.1.6)$$

In the linear theory, ρ and i_{kl} are considered to be constants. Consequently, no need arises for the equations of conservation of mass and microinertia (2.2.10-a) and (2.2.11-a). The remaining equations of balance are given by (2.2.31-a)–(2.2.33-b), i.e.,

$$t_{kl,k} + \rho(f_l - \ddot{u}_l) = 0 \quad \text{in } \mathcal{V} - \sigma,$$
$$(7.1.7\text{-a})$$
$$[\![t_{kl} - \rho v_l(v_k - u_k)]\!]\, n_k = 0 \quad \text{on } \sigma, \; (7.1.7\text{-b})$$

$$m_{klm,k} + t_{ml} - s_{ml} + \rho(l_{lm} - \sigma_{lm}) = 0, \quad (7.1.8\text{-a})$$
$$[\![m_{klm} - \rho i_{rm}v_{lr}(v_k - u_k)]\!]\, n_k = 0 \quad \text{on } \sigma, \; (7.1.8\text{-b})$$

$$-\rho\dot{\epsilon} + t_{kl}(v_{l,k} - \nu_{lk}) + s_{kl}\nu_{lk} + m_{klm}\nu_{lm,k} + q_{k,k} + \rho h = 0 \quad \text{in } \mathcal{V} - \sigma,$$
$$(7.1.9\text{-a})$$

$$\Bigg[\!\Bigg[t_{kl}v_l + m_{klm}\nu_{lm} + q_k$$
$$-\rho\left(\epsilon + \frac{1}{2}\mathbf{v}\cdot\mathbf{v} + \frac{1}{2}i_{rl}\nu_{mr}\nu_{ml}\right)(v_k - u_k)\Bigg]\!\Bigg]\, n_k = 0, \quad \text{on } \sigma.$$
$$(7.1.9\text{-b})$$

Here the second set equations (7.1.7-b), (7.1.8-b), and (7.1.9-b) are the jump conditions across a discontinuity surface that may be sweeping the body with its own velocity u_k.

Substituting from (7.1.6) into (7.1.7-a), (7.1.8-a), and (7.1.9-a), we obtain the field equations. It is simpler to employ the energy balance law (3.3.12), which is equivalent to (7.1.9-a), for the case $_D\eta \equiv 0$ considered here

$$-A_{kl}T_{,k} + A_{klmn}u_{n,mk} + (E_{klmn} - A_{klmn})\phi_{nm,k}$$

$$+F_{klmnp}\phi_{mn,pk} + \rho(f_l - \ddot{u}_l) = 0,$$

$$(B_{ml} - A_{ml})T - C_{lmk}T_{,k} + (A_{mlpq} - E_{pqml})u_{q,p} + F_{nplmk}u_{p,nk}$$
$$+(E_{mlpq} + E_{qpml} - A_{mlqp} - B_{mlpq})\phi_{pq} + (F_{mlpqr} - F_{qplmr}$$
$$+G_{pqlmr} - G_{mlpqr})\phi_{pq,r} + C_{lmknpq}\phi_{np,qk} + \rho(l_{lm} - i_{mp}\ddot{\phi}_{lp}) = 0,$$

$$\rho C_0 \dot{T} + T_0 A_{kl}\dot{u}_{l,k} + T_0(B_{kl} - A_{kl})\dot{\phi}_{lk}$$
$$+T_0 C_{klm}\dot{\phi}_{kl,m} - \frac{1}{T_0}K_{kl}T_{,kl} - \rho h = 0.$$

$$(7.1.10)$$

Thus the field equation (7.1.10) consists of 13 linear partial differential equations to determine 13 unknown functions $u_k(\mathbf{x},t)$, $\phi_{kl}(\mathbf{x},t)$, and $T(\mathbf{x},t)$. Accompanying (7.1.10) are the boundary conditions which are deduced from the jump conditions (7.1.7-b), (7.1.8-b), and (7.1.9-b). The initial conditions generally consist of Cauchy data.

Boundary Conditions:

$$u_k = \hat{u}_k \quad \text{on } \partial\mathcal{V}_{\mathbf{u}}; \qquad t_{kl}n_k = \hat{t}_{(\mathbf{n})l} \qquad \text{on} \quad \partial\mathcal{V}_{\mathbf{t}} = \partial\mathcal{V} - \partial\mathcal{V}_{\mathbf{u}},$$
$$\phi_{kl} = \hat{\phi}_{kl} \quad \text{on } \partial\mathcal{V}_{\boldsymbol{\phi}}; \qquad m_{klm}n_k = \hat{m}_{(\mathbf{n})lm} \quad \text{on} \quad \partial\mathcal{V}_{\mathbf{m}} = \partial\mathcal{V} - \partial\mathcal{V}_{\boldsymbol{\phi}},$$
$$T = \hat{T} \quad \text{on } \partial\mathcal{V}_T; \qquad q_k n_k = \hat{q}_{(\mathbf{n})} \qquad \text{on} \quad \partial\mathcal{V}_{\mathbf{q}} = \partial\mathcal{V} - \partial\mathcal{V}_T,$$

$$(7.1.11)$$

where

$$\partial\mathcal{V}_{\mathbf{u}} \cap \partial\mathcal{V}_{\mathbf{t}} = 0, \quad \partial\mathcal{V}_{\boldsymbol{\phi}} \cap \partial\mathcal{V}_{\mathbf{m}} = 0, \quad \partial\mathcal{V}_T \cap \partial\mathcal{V}_{\mathbf{q}} = 0.$$

Initial conditions:

$$\mathbf{u}(\mathbf{x},0) = \mathbf{u}^0(\mathbf{x}), \qquad \dot{\mathbf{u}}(\mathbf{x},0) = \mathbf{v}^0(\mathbf{x}),$$
$$\phi_{kl}(\mathbf{x},0) = \phi_{kl}^0(\mathbf{x}), \qquad \dot{\phi}_{kl}(\mathbf{x},0) = \nu_{kl}^0(\mathbf{x}), \qquad (7.1.12)$$
$$T(\mathbf{x},0) = T^0(\mathbf{x}), \qquad \mathbf{x} \in \mathcal{V}.$$

Quantities on the right-hand sides of (7.1.11) and (7.1.12) are prescribed in the domains of the body indicated.

Isotropic Solids

For micromorphic solids that possess various symmetry groups, constitutive equations are simplified by subjecting material moduli to the appropriate symmetry regulations. Here we obtain constitutive equations of isotropic micromorphic thermoelastic solids. In this case, odd-order material tensors vanish and even-order tensors are built by the outer products of Kronecker tensor δ_{kl}. Consequently, we have

$$A_{kl} = \beta_0 \delta_{kl}, \qquad B_{kl} = \beta_1 \delta_{kl}, \qquad C_{klm} = 0,$$
$$A_{klmn} = \lambda \delta_{kl}\delta_{mn} + (\mu+\kappa)\delta_{km}\delta_{ln} + \mu\delta_{kn}\delta_{lm},$$
$$E_{klmn} = (\lambda+\nu)\delta_{kl}\delta_{mn} + (\mu+\sigma)(\delta_{km}\delta_{ln} + \delta_{kn}\delta_{lm}),$$
$$B_{klmn} = (\lambda+2\nu+\tau)\delta_{kl}\delta_{mn} + (\mu+2\sigma+\eta)(\delta_{km}\delta_{ln} + \delta_{kn}\delta_{lm}),$$
$$F_{klmnp} = G_{klmnp} = 0,$$
$$\begin{aligned}C_{klmnpq} = &\tau_1(\delta_{kl}\delta_{mn}\delta_{pq} + \delta_{kq}\delta_{lm}\delta_{np}) + \tau_2(\delta_{kl}\delta_{mp}\delta_{nq} + \delta_{km}\delta_{lq}\delta_{np})\\
&+ \tau_3\delta_{kl}\delta_{mq}\delta_{np} + \tau_4\delta_{kn}\delta_{lm}\delta_{pq} + \tau_5(\delta_{km}\delta_{ln}\delta_{pq} + \delta_{kp}\delta_{lm}\delta_{nq})\\
&+ \tau_6\delta_{km}\delta_{lp}\delta_{nq} + \tau_7\delta_{kn}\delta_{lp}\delta_{mq} + \tau_8(\delta_{kp}\delta_{lq}\delta_{mn} + \delta_{kq}\delta_{ln}\delta_{mp})\\
&+ \tau_9\delta_{kn}\delta_{lq}\delta_{mp} + \tau_{10}\delta_{kp}\delta_{ln}\delta_{mq} + \tau_{11}\delta_{kq}\delta_{lp}\delta_{mn},\end{aligned}$$
$$K_{kl}/T_0 = K_0\delta_{kl}, \quad K_0 \geq 0 \tag{7.1.13}$$

where 21 material constants $\beta_0, \beta_1, \lambda_1, \ldots, \tau_{11}, K_0$ are functions of ρ_0, i_{kl}^0, T^0, and $\mathbf{x}$.

Using (7.1.13), the free energy reads

$$\Sigma - \Sigma_0 = \Sigma_T + U(\boldsymbol{\epsilon}, \mathbf{e}) + U(\boldsymbol{\gamma}), \tag{7.1.14}$$

where[2]

$$\Sigma_T = -\rho\eta_0 T - \frac{\rho C_0}{2T_0}T^2 - \beta_0 T\epsilon_{kk} - \beta_1 Te_{kk},$$

$$\begin{aligned}2U(\boldsymbol{\epsilon}, \mathbf{e}) = &\frac{\rho C_0}{T_0}T^2 + \lambda\epsilon_{kk}\epsilon_{ll} + (\lambda+2\nu+\tau)e_{kk}e_{ll} + 2(\lambda+\nu)\epsilon_{kk}e_{ll}\\
&+ \mu\epsilon_{kl}\epsilon_{lk} + 2(\mu+2\sigma+\eta)e_{kl}e_{lk} + 4(\mu+\sigma)\epsilon_{kl}e_{lk}\\
&+ (\mu+\kappa)\epsilon_{kl}\epsilon_{kl}\end{aligned}$$
$$\begin{aligned}2U(\boldsymbol{\gamma}) = &\tau_1(\gamma_{iij}\gamma_{jkk} + \gamma_{ijj}\gamma_{kki}) + \tau_2(\gamma_{iij}\gamma_{kjk} + \gamma_{iji}\gamma_{kkj})\\
&+ \tau_3\gamma_{iij}\gamma_{kkj} + \tau_4\gamma_{ijj}\gamma_{ikk} + \tau_5(\gamma_{ijj}\gamma_{kik} + \gamma_{iji}\gamma_{jkk})\\
&+ \tau_6\gamma_{iji}\gamma_{kjk} + \tau_7\gamma_{ijk}\gamma_{ijk} + \tau_8(\gamma_{ijk}\gamma_{jki} + \gamma_{ijk}\gamma_{kij})\\
&+ \tau_9\gamma_{ijk}\gamma_{ikj} + \tau_{10}\gamma_{ijk}\gamma_{jik} + \tau_{11}\gamma_{ijk}\gamma_{kji}. \tag{7.1.15}\end{aligned}$$

Constitutive equations for stress, entropy, and heat read

$$\begin{aligned}t_{kl} = &-\beta_0 T\delta_{kl} + [\lambda\epsilon_{rr} + (\lambda+\nu)e_{rr}]\delta_{kl} + (\mu+\kappa)\epsilon_{kl} + \mu\epsilon_{lk} + 2(\mu+\sigma)e_{kl}\\
= &-\beta_0 T\delta_{kl} + (\lambda u_{r,r} + \nu\phi_{rr})\delta_{kl} + \mu u_{k,l} + (\mu+\kappa)u_{l,k} + \sigma\phi_{kl}\\
&+ (\sigma-\kappa)\phi_{lk},\end{aligned}$$
$$s_{kl} = -\beta_1 T\delta_{kl} + [(\lambda+\nu)\epsilon_{rr} + (\lambda+2\nu+\tau)e_{rr}]\delta_{kl} + (\mu+\sigma)(\epsilon_{kl} + \epsilon_{lk})$$

---

[2]The material constants in $U(\boldsymbol{\epsilon}, \mathbf{e})$ are the same as given in Eringen and Kafadar [1976, p. 69]. The material constants τ_i are the same as given in Şuhubi and Eringen [1964, p. 397]. They are *different* from τ_i given in Eringen and Kafadar.

$$+ 2(\mu + 2\sigma + \eta)e_{kl}$$
$$= -\beta_1 T\delta_{kl} + [(\lambda + \nu)u_{r,r} + (\nu + \tau)\phi_{rr}]\delta_{kl} + (\mu + \sigma)(u_{k,l} + u_{l,k})$$
$$+ (\sigma + \eta)(\phi_{kl} + \phi_{lk}),$$
$$m_{klm} = \tau_1(\gamma_{krr}\delta_{lm} + \gamma_{rrl}\delta_{km}) + \tau_2(\gamma_{rkr}\delta_{lm} + \gamma_{rrm}\delta_{kl})$$
$$+ \tau_3\gamma_{rrk}\delta_{lm} + \tau_4\gamma_{lrr}\delta_{km} + \tau_5(\gamma_{rlr}\delta_{km} + \gamma_{mrr}\delta_{kl})$$
$$+ \tau_6\gamma_{rmr}\delta_{kl} + \tau_7\gamma_{lmk} + \tau_8(\gamma_{mkl} + \gamma_{klm}) + \tau_9\gamma_{lkm}$$
$$+ \tau_{10}\gamma_{mlk} + \tau_{11}\gamma_{kml},$$
$$\eta = \eta_0 + \frac{C_0}{T_0}T + \frac{\beta_0}{\rho_0}\epsilon_{rr} + \frac{\beta_1}{\rho_0}e_{rr}$$
$$= \eta_0 + \frac{C_0}{T_0}T + \frac{\beta_0}{\rho_0}u_{r,r} + \frac{1}{\rho_0}(\beta_1 - \beta_0)\phi_{rr},$$
$$q_k = K_0 T_{,k}. \tag{7.1.16}$$

For the case of constant material moduli $\beta_0, \beta_1, \lambda_1, \ldots, \tau_{11}, K_0$, the field equations are

$$-\beta_0 T_{,k} + (\mu + \kappa)u_{k,ll} + (\mu + \lambda)u_{l,lk} + \nu\phi_{ll,k} + \sigma\phi_{lk,l}$$
$$+ (\sigma - \kappa)\phi_{kl,l} + \rho_0\left(f_k - \frac{\partial^2 u_k}{\partial t^2}\right) = 0,$$

$$(\tau_1 + \tau_2)\phi_{kk,ml} + \tau_1\phi_{kr,rk}\delta_{lm} + \tau_2\phi_{rk,rk}\delta_{lm} + \tau_3\phi_{rr,kk}\delta_{lm}$$
$$+ (\tau_4 + \tau_9)\phi_{lk,mk} + (\tau_5 + \tau_8)(\phi_{mk,kl} + \phi_{kl,mk}) + (\tau_6 + \tau_{11})\phi_{km,kl}$$
$$+ \tau_7\phi_{lm,kk} + \tau_{10}\phi_{ml,kk} + (\beta_1 - \beta_0)T\delta_{lm} - (\nu u_{r,r} + \tau\phi_{rr})\delta_{lm}$$
$$- \sigma u_{m,l} + (\kappa - \sigma)u_{l,m} - \eta\phi_{ml} - (\kappa + \eta)\phi_{lm} + \rho_0\left(l_{lm} - i_0\frac{\partial^2\phi_{lm}}{\partial t^2}\right) = 0,$$

$$\rho_0 C_0\frac{\partial T}{\partial t} + \beta_0 T_0\left(\frac{\partial u_k}{\partial t}\right)_{,k} + (\beta_1 - \beta_0)T_0\frac{\partial\phi_{kk}}{\partial t} - K_0\nabla^2 T - \rho_0 h = 0,$$

$$\tag{7.1.17}$$

where we also wrote $i_{kl} = i_0\delta_{kl}$ (microisotropy).

A uniqueness theorem for the system (7.1.17) with suitable boundary and initial conditions (as described by (7.1.11) and (7.1.12)) may be proven, in a similar way to Section 6.2, under the conditions that $K_0 \geq 0$, $C_0 \geq 0$, and

$$U(\epsilon, e), U(\gamma) \geq 0, \tag{7.1.18}$$

for all independent variations of ϵ, e, and γ (see also Soos [1969]).

7.2 Passage to Microstretch and Micropolar Theories

The constitutive equations and field equations of the micromorphic elasticity, under special conditions, give those of microstretch and micropolar elasticities.[3] Already we have seen that this is the case. However, it is useful to obtain connections of micromorphic material moduli to those of microstretch and micropolar. Such expressions may be used to obtain extra inequalities among the material constants of the micromorphic elasticity, through those of the microstretch and micropolar theory. To this end, first we express the constitutive equations in terms of symmetric and anti-symmetric forms. From (7.1.16), we have

$$t_{(kl)} = -\beta_0 T \delta_{kl} + (\lambda u_{r,r} + \nu \phi_{rr}) \delta_{kl} + (2\mu + \kappa) u_{(k,l)} + (2\sigma - \kappa)\phi_{(kl)}$$
$$t_{[kl]} = -\kappa u_{[k,l]} + \kappa \phi_{[kl]}, \tag{7.2.1}$$

where, as usual, parentheses enclosing indices denote the symmetric parts, and brackets, the antisymmetric parts.

The constitutive equation for m_{klm} given by $(7.1.16)_3$ may be decomposed as

$$
\begin{aligned}
m_{k(lm)} &= [(\tau_1 + \tau_2)\phi_{(kr),r} + \tau_3 \phi_{rr,k}]\delta_{lm} + [\tfrac{1}{2}(\tau_4 + 2\tau_5 + \tau_6)\phi_{(lr),r} \\
&+ \tfrac{1}{2}(\tau_1 + \tau_2)\phi_{rr,l}]\delta_{km} + [\tfrac{1}{2}(\tau_4 + 2\tau_5 + \tau_6)\phi_{(mr),r} \\
&+ \tfrac{1}{2}(\tau_1 + \tau_2)\phi_{rr,m}\delta_{kl} + (\tau_7 + \tau_{10})\phi_{(lm),k} \\
&+ \tfrac{1}{2}(2\tau_8 + \tau_9 + \tau_{11})\phi_{(kl),m} + \tfrac{1}{2}(2\tau_8 + \tau_9 + \tau_{11})\phi_{(km),l} \\
&+ \tfrac{1}{2}(\tau_4 - \tau_6)\phi_{[lr],r}\delta_{km} + (\tau_1 - \tau_2)\phi_{[kr],r}\delta_{lm} \\
&+ \tfrac{1}{2}(\tau_4 - \tau_6)\phi_{[mr],r}\delta_{kl} + \tfrac{1}{2}(\tau_{11} - \tau_9)[\phi_{[kl],m} + \phi_{[km],l}], \\
m_{k[lm]} &= \tfrac{1}{2}[(\tau_4 - \tau_6)\phi_{(lr),r} + (\tau_1 - \tau_2)\phi_{rr,l}]\delta_{km} \\
&- \tfrac{1}{2}[(\tau_4 - \tau_6)\phi_{(mr),r} + (\tau_1 - \tau_2)\phi_{rr,m}]\delta_{kl} \\
&+ \tfrac{1}{2}(\tau_{11} - \tau_9)\phi_{(km),l} - \tfrac{1}{2}(\tau_{11} - \tau_9)\phi_{(kl),m} \\
&+ \tfrac{1}{2}(\tau_4 - 2\tau_5 + \tau_6)\phi_{[lr],r}\delta_{km} - \tfrac{1}{2}(\tau_4 - 2\tau_5 + \tau_6)\phi_{[mr],r}\delta_{kl} \\
&+ \tfrac{1}{2}(2\tau_8 - \tau_9 - \tau_{11})\phi_{[kl],m} - \tfrac{1}{2}(2\tau_8 - \tau_9 - \tau_{11})\phi_{[km],l} \\
&+ (\tau_7 - \tau_{10})\phi_{[lm],k}. \tag{7.2.2}
\end{aligned}
$$

The passage to microstretch theory is made by setting

$$\phi_{(kl)} = \phi \delta_{kl}, \quad \phi_{[kl]} = -\epsilon_{klm}\phi_m. \tag{7.2.3}$$

Substituting these into (7.2.1) and (7.2.2) and recalling (2.1.8), we obtain

$$
\begin{aligned}
t_{kl} &= -\beta_0 T \delta_{kl} + (\lambda u_{rr} + \lambda_0 \phi)\delta_{kl} + \mu(u_{k,l} + u_{l,k}) + \kappa(u_{l,k} - \epsilon_{klm}\phi_m), \\
m_k &= m_{kll} = a_0 \phi_{,k} - b_0 \epsilon_{krj}\phi_{j,r}, \\
m_{kl} &= \epsilon_{lmp} m_{kpm} \\
&= \alpha \phi_{r,r}\delta_{kl} + \beta \phi_{k,l} + \gamma \phi_{l,k} + \alpha_0 \epsilon_{klp}\phi_{,p}, \tag{7.2.4}
\end{aligned}
$$

[3]This section is new, not published before.

where

$$\lambda_0 = 3\nu + 2\sigma - \kappa,$$
$$a_0 = 6\tau_1 + 6\tau_2 + 9\tau_3 + \tau_4 + 2\tau_5 + \tau_6 + 3\tau_7 + 2\tau_8 + \tau_9 + 3\tau_{10} + \tau_{11},$$
$$b_0 = 3\tau_1 - 3\tau_2 + \tau_4 - \tau_6 - \tau_9 + \tau_{11},$$
$$\alpha_0 = -3\tau_1 + 3\tau_2 - \tau_4 + \tau_6 + \tau_9 - \tau_{11},$$
$$\alpha = 2\tau_8 - \tau_9 - \tau_{11},$$
$$\beta = -\tau_4 + 2\tau_5 - \tau_6,$$
$$\gamma = \tau_4 - 2\tau_5 + \tau_6 + 2\tau_7 - 2\tau_8 + \tau_9 - 2\tau_{10} + \tau_{11},$$
$$\tau_0 = -\tau_4 + \tau_6 - \tau_9 + \tau_{11},$$
$$\lambda_1 = 9\tau + 3\kappa + 6\eta. \tag{7.2.5}$$

Here we also introduced τ_0 and λ_1 which will appear in (7.2.8) and (7.2.10).

We note that neither α_0 nor b_0 contribute to the balance laws. Expressions (7.2.5) establish the expressions of the constitutive moduli of microstretch elasticity in terms of the micromorphic elastic constants. The micropolar moduli follow from these by setting

$$\lambda_0 = a_0 = b_0 = \alpha_0 = 0. \tag{7.2.6}$$

The field equations of micromorphic elasticity, given by (7.1.17) may be expressed in terms $\phi_{(kl)}$ and ϕ_k as

$$-\beta_0 T_{,k} + (\lambda + \mu)u_{l,lk} + (\mu + \kappa)u_{k,ll} + \nu\phi_{ll,k} + (2\sigma - \kappa)\phi_{(kl),l}$$
$$+\kappa\epsilon_{klm}\phi_{m,l} + \rho(f_k - \ddot{u}_k) = 0, \tag{7.2.7}$$

$$(\alpha + \beta)\phi_{l,lk} + \gamma\phi_{k,ll} + \kappa\epsilon_{klm}u_{m,l} - 2\kappa\phi_k$$
$$+\tau_0\epsilon_{klm}\phi_{(lr),rm} + \rho(l_k - j\ddot{\phi}_k) = 0, \tag{7.2.8}$$

$$[(\tau_1 + \tau_2)\phi_{(kr),rk} + \tau_3\phi_{rr,kk} + (\beta_1 - \beta_0)T - \nu u_{r,r} - \tau\phi_{rr}]\delta_{lm}$$
$$+(\tau_1 + \tau_2)\phi_{kk,lm} + \frac{1}{2}(\tau_4 + 2\tau_5 + \tau_6 + 2\tau_8 + \tau_9 + \tau_{11})(\phi_{(kl),mk} + \phi_{(km),lk})$$
$$+(\tau_7 + \tau_{10})\phi_{(lm),kk} - \frac{1}{2}\tau_0(\phi_{[lk],km} + \phi_{[mk],kl})$$
$$+(\kappa - 2\sigma)u_{(m,l)} - (\kappa + 2\eta)\phi_{(ml)} + \rho\left(l_{(lm)} - \frac{1}{2}j\ddot{\phi}_{(lm)}\right) = 0. \tag{7.2.9}$$

Field equations of microstretch elasticity are obtained by substituting (7.2.3) into (7.2.8)–(7.2.9):

$$-\beta_0 T_{,k} + (\lambda + \mu)u_{l,lk} + (\mu + \kappa)u_{k,ll} + \lambda_0\phi_{,k} + \kappa\epsilon_{klm}\phi_{m,l}$$

$$+\rho(f_k - \ddot{u}_k) = 0,$$
$$(\alpha + \beta)\phi_{l,lk} + \gamma\phi_{k,ll} + \kappa\epsilon_{klm}u_{m,l} - 2\kappa\phi_k + \rho(l_k - j\ddot{\phi}_k) = 0,$$
$$-3(\beta_0 - \beta_1)T + a_0\nabla^2\phi - \lambda_1\phi - \lambda_0 u_{r,r} + \rho\left(l - \frac{1}{2}j_0\ddot{\phi}\right) = 0. \quad (7.2.10)$$

The field equations of micropolar elasticity follow from (7.2.10), by using (7.2.6), and setting $\lambda_1 = j_0 = 0$.

Expressions (7.2.5) establish the connection of material constants of 3M continua.

7.3   Restrictions on Material Moduli[4]

Material stability requires that the strain energy be nonnegative for all variations of the strain measures. For the isotropic media, this is expressed by

$$U(\boldsymbol{\epsilon}, \mathbf{e}) \geq 0 \quad \text{for all } \boldsymbol{\epsilon} \text{ and } \mathbf{e}, \quad (7.3.1)$$
$$U(\boldsymbol{\gamma}) \geq 0 \quad \text{for all } \gamma_{ijk}. \quad (7.3.2)$$

The strain energy functions $U(\boldsymbol{\epsilon}, \mathbf{e})$ and $U(\boldsymbol{\gamma})$, given by (7.1.15), can be decomposed into several uncoupled, symmetric quadratic forms by expressing ϵ_{kl}, e_{kl} and γ_{klm} in their symmetric and antisymmetric parts. This is also achieved by taking the second partial derivatives of $U(\boldsymbol{\epsilon}, e)$ with respect to each set of the following groups of variables:

$$(\epsilon_{11}, \epsilon_{22}, \epsilon_{33}, e_{11}, e_{22}, e_{33}) = (x_1, x_2, x_3, x_4, x_5, x_6),$$
$$(\epsilon_{12}, \epsilon_{21}, e_{12}) = (y_1, y_2, y_3),$$
$$(\epsilon_{23}, \epsilon_{31}, e_{23}) = (z_1, z_2, z_3),$$
$$(\epsilon_{31}, \epsilon_{12}, e_{31}) = (\xi_1, \xi_2, \xi_3), \quad (7.3.3)$$

in the order of variables listed in each set. This enables us to express $U(\boldsymbol{\epsilon}, \mathbf{e})$ in the form

$$U(\boldsymbol{\epsilon}, \mathbf{e}) = \frac{\rho C_0}{T_0}T^2 + a_{ij}x_ix_j + b_{kl}y_ky_l + b_{kl}z_kz_l + b_{kl}\xi_k\xi_l, \quad (7.3.4)$$

where the symmetric 6×6 matrix a_{ij} and the symmetric 3×3 matrix b_{kl} are given by

$$a_{11} = a_{22} = a_{33} = \lambda + 2\mu + \kappa,$$
$$a_{12} = a_{23} = a_{31} = \lambda,$$

---

[4]This section is new.

$$a_{14} = a_{25} = a_{36} = \lambda + 2\mu + \nu + 2\sigma,$$
$$a_{15} = a_{16} = a_{24} = a_{26} = a_{34} = a_{35} = \lambda + \nu,$$
$$a_{44} = a_{55} = a_{66} = \lambda + 2\mu + 2\nu + 4\sigma + 2\eta + \tau,$$
$$a_{45} = a_{56} = a_{64} = \lambda + 2\nu + \tau. \tag{7.3.5}$$

It is clear that we only need to examine two quadratic forms, and of course $C_0 \geq 0$. This was done by Smith [1968]. However, the present constitutive moduli being different than those used by Smith, we have carried out the analysis for the conditions of positive-definiteness of two quadratic forms. To this end, we need to obtain the eigenvalues of matrices (a_{ij}) and (b_{kl}), i.e.,

$$\det(a_{ij} - a\delta_{ij}) = 0, \quad i,j = 1,2,\ldots,6, \tag{7.3.6}$$
$$\det(b_{kl} - b\delta_{kl}) = 0, \quad k,l = 1,2,3. \tag{7.3.7}$$

We examine (7.3.7) first. Expansion of this determinant gives for b

$$b = b_{11} - b_{12}, \qquad b^2 - (b_{11} + b_{12} + b_{33})b + (b_{11} + b_{12})b_{33} - 2b_{13}^2 = 0. \tag{7.3.8}$$

For the roots b_k to be positive, we obtain

$$b_{11} - b_{12} > 0, \qquad (b_{11} + b_{12})b_{33} - 2b_{13}^2 > 0.$$

Substituting from (7.3.5), we obtain

$$\kappa > 0, \qquad 6\mu + \kappa + 8\sigma + 4\eta > 0,$$
$$(\mu + 2\sigma + 2\eta)(2\mu + \kappa) - (2\mu + 2\sigma)^2 > 0. \tag{7.3.9}$$

The determination of the eigenvalues a_i of the matrix (a_{ij}) is somewhat more tedious. However, several elements of (a_{ij}) are identical. By adding and subtracting some rows and columns, the expansion of the 6×6 determinant (7.3.6) is carried out leading to two quadratic equations:

$$a^2 - (a_{11} - \lambda + a_{44} - a_{45})a + (a_{11} - \lambda)(a_{44} - a_{45}) - (a_{14} - a_{15})^2 = 0,$$
$$a^2 - (2a_{45} + 2\lambda + a_{11} + a_{44})a + (2a_{45} + a_{44})(a_{11} + 2\lambda) - (2a_{15} + a_{14})^2 = 0. \tag{7.3.10}$$

The roots a_i of these equations will be positive, if and only if,

$$\mu + 2\sigma + \eta > 0, \qquad 2\mu + \kappa > 0,$$
$$(2\mu + \kappa)(\mu + 2\sigma + \eta) - 2(\mu + \sigma)^2 > 0,$$
$$3\lambda + 2\mu + 6\nu + 3\tau + 4\sigma + 2\eta > 0,$$
$$3\lambda + 2\mu + \kappa > 0,$$
$$(3\lambda + 2\mu + \kappa)(3\lambda + 2\mu + 6\nu + 3\tau + 4\sigma + 2\eta)$$
$$- (3\lambda + 2\mu + 3\nu + 2\sigma)^2 > 0. \tag{7.3.11}$$

Thus, we have proved:

Theorem. The necessary and sufficient conditions, for the strain energy $U(\epsilon, \mathbf{e})$ to be positive-definite are $C_0 > 0$, (7.3.9), and (7.3.11).

Smith decomposes the strain energy $U(\boldsymbol{\gamma})$ into two symmetric second-order polynomials involving one 6×6 symmetric matrix (c_{ij}) and another 7×7 symmetric matrix (d_{ij}): by taking second partial derivatives of $U(\boldsymbol{\gamma})$ with respect to two groups of variables

$$(\gamma_{123}, \gamma_{231}, \gamma_{312}, \gamma_{132}, \gamma_{321}, \gamma_{213}) = (x_1, x_2, x_3, x_4, x_5, x_6),$$
$$(\gamma_{111}, \gamma_{122}, \gamma_{133}, \gamma_{212}, \gamma_{313}, \gamma_{221}, \gamma_{331}) = (y_1, y_2, y_3, y_4, y_5, y_6, y_7),$$
$$(7.3.12)$$

in the order in which they are listed in each group. Hence,

$$U(\boldsymbol{\gamma}) = \sum_{i=1}^{6} c_{ij} x_i x_j + \sum_{k=1}^{7} d_{kl} y_k y_l. \qquad (7.3.13)$$

The eigenvalues of c_i of (c_{ij}) and d_k of (d_{kl}) are then obtained with a procedure similar to that presented above. Here

$$c_{11} = c_{22} = c_{33} = c_{44} = c_{55} = c_{66} = \tau_7,$$
$$c_{12} = c_{13} = c_{23} = c_{45} = c_{46} = c_{56} = \tau_8,$$
$$c_{14} = c_{35} = c_{26} = \tau_9,$$
$$c_{24} = c_{15} = c_{36} = \tau_{11},$$
$$c_{34} = c_{25} = c_{16} = \tau_{10}, \qquad (7.3.14)$$

$$d_{11} = 2\tau_1 + 2\tau_2 + \tau_3 + \tau_4 + 2\tau_5 + \tau_6 + \tau_7 + 2\tau_8 + \tau_9 + \tau_{10} + \tau_{11},$$
$$d_{12} = d_{13} = \tau_1 + \tau_4 + \tau_5, \qquad d_{14} = d_{15} = \tau_2 + \tau_5 + \tau_6,$$
$$d_{16} = d_{17} = \tau_1 + \tau_2 + \tau_3, \qquad d_{22} = d_{33} = \tau_4 + \tau_7 + \tau_9,$$
$$d_{23} = \tau_4, \qquad d_{24} = d_{35} = \tau_5 + \tau_8 + \tau_{10},$$
$$d_{25} = d_{34} = \tau_5, \qquad d_{26} = d_{37} = \tau_1 + \tau_8 + \tau_{11},$$
$$d_{27} = d_{36} = \tau_1, \qquad d_{44} = d_{55} = \tau_6 + \tau_7 + \tau_{11},$$
$$d_{45} = \tau_6, \qquad d_{46} = d_{57} = \tau_2 + \tau_8 + \tau_9,$$
$$d_{47} = d_{56} = \tau_2, \qquad d_{66} = d_{77} = \tau_3 + \tau_7 + \tau_{10},$$
$$d_{67} = \tau_8. \qquad (7.3.15)$$

The restrictions on the material moduli τ_i arising from the positive eigenvalues of c_{ij} and d_{kl} are then given by Smith as follows:

$$\tau_7 + 2\tau_8 > |\tau_9 + \tau_{10} + \tau_{11}|,$$
$$\tau_7 - \tau_8 > \frac{1}{\sqrt{2}}[(\tau_9 - \tau_{10})^2 + (\tau_{10} - \tau_{11})^2 + (\tau_{11} - \tau_9)^2]^{1/2}$$
$$\text{tr } T > 0, \quad tr(\text{COT}) > 0, \quad \det T > 0, \qquad (7.3.16)$$

where T is the following 3×3 matrix and COT denotes the cofactor of the matrix of T:

$$T = \begin{pmatrix} \tau_1 + \tau_2 + 3\tau_3 + \tau_7 + \tau_{10} & 3\tau_1 + \tau_4 + \tau_5 + \tau_8 + \tau_{11} & 3\tau_2 + \tau_5 + \tau_6 + \tau_8 + \tau_9 \\ 3\tau_1 + \tau_2 + \tau_3 + \tau_8 + \tau_{11} & \tau_1 + 3\tau_4 + \tau_5 + \tau_7 + \tau_9 & \tau_2 + 3\tau_5 + \tau_6 + \tau_8 + \tau_{10} \\ \tau_1 + 3\tau_2 + \tau_3 + \tau_8 + \tau_9 & \tau_1 + \tau_4 + 3\tau_5 + \tau_8 + \tau_{10} & \tau_2 + \tau_5 + 3\tau_6 + \tau_7 + \tau_{11} \end{pmatrix}.$$

$$(7.3.17)$$

Thus we have:

Theorem. The isotropic micromorphic thermoelastic solid will be stable if and only if the material moduli do not violate the inequalities (7.3.9), (7.3.11), (7.3.16), and $C_0 \geq 0$.

Special Restrictions

Some special restrictions on the material constants of the micromorphic continua are obtained through those of microstretch continua (6.1.24). These are

$$a_0 = 6\tau_1 + 6\tau_2 + 9\tau_3 + \tau_4 + 2\tau_5 + \tau_6 + 3\tau_7 + 2\tau_8 + \tau_9 + 3\tau_{10} + \tau_{11} > 0,$$
$$\lambda_1 = 9\tau + 3\kappa + 6\eta > 0,$$
$$\gamma = \tau_4 - 2\tau_5 + \tau_6 + 2\tau_7 - 2\tau_8 + \tau_9 - 2\tau_{10} + \tau_{11} > 0. \qquad (7.3.18)$$

Other restrictions may be obtained from the conditions of existence of the various *real* acoustic and optical modes.

7.4 Plane Harmonic Waves[5]

For the isothermal case and no applied loads, we introduce the usual decomposition

$$\mathbf{u} = \nabla\sigma + \nabla \times \mathbf{U}, \quad \nabla \cdot \mathbf{U} = 0,$$
$$\boldsymbol{\phi} = \nabla\tau + \nabla \times \boldsymbol{\Phi}, \quad \nabla \cdot \boldsymbol{\Phi} = 0, \qquad (7.4.1)$$

into the field equations (7.2.7) and (7.2.8)

$$(\lambda + 2\mu + \kappa)\sigma_{,kll} + (\mu + \kappa)\epsilon_{kij}U_{j,ill} + (2\sigma - \kappa)\phi_{(kl),l} + \nu\phi_{ll,k} - \kappa\Phi_{k,ll}$$
$$-\rho\ddot{\sigma}_{,k} - \rho\epsilon_{kij}\ddot{U}_{j,i} = 0,$$
$$(\alpha + \beta + \gamma)\tau_{,llk} + \gamma\epsilon_{kij}\Phi_{j,ill} - \kappa U_{k,ll} - 2\kappa\tau_{,k}$$
$$-2\kappa\epsilon_{kij}\Phi_{j,i} + \tau_0\epsilon_{klm}\phi_{(lr),rm} - \rho j(\ddot{\tau}_{,k} + \epsilon_{kij}\ddot{\Phi}_{j,i}) = 0. \qquad (7.4.2)$$

---

[5]This section is new, not published before.

In addition to these, (7.2.9) is valid.

Plane harmonic waves are expressed by

$$\{\sigma, \tau, \mathbf{U}, \boldsymbol{\Phi}, \phi_{(kl)}\} = \{\sigma^0, \tau^0, \mathbf{U}^0, \boldsymbol{\Phi}^0, \phi^0_{(kl)}\} \exp[i(\mathbf{k} \cdot \mathbf{x} - \omega t)], \qquad (7.4.3)$$

where $\{\sigma^0, \tau^0, \mathbf{U}^0, \boldsymbol{\Phi}^0, \phi^0_{kl}\}$ are independent of $\mathbf{x}$ and t. Substituting (7.4.3) into (7.4.2) and (7.2.9), we have

$$[\rho\omega^2 - (\lambda + 2\mu + \kappa)k^2]\sigma^0\mathbf{k} + [\rho\omega^2 - (\mu + \kappa)k^2]\mathbf{k} \times \mathbf{U}^0 - i\kappa k^2 \boldsymbol{\Phi}^0$$
$$+\nu\varpi^0\mathbf{k} + (2\sigma - \kappa)\mathbf{P}^0 = \mathbf{0}, \qquad (7.4.4)$$

$$[\rho j\omega^2 - 2\kappa - (\alpha + \beta + \gamma)k^2]\tau^0\mathbf{k} - i\kappa k^2\mathbf{U}^0 + (\rho j\omega^2 - 2\kappa - \gamma k^2)\mathbf{k} \times \boldsymbol{\Phi}^0$$
$$+i\tau_0\mathbf{P}^0 \times \mathbf{k} = \mathbf{0}, \qquad (7.4.5)$$

$$[(\tau_1 + \tau_2)Q^0 + (\tau_3 k^2 + \tau)\varpi^0 - \nu k^2\sigma^0]\delta_{lm}$$
$$+\frac{1}{2}(\tau_4 + 2\tau_5 + \tau_6 + 2\tau_8 + \tau_9 + \tau_{11})(P^0_l k_m + P^0_m k_l) + (\tau_1 + \tau_2)\varpi^0 k_l k_m$$
$$-[\frac{1}{2}\rho j\omega^2 - \kappa - 2\eta - (\tau_7 + \tau_{10})k^2]\phi^0_{(lm)} + \frac{\kappa - 2\sigma}{2}[2\sigma_0 k_l k_m + (\mathbf{k} \times \mathbf{U}^0)_m k_l$$
$$+(\mathbf{k} \times \mathbf{U}^0)_l k_m] = 0, \qquad (7.4.6)$$

where we put

$$\phi^0_{(ll)} \equiv \varpi^0, \quad \phi^0_{(kl)}k_l \equiv P^0_k, \quad \phi^0_{(kl)}k_k k_l = P^0_k k_k = Q^0. \qquad (7.4.7)$$

From (7.4.6), we solve $\phi^0_{(lm)}$:

$$D\phi^0_{(lm)} = (d_{11}Q^0 + d_{12}\varpi^0 - d_{13}\sigma^0)\delta_{lm} + d_2(P^0_l k_m + P^0_m k_l) + d_3\varpi^0 k_l k_m$$
$$+ d_4[2\sigma_0 k_l k_m + (\mathbf{k} \times \mathbf{U}^0)_m k_l + (\mathbf{k} \times \mathbf{U}^0)_l k_m]. \qquad (7.4.8)$$

This enables us to express ϖ^0, Q^0, and P^0_k given by (7.4.7). Here

$$d_3 = d_{11} \equiv \tau_1 + \tau_2, \quad d_{12} = \tau_3 k^2 + \tau, \quad d_{13} = \nu k^2,$$
$$d_2 \equiv \tfrac{1}{2}(\tau_4 + 2\tau_5 + \tau_6 + 2\tau_8 + \tau_9 + \tau_{11}),$$
$$D \equiv \tfrac{1}{2}\rho j\omega^2 - \kappa - 2\eta - (\tau_7 + \tau_{10})k^2, \quad d_4 = \tfrac{\kappa - 2\sigma}{2}, \qquad (7.4.9)$$

The scalar and vector products of (7.4.4) and (7.4.5), by $\mathbf{k}$, give

$$k^2 A\sigma^0 + a_1\varpi^0 + a_2 Q^0 = 0, \qquad (7.4.10\text{-a})$$
$$k^2 B\mathbf{U}^0 - i\kappa k^2\boldsymbol{\Phi}^0 \times \mathbf{k} + a_2\mathbf{P}^0 \times \mathbf{k} = \mathbf{0}, \qquad (7.4.10\text{-b})$$
$$[\rho j\omega^2 - 2\kappa + (\alpha + \beta + \gamma)k^2]\tau^0 k^2 = 0, \qquad (7.4.11\text{-a})$$
$$\kappa k^2\mathbf{k} \times \mathbf{U}^0 - ik^2 C\boldsymbol{\Phi}^0 - \tau_0 k^2\mathbf{P}^0 + \tau_0 Q^0\mathbf{k} = \mathbf{0}, \qquad (7.4.11\text{-b})$$

where

$$A \equiv \rho\omega^2 - (\lambda + 2\mu + \kappa)k^2, \quad a_1 \equiv \nu k^2,$$
$$B \equiv \rho\omega^2 - (\mu + \kappa)k^2, \quad a_2 \equiv 2\sigma - \kappa, \qquad (7.4.12)$$
$$C \equiv \rho j\omega^2 - 2\kappa - \gamma k^2.$$

Using (7.4.8), we calculate ϕ_{ll}^0, $\mathbf{P}^0$, and Q^0:

$$(2d_4k^2 - 3d_{13})\sigma^0 + (3d_{12} + d_3k^2 - D)\varpi^0 + (3d_{11} + 2d_2)Q^0 = 0,$$
(7.4.13)

$$d_4k^2\mathbf{k} \times \mathbf{U}^0 + (d_2k^2 - D)\mathbf{P}^0$$
$$+[(2k^2d_4 - d_{13})\sigma^0 + (d_{12} + d_3k^2)\varpi^0 + (d_{11} + d_2)Q^0]\mathbf{k} = \mathbf{0},$$
(7.4.14)

$$k^2(2k^2d_4 - d_{13})\sigma^0 + k^2(d_{12} + d_3k^2)\varpi^0 + [(d_{11} + 2d_2)k^2 - D]Q^0 = 0.$$
(7.4.15)

The cross products of (7.4.11-b) and (7.4.14), by $\mathbf{k}$, obtain

$$\kappa k^4\mathbf{U}^0 + ik^2 C\mathbf{k} \times \mathbf{\Phi}^0 + \tau_0 k^2\mathbf{k} \times \mathbf{P}^0 = \mathbf{0}, \qquad (7.4.16)$$
$$k^4 d_4\mathbf{U}^0 + (D - d_2k^2)\mathbf{k} \times \mathbf{P}^0 = \mathbf{0}. \qquad (7.4.17)$$

We now have a set of three scalar equations, (7.4.10-a), (7.4.13) and (7.4.15)

$$k^2 A\sigma^0 + a_1\varpi^0 + a_2 Q^0 = 0,$$
$$\alpha_1\sigma^0 + \alpha_2\varpi^0 + \alpha_3 Q^0 = 0,$$
$$\beta_1\sigma^0 + \beta_2\varpi^0 + \beta_3 Q^0 = 0, \qquad (7.4.18)$$

and a set of three vector equations, (7.4.10-b), (7.4.16), and (7.4.17),

$$k^2 B\mathbf{U}^0 - i\kappa k^2\mathbf{\Phi}^0 \times \mathbf{k} + a_2\mathbf{P}^0 \times \mathbf{k} = \mathbf{0},$$
$$\kappa k^4\mathbf{U}^0 - ik^2 C\mathbf{\Phi}^0 \times \mathbf{k} - \tau_0 k^2\mathbf{P} \times \mathbf{k} = \mathbf{0},$$
$$k^4 d_4\mathbf{U}^0 + (d_2k^2 - D)\mathbf{P}^0 \times \mathbf{k} = \mathbf{0}, \qquad (7.4.19)$$

where

$$\alpha_1 = 2d_4k^2 - 3d_{13}, \qquad \alpha_2 \equiv 3d_{12} + d_3k^2 - D,$$
$$\alpha_3 \equiv 3d_{11} + 2d_2, \qquad \beta_1 \equiv (2k^2d_4 - d_{13})k^2,$$
$$\beta_2 \equiv (d_{12} + d_3k^2)k^2, \qquad \beta_3 \equiv (d_{11} + 2d_2)k^2 - D.$$
(7.4.20)

Setting the determinant of the coefficient equal to zero in each set, we obtain the dispersion relations. Thus, for the scalar waves, we have

$$AD^2 - 2p_1 AD + p_2 A + p_3 k^3 D + p_4 k^3 = 0, \qquad (7.4.21)$$

where

$$2p_1 \equiv 3\tau + k^2(2\tau_1 + 2\tau_2 + 3\tau_3 + 2d_2),$$
$$p_2 \equiv 2d_2\tau + 2k^2\tau_3 d_2 - k^2(\tau_1 - \tau_2)^2,$$
$$p_3 \equiv 4\sigma(\kappa - \nu - \sigma) - \kappa^2 - 3\nu^2,$$
$$p_4 \equiv (\kappa - \sigma)[-2\tau(2\sigma - \kappa) + 2k^2(\tau_1 + \tau_2)(2\sigma - \kappa - 2\nu)$$
$$- 2k^2 d_2\nu - 2k^2\tau_3(2\sigma - \kappa) - 2\nu^2 k^4]$$
$$+ \nu[-6\tau(2\sigma + \kappa) + 4k^2(\kappa - 2\sigma)(\tau_1 + \tau_2) - 2k^2\nu d_2$$
$$- 6k^2\tau_3(2\sigma - \kappa) - 6k^4\nu^2].$$
(7.4.22)

For the vector waves, the dispersion relations read

$$BCD - k^2 d_2 BC - k^2 q_1 C - k^2 \kappa^2 D + k^4 q_2 = 0, \qquad (7.4.23)$$

where

$$q_1 \equiv \frac{1}{2}(2\sigma - \kappa)^2, \quad q_2 = \frac{1}{2}\kappa(\kappa - 2\sigma)\tau_0. \qquad (7.4.24)$$

In addition to the dispersion relations (7.4.21) and (7.4.23), we have (7.4.11-a), namely,

$$\rho j \omega^2 - 2\kappa + (\alpha + \beta + \gamma)k^2 = 0. \qquad (7.4.25)$$

Equations (7.4.21) and (7.4.23) each constitute a third-degree polynomial in ω^2. In view of the fact that many material constants are not measured to date, the solution ω^2 as a function of k^2 will not provide any useful information. However, the fundamental nature of various branches, as to whether they are acoustic or optic type, may be identified near the origin of the k-axis, near $k = 0$. From (7.4.21), we have

$$A \equiv \rho \omega^2 - (\lambda + 2\mu + \kappa)k^2 = 0 \quad \text{LA-branch,}$$
$$D \equiv \tfrac{1}{2}\rho j \omega^2 - \kappa - 2\eta - (\tau_7 + \tau_{10})k^2 - \tfrac{3}{2}\tau$$
$$- \tfrac{1}{2}(9\tau^2 - 4d_2\tau)^{1/2} = 0 \quad \text{O-branch,}$$
$$D = \tfrac{1}{2}\rho j \omega^2 - \kappa - 2\eta - (\tau_7 + \tau_{10})k^2 - p_1 - \tfrac{3}{2}\tau$$
$$+ \tfrac{1}{2}(9\tau^2 - 4d_2\tau)^{1/2} = 0 \quad \text{O-branch.} \qquad (7.4.26)$$

For the two optical branches, the cut-off frequencies are given by

$$\left.\begin{array}{c} \omega_{c1} \\ \omega_{c2} \end{array}\right\} = \left\{ \frac{2}{\rho j}[\kappa + 2\eta \pm p_1 + (p_1^2 - p_2^2)^{1/2}] \right\}^{1/2}. \qquad (7.4.27)$$

Similarly, for the vector waves from (7.4.23) near $k = 0$, we deduce

$$B \equiv \rho \omega^2 - (\mu + \kappa)k^2 + \cdots = 0 \quad \text{TA-branch,}$$
$$C \equiv \rho j \omega^2 - 2\kappa - \gamma k^2 + \cdots = 0 \quad \text{O-branch,}$$
$$D = \frac{1}{2}\rho j \omega^2 - \kappa - 2\eta - (\tau_7 + \tau_{10})k^2 - k^2 d_2 = 0 \quad \text{O-branch.}$$
$$(7.4.28)$$

Here, the cut-off frequencies for the optical branches are

$$\omega_{c3} = \left(\frac{2\kappa}{\rho j}\right)^{1/2}, \quad \omega_{c4} = \left(\frac{2\kappa + 4\eta}{\rho j}\right)^{1/2}. \qquad (7.4.29)$$

Recalling the remaining dispersion relation (7.4.25) here, we have another optical branch with the same cut-off frequency given by ω_{c3}.

This exercise indicates that plane harmonic waves in micromorphic elasticity give rise to two acoustic and five optical branches. Some of these

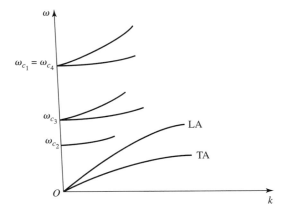

FIGURE 7.4.1. Sketch of Dispersion Curves Near $k = 0$ in Micromorphic Elastic Solids

branches may interfere with each other. In fact, (7.4.21) indicates the coupling of the branches A and D, and (7.4.23) those of B, C, and D. To have a clear picture of their behavior with increasing k more elaborate calculations are needed. Unfortunately, such calculations will have to be left to a future date, when material constants are determined experimentally.

7.5 Other Contributions

Because of its complexity, contributions to micromorphic mechanics have been rare. Soos [1969] gave a uniqueness theorem which states that:

Theorem. Coupled thermoelasticity problem having balance laws (7.1.7-a)–(7.1.9-a); constitutive equations (7.1.6) with strain measures given by (7.1.3) under the boundary conditions of the type (7.1.11), and initial conditions (7.1.12) have unique solutions if the strain energy $\Sigma(T = 0)$ and the dissipation potential Φ are nonnegative for all strain measures and temperature gradients.

The proof of this theorem is similar to those already given for the micropolar and microstretch solids.

Maugin [1970] gave a variational principle for the nondissipative micromorphic solids. A theory of micromorphic media with memory was established by Eringen [1972a]. In two papers, Twiss and Eringen [1971, 1972] have established the theory of mixtures of micromorphic materials (solids and fluids). This theory and its constitutive equations were specialized to micropolar media. Application of the theory is made to propagation of waves in two constituent micropolar mixtures. Dispersion curves are studied in their paper [1972]. Other applications of micromorphic mechanics to earthquake problems were suggested by Teisseyre [1982], Dresen *et al.* [1984], and Teisseyre *et al.* [1985].

Balance laws of micromorphic mechanics were generalized to higher-grade materials by Eringen [1970b] to include an arbitrary number of deformable directors. The problems of dislocations and disclinations were discussed by Eringen and Claus in [1970]. In spite of its age, since its inception in 1964, micromorphic mechanics remains a field that is not cultivated. Its potential is yet to be discovered by research workers. This is probably due to the complexity of the field equations. Nevertheless, this field of mechanics offers great potential in dealing with deformable microstructure. For example, the important field of polymers with deformable molecules, granular and porous solids with deformable grains and pores, biological tissues, e.g., animal bones and muscles, and composites with flexible cords, are but a few deformable bodies that involve the degrees of freedom offered by the micromorphic continuum theory. Contrary to solids, there are more studies in the area of micromorphic fluids. These are discussed in Volume II.

Chapter 7 Problems

7.1. Give the detailed calculations leading to the restriction (7.3.16) and (7.3.17).

7.2. Using the restrictions in Section 7.3 for the material constants, obtain the restrictions of the material constants of the microstretch elastic solids.

7.3. Determine the reflected micromorphic waves, from the free boundary of a half-space.

7.4. Determine the dispersion relations for the micromorphic surface waves.

7.5. Express the field equations of the micromorphic elastic solids in spherical coordinates.

7.6. (Short-Term Paper) Obtain the fundamental solutions for the micromorphic elastic solids,

7.7. (Short-Term Paper) Study the literature and construct a micromorphic theory for:

(a) the binary mixture of two solids; and

(b) the binary mixture of a porous solid with fluid.

8
Electromagnetic Interactions

8.0 Scope

The chapter is devoted to the development of the basic equations of the E-M theory of thermo-micropolar elasticity.[1] The linear constitutive equations for the anisotropic solids are established in Section 8.1, where passage is also made to isotropic solids. In Section 8.2 we study material stability conditions. This leads to some inequalities among the thermomechanical and E-M material moduli for the isotropic solids. Section 8.3 displays the balance laws and jump conditions. Field equations, boundary and initial conditions are obtained in Section 8.4. Special cases of these equations, the micropolar piezoelectricity and magnetoelasticity, are formulated in the same section. In Section 8.5 we discuss the nature of piezoelectric waves and new coupling presented by the micropolar effects, e.g., the microrotations due to the electric field and, conversely, the electric field generated by the microrotations in the body.

8.1 Linear Constitutive Equations

Constitutive equations of micropolar electromagnetic (E-M) elastic solids were obtained in Section 4.2. Here we present the linear theory.

---

[1]The material presented in this chapter is new.

Carrying out the linear approximation in the sense expressed by (5.1.4), with the inclusion of E-M fields $\mathcal{E}$ and $\mathbf{B}$ among the independent variables, the free energy Σ, and the dissipation potential Φ, may be expressed as

$$\Sigma = \Sigma^M + \Sigma^E, \qquad \Phi = \Phi^M + \Phi^E, \tag{8.1.1}$$

where (Σ^M, Φ^M) denote purely mechanical parts, given by $(5.1.12)_1$ and $(5.1.13)_1$, with $M_k = N = 0$, namely,

$$
\begin{aligned}
\Sigma^M - \Sigma_0 &= -\rho\eta_0 T - (\rho C_0/2T_0)T^2 - A_{kl}T\epsilon_{kl} - B_{kl}T\gamma_{kl} \\
&\quad + \tfrac{1}{2}A_{klmn}\epsilon_{kl}\epsilon_{mn} + \tfrac{1}{2}B_{klmn}\gamma_{kl}\gamma_{mn} + C_{klmn}\epsilon_{kl}\gamma_{mn}, \\
\Phi^M &= \frac{1}{2T_0^2}K_{kl}T_{,k}T_{,l},
\end{aligned} \tag{8.1.2}
$$

and (Σ^E, Φ^E) the E-M interactions, which may be expressed as

$$
\begin{aligned}
\Sigma^E &= -\rho\lambda_k^{TE}T\mathcal{E}_k - \rho\lambda_k^{TB}TB_k + \tfrac{1}{2}\chi_{kl}^E \mathcal{E}_k\mathcal{E}_l + \tfrac{1}{2}\chi_{kl}^B B_k B_l \\
&\quad + \lambda_{kl}^{EB}\mathcal{E}_k B_l + \lambda_{klm}^{\epsilon E}\epsilon_{kl}\mathcal{E}_m + \lambda_{klm}^{\epsilon B}\epsilon_{kl}B_m + \lambda_{klm}^{\gamma E}\gamma_{lk}\mathcal{E}_m \\
&\quad + \lambda_{klm}^{\gamma B}\gamma_{lk}B_m, \\
\Phi^E &= \tfrac{1}{2}\sigma_{kl}^E \mathcal{E}_k\mathcal{E}_l + \sigma_{kl}^{EB}\mathcal{E}_k B_l + \tfrac{1}{T_0}\sigma_{kl}^{ET}\mathcal{E}_k T_{,l} + \sigma_{klm}^{\epsilon E}\mathcal{E}_k\epsilon_{lm} \\
&\quad + \frac{1}{T_0}\sigma_{kl}^{BT}B_k T_{,l},
\end{aligned} \tag{8.1.3}
$$

where $\lambda_k^{TE}, \lambda_k^{TB}, \dots, \lambda_{klm}^{\gamma B}$ are material constants that depend on ρ_0, T_0, I_{KL}, and $\mathbf{X}$ at the natural state. The conduction moduli $\sigma_{kl}^E, \dots, \sigma_{kl}^{BT}$ are also functions of these natural state variables, as described in Section 4.2.

From (4.2.16) and (4.2.17), there follow the constitutive equations:

$$
\begin{aligned}
\eta &= \eta^M + \eta^E, \quad \mathbf{t} = \mathbf{t}^M + \mathbf{t}^E, \quad \mathbf{m} = \mathbf{m}^M + \mathbf{m}^E, \\
\mathbf{q} &= \mathbf{q}^M + \mathbf{q}^E, \quad \mathbf{P} = \mathbf{P}^E, \quad \mathbf{M} = \mathbf{M}^E, \quad \mathbf{J} = \mathbf{J}^E,
\end{aligned} \tag{8.1.4}
$$

where

$$
\begin{aligned}
\eta^M &= \eta_0^M + \frac{C_0}{T_0}T + \frac{1}{\rho}A_{kl}\epsilon_{kl} + \frac{1}{\rho}B_{kl}\gamma_{kl}, \\
t_{kl}^M &= -A_{kl}T + A_{klmn}\epsilon_{mn} + C_{klmn}\gamma_{mn}, \\
m_{kl}^M &= -B_{lk}T + C_{mnlk}\epsilon_{mn} + B_{lkmn}\gamma_{mn}, \\
q_k^M &= \frac{1}{T_0}K_{kl}T_{,l},
\end{aligned} \tag{8.1.5}
$$

and

$$
\begin{aligned}
\eta^E &= \lambda_k^{TE}\mathcal{E}_k + \lambda_k^{TB}B_k, \\
t_{kl}^E &= \lambda_{klm}^{\epsilon E}\mathcal{E}_m + \lambda_{klm}^{EB}B_m,
\end{aligned}
$$

$$m_{kl}^E = \lambda_{klm}^{\gamma E} \mathcal{E}_m + \lambda_{klm}^{\gamma B} B_m,$$
$$q_k^E = \sigma_{lk}^{ET} \mathcal{E}_l + \sigma_{lk}^{BT} B_l,$$
$$P_k^E = \rho \lambda_k^{TE} T - \chi_{kl}^E \mathcal{E}_l - \chi_{kl}^{EB} B_l - \lambda_{lmk}^{\epsilon E} \epsilon_{lm} - \lambda_{lmk}^{\gamma E} \gamma_{ml},$$
$$\mathcal{M}_k^E = \rho \lambda_k^{TB} T - \chi_{kl}^B B_l - \lambda_{lk}^{EB} \mathcal{E}_l - \lambda_{lmk}^{\epsilon B} \epsilon_{lm} - \lambda_{lmk}^{\gamma B} \gamma_{lm},$$
$$\mathcal{J}_k^E = \sigma_{kl}^E \mathcal{E}_l + \sigma_{kl}^{EB} B_l + \frac{1}{T_0} \sigma_{kl}^{ET} T_{,l} + \sigma_{klm}^{\epsilon E} \epsilon_{lm}. \qquad (8.1.6)$$

The strain measure ϵ_{kl} and γ_{kl} have their usual expressions

$$\epsilon_{kl} = u_{l,k} + \epsilon_{lkm} \phi_m, \quad \gamma_{kl} = \phi_{k,l}. \qquad (8.1.7)$$

Constitutive equations, obtained above, are for anisotropic micropolar E-M thermoelastic solids. As such, the material moduli depend on the material symmetry group. We recall that $\mathbf{B}$ and $\mathbf{H}$ are not *time-symmetric* vectors, i.e., upon the reversal of time (e.g., the direction of electronic spin), $\mathbf{B}$ and $\mathbf{H}$ reverse their signs. Consequently, the material moduli λ_k^{TB}, λ_{kl}^{EB}, λ_{klm}^{EB}, $\lambda_{klm}^{\gamma B}$, σ_{kl}^{EB}, and σ_{kl}^{BT}, which are the coefficients of terms involving $\mathbf{B}$, must produce time symmetry in the constitutive equations. Crystals having this property have been discussed in a book by Kiral and Eringen [1990].

Equations (8.1.6) display some new physical phenomena over the classical E-M theory:

(a) Piezoelectricity represented by $\lambda_{klm}^{\epsilon E}$ is not symmetric in indices k and l.

(b) Piezomagnetism represented by $\lambda_{klm}^{\epsilon B}$ is not symmetric also in indices k and l.

(c) The presence of the material constant $\lambda_{klm}^{\gamma E}$ in the expression of $\mathbf{P}^E$ indicates that the wryness tensor γ_{kl} will produce polarization. This is a new phenomena, which we shall name the *piezowryness* effect.

(d) Heat conduction due to E-field (Peltier effect) and the current due to temperature gradient (Seebeck effect) are well known in classical E-M theory.

(e) Besides the classical pyroelectric effect (λ_{kl}^{TE}) and the magnetoelectric effect (λ_{kl}^{EB}), new effects, denoted by $\lambda_{lmk}^{\gamma B}$, appear. This shows that wryness can cause magnetization. We name this phenomena the *magnetowryness* effect.

Thus, micropolar theory produces not only a larger number of coefficients for the classical effects by eliminating symmetries due to strain tensor, but also brings to focus the importance of the rotational degrees of freedom for a set of new physical phenomena. The importance of these effects depend on the nature and magintudes of the microrotation gradient.

Isotropic Solids

For the isotropic micropolar E-M solids, the material tensors acquire their isotropic forms. The mechanical part of the constitutive equations were given by (5.3.3), i.e.,

$$\Sigma^M = \Sigma_0 - \rho[\eta_0 T + (C_0/2T_0)T^2] - \beta_0 T \epsilon_{mm} \qquad (8.1.8)$$
$$+ \tfrac{1}{2}[\lambda\epsilon_{kk}\epsilon_{ll} + (\mu+\kappa)\epsilon_{kl}\epsilon_{kl} + \mu\epsilon_{kl}\epsilon_{lk}]$$
$$+ \tfrac{1}{2}(\alpha\gamma_{kk}\gamma_{ll} + \beta\gamma_{kl}\gamma_{lk} + \gamma\gamma_{kl}\gamma_{kl}),$$

$$\Phi = \frac{1}{2T_0^2}KT_{,k}T_{,k},$$

$$\eta = \eta_0 + (C_0/T_0)T + (\beta_0/\rho)\epsilon_{mm},$$

$$t_{kl} = -\beta_0 T \delta_{kl} + \lambda\epsilon_{mm}\delta_{kl} + (\mu+\kappa)\epsilon_{kl} + \mu\epsilon_{lk},$$

$$m_{kl} = \alpha\gamma_{mm}\delta_{kl} + \beta\gamma_{kl} + \gamma\gamma_{lk},$$

$$q_k = \frac{K}{T_0}T_{,k}, \qquad (8.1.9)$$

and the E-M parts reduce to

$$\Sigma^E = \tfrac{1}{2}\chi^E \boldsymbol{\mathcal{E}}\cdot\boldsymbol{\mathcal{E}} + \tfrac{1}{2}\chi^B \mathbf{B}\cdot\mathbf{B} + \lambda^{\gamma E}\epsilon_{klm}\gamma_{lk}\mathcal{E}_m,$$

$$\Phi^E = \tfrac{1}{2}\sigma^E \boldsymbol{\mathcal{E}}\cdot\boldsymbol{\mathcal{E}} + \frac{1}{T_0}\sigma^{ET}\boldsymbol{\mathcal{E}}\cdot\boldsymbol{\nabla}T,$$

$$t_{kl}^E = m_{kl}^E = 0, \quad \eta^E = 0,$$

$$q_k^E = \sigma^{ET}\mathcal{E}_k,$$

$$P_k^E = -\chi^E\mathcal{E}_k - \lambda^{\gamma E}\epsilon_{lmk}\gamma_{ml},$$

$$\mathcal{M}_k^E = -\chi^B B_k,$$

$$\mathcal{J}_k^E = \sigma^E\mathcal{E}_k + \frac{1}{T_0}\sigma^{ET}T_{,k}. \qquad (8.1.10)$$

Note that the piezowryness effect remains intact, in isotropic micropolar solids. However, the magnetowryness effect disappears due to the condition of time symmetry on γ_{kl}.

8.2 Material Stability

The material stability requires that the strain energy be nonnegative. Thermodynamic restrictions imply that the dissipation potential Φ be nonnegative

$$U^M + \Sigma^E \geq 0, \quad \Phi \geq 0. \qquad (8.2.1)$$

Hence, the ensuing quadratic polynomials must be positive semidefinite for all independent variations of $\mathbf{E}$, $\mathbf{B}$, ϵ, γ, and ∇T. These conditions require that several independent quadratic forms must be positive semidefinite

leading to inequalities for the material moduli. These restrictio~
obtained as described in Section 5.2. Here, we obtain these inequa~
for the isotropic media.

For vanishing E-M effects, the inequalities given by (5.3.12) remain va...

$$3\lambda + 2\mu + \kappa \geq 0, \qquad 2\mu + \kappa \geq 0, \qquad \kappa \geq 0,$$
$$3\alpha + \beta + \gamma \geq 0, \qquad \alpha + \beta \geq 0, \qquad \gamma - \beta \geq 0, \qquad (8.2.2)$$
$$C_0 \geq 0, \qquad K \geq 0.$$

From the expressions of Σ^E and Φ^E given by (8.1.10), it is immediately clear that

$$\chi^E \geq 0, \quad \chi^B \geq 0, \quad \sigma^E \geq 0, \qquad (8.2.3)$$

$$\frac{1}{T_0}\sigma^{ET}\mathcal{E}_k T_{,k} + \frac{K}{2T_0^2}T_{,k}T_{,k} + \tfrac{1}{2}\sigma^E \mathcal{E}_k \mathcal{E}_k \geq 0. \qquad (8.2.4)$$

The quadratic form (8.2.4) will be nonnegative if

$$K \geq 0, \quad K\sigma^E - (\sigma^{ET})^2 \geq 0. \qquad (8.2.5)$$

In the expression of $U^M + \Sigma^E$, γ_{kl}, and $\mathcal{E}_k$ are coupled by the quadratic form

$$\tfrac{1}{2}(\alpha\gamma_{kk}\gamma_{ll} + \beta\gamma_{kl}\gamma_{lk} + \gamma\gamma_{kl}\gamma_{kl}) + \lambda^{\gamma E}\epsilon_{klm}\gamma_{lk}\mathcal{E}_m + \tfrac{1}{2}\chi^E\mathcal{E}_k\mathcal{E}_k \geq 0. \; (8.2.6)$$

This quadratic form shows that only the antisymmetric part $\gamma_{[lk]} = (\gamma_{lk} - \gamma_{kl})/2$ is coupled with $\mathcal{E}_k$. Hence setting $\gamma_{(lk)} = (\gamma_{lk}+\gamma_{kl})/2 = 0$, we obtain a quadratic form in $\gamma_{[lk]}$ and $\mathcal{E}_k$, which will be nonnegative if

$$\lambda^{\gamma E} \geq 0, \quad (\gamma - \beta)\chi^E - 2(\lambda^{\gamma E})^2 \geq 0. \qquad (8.2.7)$$

Collecting, we have for the material stability regulations (8.2.2) and

$$\chi^E \geq 0, \quad \chi^B \geq 0, \quad \sigma^E \geq 0, \lambda^{\gamma E} \geq 0,$$
$$K \geq 0, \quad T_0 K\sigma^E - (\sigma^{ET})^2 \geq 0, \quad (\gamma - \beta)\chi^E - 4(\lambda^{\gamma E})^2 \geq 0, \; (8.2.8)$$

for the isotropic micropolar E-M solids.

8.3 Balance Laws

Balance laws of thermoelastic E-M solids consist of the union of the mechanical balance laws (4.1.6)–(4.1.10) and Maxwell's equations (4.1.1)–(4.1.4). For the linear theory, equations of conservation of mass and microinertia are dropped. For micropolar solids then, we have

Mechanical Balance Laws::

$$t_{kl,k} + \rho(f_l - \ddot{u}_l) = 0 \quad \text{in } \mathcal{V} - \sigma, \quad (8.3.1\text{-a})$$

$$[[t_{kl} + t_{kl}^E + u_k G_l - \rho(v_k - u_k)]]n_k = 0 \quad \text{on } \sigma, \quad (8.3.1\text{-b})$$

$$m_{kl,k} + \epsilon_{lmn}t_{mn} + \rho(l_l - j\ddot{\phi}_l) = 0 \quad \text{in } \mathcal{V} - \sigma, \quad (8.3.2\text{-a})$$

$$[[m_{kl} - \rho j_{pl}\nu_p(v_k - u_k)]]\,n_k = 0 \quad \text{on } \sigma, \quad (8.3.2\text{-b})$$

$$\rho\theta\dot{\eta} - \nabla \cdot \mathbf{q} - \boldsymbol{\mathcal{J}} \cdot \boldsymbol{\mathcal{E}} - \rho h = 0 \quad \text{in } \mathcal{V} - \sigma, \quad (8.3.3\text{-a})$$

$$[[(t_{kl} + t_{kl}^E + u_k G_l)v_l + m_{kl}\nu_l + q_k - \mathcal{S}_k$$

$$-(\rho\epsilon + \frac{1}{2}\rho\mathbf{v} \cdot \mathbf{v} + \frac{1}{2}\rho j\boldsymbol{\nu} \cdot \boldsymbol{\nu} + \frac{1}{2}\mathbf{E} \cdot \mathbf{E} + \frac{1}{2}\mathbf{B} \cdot \mathbf{B})$$

$$(v_k - u_k)]]n_k = 0 \quad \text{on } \sigma. \quad (8.3.3\text{-b})$$

Maxwell's Equations::

$$\nabla \cdot \mathbf{D} = q_e \quad \text{in } \mathcal{V} - \sigma, \quad (8.3.4\text{-a})$$

$$\mathbf{n} \cdot [[\mathbf{D}]] = w_e \quad \text{on } \sigma, \quad (8.3.4\text{-b})$$

$$\nabla \times \mathbf{E} + \frac{1}{c}\frac{\partial \mathbf{B}}{\partial t} = \mathbf{0} \quad \text{in } \mathcal{V} - \sigma \quad (8.3.5\text{-a})$$

$$\mathbf{n} \times \left[\left[\mathbf{E} + \frac{1}{c}\mathbf{u} \times \mathbf{B}\right]\right] = \mathbf{0} \quad \text{on } \sigma, \quad (8.3.5\text{-b})$$

$$\nabla \cdot \mathbf{B} = 0 \quad \text{in } \mathcal{V} - \sigma, \quad (8.3.6\text{-a})$$

$$\mathbf{n} \cdot [[\mathbf{B}]] = 0 \quad \text{on } \sigma, \quad (8.3.6\text{-b})$$

$$\nabla \times \mathbf{H} - \frac{1}{c}\frac{\partial \mathbf{D}}{\partial t}, = \frac{1}{c}\mathbf{J} \quad \text{in } \mathcal{V} - \sigma, \quad (8.3.7\text{-a})$$

$$\mathbf{n} \times \left[\left[\mathbf{H} - \frac{1}{c}\mathbf{u} \times \mathbf{D}\right]\right] = \mathbf{0} \quad \text{on } \sigma, \quad (8.3.7\text{-b})$$

where we replaced the energy balance law (4.1.10-a) by its short form (4.2.18) (with $_D\dot{\eta} \equiv 0$).

Applied loads **f** and **l** are given by

$$\rho\mathbf{f} = \rho\mathbf{f}^M + q_e\mathbf{E} + \frac{1}{c}\mathbf{J} \times \mathbf{B} + (\nabla\mathbf{E}) \cdot \mathbf{P} + (\nabla\mathbf{B}) \cdot \mathbf{M} + \frac{1}{c}[(\mathbf{P} \times \mathbf{B})v_k]_{,k}$$

$$+ \frac{1}{c}\frac{\partial}{\partial t}(\mathbf{P} \times \mathbf{B}),$$

$$\rho\mathbf{l} = \rho\mathbf{l}^M + \mathbf{P} \times \boldsymbol{\mathcal{E}} + \boldsymbol{\mathcal{M}} \times \mathbf{B}, \quad (8.3.8)$$

the E-M stress tensor t_{kl}^E, the E-M momentum G_k, and the Poynting vector $\mathbf{S}$ are expressed by

$$\mathbf{t}^E = \mathbf{P} \otimes \boldsymbol{\mathcal{E}} - \mathbf{B} \otimes \boldsymbol{\mathcal{M}} + \mathbf{E} \otimes \mathbf{E} + \mathbf{B} \otimes \mathbf{B} - \tfrac{1}{2}(E^2 + B^2 - 2\boldsymbol{\mathcal{M}} \cdot \mathbf{B})\mathbf{1},$$

$$\mathbf{G} = \frac{1}{c}\mathbf{E} \times \mathbf{B},$$

$$\mathbf{S} = c\boldsymbol{\mathcal{E}} \times \boldsymbol{\mathcal{H}}. \tag{8.3.9}$$

E-M vectors in a comoving frame, represented by majuscule capital letters, are related to those at the fixed laboratory frame as follows:

$$\boldsymbol{\mathcal{E}} = \mathbf{E} + \frac{1}{c}\mathbf{v} \times \mathbf{B}, \qquad \boldsymbol{\mathcal{H}} = \mathbf{H} - \frac{1}{c}\mathbf{v} \times \mathbf{D},$$

$$\boldsymbol{\mathcal{M}} = \mathbf{M} + \frac{1}{c}\mathbf{v}, \times \mathbf{P}, \qquad \boldsymbol{\mathcal{J}} = \mathbf{J} - q_e\mathbf{v}, \tag{8.3.10}$$

$$\mathbf{D} = \mathbf{E} + \mathbf{P}, \qquad \mathbf{B} = \mathbf{H} + \mathbf{M}.$$

Note that the second set of equations, marked by b after the equation number, are the jump conditions at a discontinuity surface σ that is moving in the positive direction of its normal with the velocity $\mathbf{u}$.

8.4   Field Equations

Field equations for micropolar thermoelastic E-M solids are obtained by substituting the constitutive equations into balance laws.

A. Anisotropic Solids

$$[-A_{kl}T + A_{klmn}(u_{n,m} + \epsilon_{nmr}\phi_r) + C_{klmn}\phi_{m,n} + \lambda_{klm}^{\epsilon E}\mathcal{E}_m$$
$$+\lambda_{klm}^{EB}B_m]_{,k} + \rho(f_l - \ddot{u}_l) = 0, \tag{8.4.1}$$

$$[-B_{kl}T + B_{lkmn}\phi_{m,n} + C_{mnlk}(u_{n,m} + \epsilon_{nmr}\phi_r) + \lambda_{klm}^{\gamma E}\mathcal{E}_m + \lambda_{klm}^{\gamma B}B_m]_{,k}$$
$$+\epsilon_{lmn}[-A_{mn}T + A_{mnpq}(u_{q,p} - \epsilon_{pqr}\phi_r) + C_{mnpq}\phi_{p,q} + \lambda_{mnp}^{\gamma E}\mathcal{E}_p + \lambda_{mnp}^{\gamma B}B_p]$$
$$+\rho(l_l - j_{kl}\ddot{\phi}_k) = 0, \tag{8.4.2}$$

$$\rho C_0\dot{T} + \rho\lambda_k^{TE}\dot{\mathcal{E}}_k + \rho\lambda_k^{TB}\dot{B} + T_0 A_{kl}(\dot{u}_{l,k} - \epsilon_{klr}\dot{\phi}_r) + T_0 B_{kl}\dot{\phi}_{k,l}$$
$$- \left[\frac{K_{kl}}{T_0}T_{,l} + \sigma_{lk}^{ET}\mathcal{E}_l + \sigma_{lk}^{BT}B_l\right]_{,k} - \boldsymbol{\mathcal{J}} \cdot \boldsymbol{\mathcal{E}} - \rho h^M = 0, \tag{8.4.3}$$

$$\nabla \cdot \mathbf{E} + [\rho\lambda_k^{TE}T - \chi_{kl}^E\mathcal{E}_l - \chi_{kl}^{EB}B_l - \lambda_{lmk}^{\epsilon E}(u_{m,l} + \epsilon_{mlr}\phi_r)$$

$$-\lambda_{lmk}^{\gamma E}\phi_{m,l}]_{,k} - q_e = 0,$$
$$(8.4.4)$$

$$\nabla \times \mathbf{E} + \frac{1}{c}\frac{\partial \mathbf{B}}{\partial t} = \mathbf{0}, \quad \nabla \cdot \mathbf{B} = 0,$$
$$(8.4.5)$$

$$\epsilon_{klm}[B_m - \rho\lambda_m^{TB}T + \chi_{mr}^B B_r + \lambda_{rm}^{EB}\mathcal{E}_r + \lambda_{ijm}^{\epsilon B}(u_{j,i} + \epsilon_{jip}\phi_p)$$
$$+\lambda_{ijm}^{\gamma B}\phi_{i,j} + \frac{1}{c}\epsilon_{mij}v_iP_j]_{,l} - \frac{1}{c}\frac{\partial}{\partial t}[E_k + \rho\lambda_k^{TE}T - \chi_{kl}^E E_l - \chi_{kl}^{EB}B_l$$
$$-\lambda_{lmk}^{\epsilon E}(u_{m,l} + \epsilon_{mlr}\phi_r) - \lambda_{lmk}^{\gamma E}\phi_{m,l}] - \frac{1}{c}[\sigma_{kl}^E\mathcal{E}_l + \sigma_{kl}^{EB}B_l + \frac{1}{T_0}\sigma_{kl}^{ET}T_{,l}$$
$$+\sigma_{klm}^{\epsilon E}(u_{m,l} + \epsilon_{mlr}\phi_r) - q_ev_k] = 0 \quad \text{on } \bar{\mathcal{V}} \times T^+,$$
$$(8.4.6)$$

where the applied loads $\mathbf{f}$ and $\mathbf{l}$ are given by (8.3.8). In a fully linear theory, terms containing products of fields are dropped, simplifying the field equations.

Given the external loads (which involve E-M fields as well), the system of equations (8.4.1)–(8.4.6) must be solved under appropriate boundary conditions selected from the jump conditions given in Section 4.1, and initial conditions (generally Cauchy data), to determine the fields $u_k(\mathbf{x},t), \phi_k(\mathbf{x},t), T(\mathbf{x},t), E_k(\mathbf{x},t)$, and $B_k(\mathbf{x},t)$.

Boundary Conditions:

In the absence of a discontinuity surface:

$$\begin{aligned}
\mathbf{u} &= \hat{\mathbf{u}} &&\text{on } \bar{S}_1 \times T^+, & \mathbf{t}^M &= t_{kl}^M n_k \mathbf{i}_l = \hat{\mathbf{t}}^M &&\text{on } S_2 \times T^+ \\
\boldsymbol{\phi} &= \hat{\boldsymbol{\phi}} &&\text{on } \bar{S}_3 \times T^+, & \mathbf{m}^M &= m_{kl}^M n_k \mathbf{i}_l = \hat{\mathbf{m}}^M &&\text{on } S_4 \times T^+, \\
T &= \hat{T} &&\text{on } \bar{S}_5 \times T^+, & q &= \mathbf{q} \cdot \mathbf{n} = \hat{q} &&\text{on } S_6 \times T^+, \\
\mathbf{E} &= \hat{\mathbf{E}} &&\text{on } \bar{S}_7 \times T^+, & \mathbf{B} &= \hat{\mathbf{B}} &&\text{on } \bar{S}_8 \times T^+,
\end{aligned}$$
$$(8.4.7)$$

where $\hat{\mathbf{u}}, \hat{\mathbf{t}}^M, \hat{\boldsymbol{\phi}}, \hat{\mathbf{m}}^M, \hat{T}, \hat{q}, \hat{\mathbf{E}}$, and $\hat{\mathbf{B}}$ are prescribed in nonoverlapping surfaces $\bar{S}_i (i = 1, \ldots, 8)$, of subsurfaces of $\partial\mathcal{V}$ of the body, and T^+ is the time interval. Some of these subsurfaces may vanish, excluding the corresponding boundary conditions.

Initial Conditions:

These consist of the Cauchy data:

$$\begin{aligned}
\mathbf{u}(\mathbf{x},0) &= \mathbf{u}^0(\mathbf{x}), & \dot{\mathbf{u}}(\mathbf{x},0) &= \mathbf{v}^0(\mathbf{x}), \\
\boldsymbol{\phi}(\mathbf{x},0) &= \boldsymbol{\phi}^0(\mathbf{x}), & \dot{\boldsymbol{\phi}}(\mathbf{x},0) &= \boldsymbol{\nu}^0(\mathbf{x}), \\
T(\mathbf{x},0) &= T^0(\mathbf{x}), & \mathbf{E}(\mathbf{x},0) &= \mathbf{E}^0(\mathbf{x}), \\
\mathbf{B}(\mathbf{x},0) &= \mathbf{B}^0(\mathbf{x}), & \mathbf{x} &\in \bar{\mathcal{V}}.
\end{aligned}$$
$$(8.4.8)$$

Of course, a well-posed boundary value problem must obey the existence theorem which is not given. For the linear piezoelectric media, a uniqueness theorem was given by Cracium [1995]. However, Cracium's constitutive equations lack several terms, e.g., in Fourier's law, $\mathcal{E}$ is not present. The proof is similar to that given in Section 5.7.

B. Isotropic Solids

For the isotropic solids, a great deal of simplifications are achieved

$$-\beta_0 T_{,k} + (\lambda + \mu)u_{l,lk} + (\mu + \kappa)u_{k,ll} + \kappa\epsilon_{klm}\phi_{m,l} + \rho(f_k - \ddot{u}_k) = 0,$$
$$(\alpha + \beta)\phi_{l,lk} + \gamma\phi_{k,ll} + \kappa\epsilon_{klm}u_{m,l} - 2\kappa\phi_k + \rho(l_k - j_{kl}\dot{\phi}_l) = 0,$$
$$(8.4.9)$$

$$\rho C_0 \dot{T} + \beta_0 T_0 \nabla \cdot \dot{\mathbf{u}} - \frac{k}{T_0}\nabla^2 T - \sigma^{ET}\nabla \cdot \left(\mathbf{E} + \frac{1}{c}\mathbf{v} \times \mathbf{B}\right)$$
$$-\sigma^E\left(\mathbf{E} + \frac{1}{c}\mathbf{v} \times \mathbf{B}\right) \cdot \mathbf{E} - \frac{1}{T_0}\sigma^{ET}\mathbf{E} \cdot \nabla T - \rho h^M = 0,$$
$$(8.4.10)$$

$$(1 - \chi^E)\nabla \cdot \mathbf{E} + \frac{\chi^E}{c}\mathbf{v} \cdot (\nabla \times \mathbf{B}) - \frac{\chi^E}{c}(\nabla \times \mathbf{v}) \cdot \mathbf{B} - q_e = 0,$$
$$(8.4.11)$$

$$\nabla \times \mathbf{E} + \frac{1}{c}\frac{\partial \mathbf{B}}{\partial t} = \mathbf{0}, \quad \nabla \cdot \mathbf{B} = 0,$$
$$(8.4.12)$$

$$(1 + \chi^B)\nabla \times \mathbf{B} + \frac{1}{c}\mathbf{v} \times [\chi^E(\mathbf{E} + \frac{1}{c}\mathbf{v} \times \mathbf{B}) + \lambda^{\gamma E}\nabla \times \phi]$$
$$-\frac{1}{c}\frac{\partial}{\partial t}\left[(1 - \chi^E)\mathbf{E} - \frac{\chi^E}{c}\mathbf{v} \times \mathbf{B} - \lambda^{\gamma E}\nabla \times \phi\right]$$
$$= \frac{1}{c}\left[\sigma^E\left(\mathbf{E} + \frac{1}{c}\mathbf{v} \times \mathbf{B}\right) + q_e\mathbf{v} + \frac{1}{T_0}\sigma^{ET}\nabla T\right], \quad (8.4.13)$$

where coefficients of $1/c^2$ may also be dropped, providing further simplification.

Here we see that the only coupling in the field equations (excluding the boundary conditions) among mechanical and E-M fields is through the temperature T and the microrotation ϕ. Particularly, the new element of coupling is through $\nabla \times \phi$.

In fully linear theory, we may also drop nonlinear terms involving products of $\mathbf{v}$ and $\mathbf{B}$.

C. Piezoelectricity

For piezoelectric materials at uniform temperature, the effect of T and the **B**-field may be ignored. Dropping also the free charge which is absent, and the nonlinear terms in the expressions of $\mathbf{f}$, $\mathbf{l}$, and $\mathcal{E}$, we arrive at the following field equations:

$$A_{klmn}(u_{n,mk} + \epsilon_{nmr}\phi_{r,k}) + C_{klmn}\phi_{m,nk} - \lambda_{klm}^{\epsilon E}\psi_{,mk} + \rho f_l^M - \rho \ddot{u}_l = 0, \tag{8.4.14}$$

$$B_{lkmn}\phi_{m,nk} + C_{mnlk}(u_{n,mk} + \epsilon_{nmr}\phi_{r,k}) - \lambda_{klm}^{\gamma E}\psi_{,mk}$$
$$+\epsilon_{lmn}[A_{mnpq}(u_{q,p} + \epsilon_{qpr}\phi_r) + C_{mnpq}\phi_{p,q} - \lambda_{mnp}^{\gamma E}\psi_{,p}] + \rho l_l^M - \rho j_{kl}\ddot{\phi}_k = 0, \tag{8.4.15}$$

$$-\nabla^2\psi + \chi_{kl}^E\psi_{,lk} - \lambda_{lmk}^{\epsilon E}(u_{m,lk} + \epsilon_{mlr}\phi_{r,k}) - \lambda_{lmk}^{\gamma E}\phi_{m,lk} - q_e = 0, \tag{8.4.16}$$

where we introduced electrostatic potential ψ by $\mathbf{E} = -\nabla\psi$ which satisfies $\nabla \times \mathbf{E} = \mathbf{0}$.

Equations (8.4.10) constitute seven partial differential equations to determine seven unknowns $u_k(\mathbf{x},t)$, $\phi_k(\mathbf{x},t)$, and $\psi(\mathbf{x},t)$. Boundary and initial conditions are given by (8.4.7) and (8.4.8) with $T = 0$.

D. Magneto-elasticity

For nonferrous materials, effects of charge, displacement current, and polarization can be neglected. With this provision, the field equations of anisotropic media can be deduced from the balance laws and constitutive equations. Here, we give the field equations for the isotropic solids.

Balance laws reduce to

$$\nabla \times \mathbf{E} + \frac{1}{c}\frac{\partial \mathbf{B}}{\partial t} = \mathbf{0}, \quad \nabla \cdot \mathbf{B} = 0$$
$$\nabla \times (\mathbf{B} - \mathbf{M}) = \mathbf{J}/c,$$
$$\nabla \cdot \mathbf{t}^M + \frac{1}{c}\mathbf{J} \times \mathbf{B} + (\nabla\mathbf{B}) \cdot \mathbf{M} + \rho\mathbf{f}^M - \rho\ddot{\mathbf{u}} = \mathbf{0},$$
$$m_{kl,k}^M + \epsilon_{lmn}t_{mn}^M + (\mathbf{M} \times \mathbf{B})_l + \rho l_l^M - \rho j\ddot{\phi}_l = 0,$$
$$\rho T_0\dot{\eta} - \nabla \cdot \mathbf{q} - \mathbf{J} \cdot (\mathbf{E} + \frac{1}{c}\mathbf{v} \times \mathbf{B}) - \rho h^M = 0. \tag{8.4.17}$$

Constitutive equations are required to close the system:

$$\mathbf{M} = -\chi^B\mathbf{B},$$
$$\mathbf{J} = \sigma^E\left(\mathbf{E} + \frac{1}{c}\mathbf{v} \times \mathbf{B}\right) + \frac{1}{T_0}\sigma^{ET}\nabla T,$$

$$t_{kl}^M = -\beta_0 T \delta_{kl} + \lambda u_{r,r} \delta_{kl} + (\mu + \kappa)(u_{l,k} + \epsilon_{lkm}\phi_m)$$
$$+ \mu(u_{k,l} + \epsilon_{klm}\phi_m),$$
$$m_{kl}^M = \alpha\phi_{r,r}\delta_{kl} + \beta\phi_{k,l} + \gamma\phi_{l,k},$$
$$\mathbf{q} = \frac{K}{T_0}\nabla T + \sigma^{ET}\left(\mathbf{E} + \frac{1}{c}\mathbf{v}\times\mathbf{B}\right),$$
$$\eta = \eta_0 + (C_0/T_0)T + (\beta_0/\rho)\nabla\cdot\mathbf{u}. \tag{8.4.18}$$

Substituting these into (8.4.16), we obtain the field equations of magnetothermo-micropolar elasticity

$$\nabla\times\mathbf{E} + \frac{1}{c}\frac{\partial\mathbf{B}}{\partial t} = \mathbf{0}, \quad \nabla\cdot\mathbf{B} = 0, \tag{8.4.19}$$

$$(1+\chi^B)\nabla\times\mathbf{B} + \frac{\sigma^E}{c}\left(\mathbf{E}+\frac{1}{c}\mathbf{v}\times\mathbf{B}\right) - \frac{1}{cT_0}\sigma^{ET}\nabla T = \mathbf{0}, \tag{8.4.20}$$

$$-\beta_0\nabla T + (\lambda+2\mu+\kappa)\boldsymbol{\nabla}\boldsymbol{\nabla}\cdot\mathbf{u} - (\mu+\kappa)\nabla\times\nabla\times\mathbf{u} + \kappa\nabla\times\boldsymbol{\phi}$$
$$+\frac{\sigma^E}{c}\left(\mathbf{E}+\frac{1}{c}\mathbf{v}\times\mathbf{B}\right)\times\mathbf{B} + \frac{\sigma^{ET}}{cT_0}\nabla T\times\mathbf{B} - \chi^B(\nabla\mathbf{B})\cdot\mathbf{B}$$
$$+\rho\mathbf{f}^M - \rho j\ddot{\mathbf{u}} = \mathbf{0}, \tag{8.4.21}$$

$$(\alpha+\beta+\gamma)\boldsymbol{\nabla}\boldsymbol{\nabla}\cdot\boldsymbol{\phi} - \gamma\nabla\times\nabla\times\boldsymbol{\phi} + \kappa\nabla\times\mathbf{u} - 2\kappa\boldsymbol{\phi} + \rho\mathbf{l}^M - \rho j\ddot{\boldsymbol{\phi}} = \mathbf{0}, \tag{8.4.22}$$

$$\rho C_0\dot{T} + \beta_0 T_0\nabla\cdot\dot{\mathbf{u}} - \frac{K}{T_0}\nabla^2 T - \sigma^{ET}[\nabla\cdot\mathbf{E} + \frac{1}{c}\nabla\cdot(\mathbf{v}\times\mathbf{B})]$$
$$-\frac{1}{T_0}\sigma^{ET}\nabla T\cdot\left(\mathbf{E}+\frac{1}{c}\mathbf{v}\times\mathbf{B}\right) - \sigma^E\left(\mathbf{E}+\frac{1}{c}\mathbf{v}\times\mathbf{B}\right)\cdot\left(\mathbf{E}+\frac{1}{c}\mathbf{v}\times\mathbf{B}\right) = 0. \tag{8.4.23}$$

This is a system of partial differential equations to determine $\mathbf{u}(\mathbf{x},t)$, $\boldsymbol{\phi}(\mathbf{x},t)$, $T(\mathbf{x},t)$, and $\mathbf{B}(\mathbf{x},t)$ under boundary and initial conditions specified by (8.4.7) and (8.4.8).

Further reductions of these equations are possible by elimination of $\mathbf{E}$ from these equations.

8.5   Piezoelectric Waves

Field equation (8.4.14–8.4.16) for piezoelectricity shows that micropolar displacement and rotation fields are coupled with the electric field. For simplicity, we consider centrosymmetric crystals for which $C_{klmn} = 0$. Then, for plane-harmonic waves, we write

$$(u_k, \phi_k, \psi) = (U_k, \Phi_k, \Psi) \exp[i(\mathbf{k} \cdot \mathbf{x} - \omega t)], \tag{8.5.1}$$

where U_k, Φ_k, and Ψ are constant wave amplitudes, $\mathbf{k}$ is the wave-vector, and ω is the circular frequency. Substituting (8.5.1) into (8.4.14–8.4.16), we obtain a system of linear homogeneous equations

$$A_{lr}^i U_r + B_{lr}^i \Phi_r + C_l^i \Psi = 0, \quad A_r^3 U_r + B_r^3 \Phi_r + C^3 \Psi = 0, \tag{8.5.2}$$

where

$$
\begin{aligned}
A_{lr}^1 &\equiv (-A_{klmr} k_m k_k + \rho \omega^2 \delta_{lr}, \\
B_{lr}^1 &\equiv (A_{klmn} i k_k \epsilon_{nmr}), \quad C_l^1 \equiv \lambda_{klm}^{\epsilon E} k_m k_k, \\
A_{lr}^2 &\equiv \epsilon_{lmn} i k_p A_{mnpr}, \quad B_{lr}^2 \equiv (-B_{lkrn} k_n k_k + \epsilon_{lmn} A_{mnpq} \epsilon_{qpr} + \rho j_{rl} \omega^2), \\
C_l^2 &\equiv \lambda_{klm}^{\gamma E} k_m k_k - \lambda_{mnp}^{\gamma E} i k_p \epsilon_{lmn}, \\
A_r^3 &\equiv -\lambda_{lrk}^{\epsilon E} k_l k_k, \quad B_r^3 \equiv (-\lambda_{lmk}^{\epsilon E} i k_k \epsilon_{mlr} + \lambda_{mrk}^{\gamma E} k_m k_k), \\
C^3 &= k^2 - \chi_{kl}^E k_k k_l.
\end{aligned} \tag{8.5.3}
$$

For the solution U_r, Φ_r, and Ψ to exist, the coefficient determinant of (8.5.2) must vanish

$$
\begin{vmatrix}
A_{11}^1 & A_{12}^1 & A_{13}^1 & B_{11}^1 & B_{12}^1 & B_{13}^1 & C_1^1 \\
A_{21}^1 & A_{22}^1 & A_{23}^1 & B_{21}^1 & B_{22}^1 & B_{23}^1 & C_2^1 \\
\ldots & \ldots & \ldots & \ldots & \ldots & \ldots & \ldots \\
A_1^3 & A_2^3 & A_3^3 & B_1^3 & B_2^3 & B_3^3 & C^3
\end{vmatrix} = 0. \tag{8.5.4}
$$

This is a 7×7 determinant whose roots give the dispersion relations

$$D(k_i, \omega^2) = 0. \tag{8.5.5}$$

This is a polynomial of 14 degrees in the products of k_i and 6 degrees in ω^2. Thus, the determination of ω^2 as a function of k_i requires finding the roots of a sixth-degree polynomial. In general, this is not possible to obtain in explicit form.

For the system (8.5.2), it is clear that the displacement and rotation fields produce an electric field and, conversely, the electric field will affect the displacement and rotations. Excluding the Peltier effect (σ^{ET}), the interaction of the electric field with the motion is through third-degree material tensors $\lambda_{klm}^{\epsilon E}$ and $\lambda_{klm}^{\gamma E}$. Thus, for isotropic media, there will be no coupling (with $\sigma^{ET} = 0$). For the coupling to exist, some components of these third-order tensors must be admitted by the material symmetry group. For example, this will be the case for the Rochelle Salt (34°C) which shows that

$$
\begin{aligned}
D_1 &= E_1 + P_1 = d_{14} t_{23}, \quad D_2 = E_2 + P_2 = d_{25} t_{31}, \\
D_3 &= E_3 + P_3 = d_{36} t_{12},
\end{aligned} \tag{8.5.6}
$$

where

$$d_{14} = 345, \quad d_{25} = 54, \quad d_{36} = 12 \quad \text{in units } 10^{-12}\text{C/N},$$

cf. Berlincourt et al. [1964]. From (8.5.6), it is clear that, at least, $\lambda_{231}^{\epsilon E}$, $\lambda_{312}^{\epsilon E}$, $\lambda_{123}^{\epsilon E}$ do not vanish for the Rochelle Salt. For this material then, there will be coupling among the displacement $\mathbf{u}$, rotation $\boldsymbol{\phi}$, and the electric field $\mathbf{E}$. Of interest here is the coupling between $\boldsymbol{\phi}$ and $\mathbf{E}$. The electric field in a micropolar solid not only produces a displacement, but also a rotation. Conversely, rotation can produce an electric field. Thus, we have a new physical phenomena due to the coupling of microrotation with the electric field. This effect may be used to determine new piezoelectric material constants.

Chapter 8 Problems

8.1. Obtain the dispersion relations for the micropolar piezoelectric solids.

8.2. Determine the dispersion relations for the magneto-micropolar elastic waves in an infinite solid having a constant bias magnetic field.

8.3. Determine the dispersion relations for the piezoelectric quartz.

8.4. Determine static stress concentration around a circular hole in an infinite isotropic micropolar plate subject to a constant bias magnetic field.

8.5. (Short-Term Paper) Construct a plate theory for the magneto-micropolar elastic solids.

References

[1] Akiyama, K., Morioka, Y., and Nakagawa, I. [1980] Raman and Infrared Spectra and Lattice Vibrations, *J. Phys. Soc. Japan* **48**, 898.

[2] Anderson, R.W. [1960] in *Fizika Dielektrikov*, (G.I. Skamavi, ed.) (*Akad Nauk SSSR*), Fizikeskii Inst., Lebedevn, Moscow.

[3] Anthony, K.H. [1970] Die Theorie der Dislocationen, *Arch. Rational Mech. Anal.* **39**, 1, 43.

[4] Ariman, T., Turk, M.A., and Sylvester, N.D. [1973] Microcontinuum Fluid Mechanics—A Review, *Int. J. Engng. Sci.* **11**, 905.

[5] Ariman, T., Turk, M.A., and Sylvester, N.D. [1974] Applications of Microcontinuum Fluid Mechanics, *Int. J. Engng. Sci.* **12**, 273.

[6] Aşkar, A. [1972] Molecular Crystals and the Polar Theories of Continua. Experimental Values of Material Coefficients for KNO_3, *Int. J. Engng. Sci.* **10**, 293–300.

[7] Aşkar, A. [1985] *Lattice Dynamical Foundations of Continuum Theories*, World Scientific, Singapore.

[8] Aşkar, A. and Çakmak, A.S. [1968] A Structural Model of Micropolar Continuum, *Int. J. Engng. Sci.* **6**, 583–589.

[9] Balkanski, M., Teng, M.K., and Musimovici, M. [1968] Raman Scattering in KNO_3 Phases I, II, and III, *Phys. Res.* **176**, 1098.

[10] Balkanski, M. and Teng, M.K. [1969] *Physics of the Solid State* (Balakrishna, S., Krishnamurthy, B., and Ramachandra Rao, B., eds.), Academic Press, New York.

[11] Barret, H.H. [1970] Acoustic Properties of Material of the Perovskite Structure, *Phys. Acoustic* **V1**, edited by W.P. Mason and R.N. Thurston, Academic Press 65.

[12] Barrow, J.D. [1995] *The Artful Universe*, Clarendon Press, Oxford.

[13] Bateman, H. [1953] *Higher Transcendental Functions*, Vol. II, McGraw-Hill, New York, p. 92.

[14] Bažant, Z.P. and Christensen, M. [1972] Analogy Between Micropolar Continuum and Grid Frameworks Under Initial Stress, *Int. J. Solids Structures* **8**, 327–346.

[15] Bedford, A. and Drumheller, D.S. [1983] Theories of Immiscible and Structured Mixtures, *Int. J. Engng. Sci.* **21**, 863–960.

[16] Berglund, K. [1977] Investigation of a Two-Dimensional Model of a Micropolar Continum, *Arch. Mech.* **29**, 383-392.

[17] Berglund, K. [1982] Structural Models of Micropolar Media, in *Mechanics of Micropolar Continua*, (Brulin, O. and Tsich, R.K., eds.), World Scientific, Singapore, pp. 35–85.

[18] Bhargava, R.D. and Ghosh, S.K. [1975] On the Stress Concentration Problem Around a Circular Hole in Plane Asymmetric Elasticity, *Acta Mech.* **21**, 127–140.

[19] Bofill, F. and Quintanilla, R. [1995] Some Qualitiative Results for the Linear Theory of Thermo-Microstretch Elastic Solids, *Int. J. Engng. Sci.* **35**, 2115–2125.

[20] Born, M. and Huang, K. [1954] *Dynamical Theory of Crystal Lattices*, Oxford University Press, London.

[21] Carbonaro, B. and Russo, R. [1985] On Some Classical Theorems in Micropolar Elasticity, *Int. J. Engng. Sci.* **23**, 119–129.

[22] Chadha, T.K. and Kumar, R. [1988] Steady-State Axisymmetric Problem in Micropolar Thermoelastic Half-Space with Stretch, *Int. J. Engng. Sci.* **26**, 663–672.

[23] Chandrasekharaiah, D.S. [1987] Variational and Reciprocal Principles in Micropolar Thermoelasticity, *Int. J. Engng. Sci.* **25**, 55–63.

[24] Cheng, Z.Q. and He, L. [1995] Micropolar Elastic Fields Due to a Spherical Inclusion, *Int. J. Engng. Sci.* **33**, 389–398.

[25] Ciarletta, M. [1998] On the Bending of Micro-Stretch Elastic Plates, submitted to *Int. J. Engng. Sci.*

[26] Claus, W.D. and Eringen, A.C. [1971] Dislocation Dispersion of Elastic Waves, *Int. J. Engng. Sci.* **9**, 605–610.

[27] Claus, W.D. Jr. and Eringen, A.C. [1969] Three Dislocation Concepts and Micromorphic Mechanics, in *Development in Mechanics*, Vol. **6**, Proc. of the 12th Midwestern Mech. Conf., pp. 349–358.

[28] Cochran, W. [1960] *Adv. Phys.* Taylor and Francis, London, **9**, 387 edited by Mott N.F.

[29] Constanda, C. [1974] On the Bending of Micropolar Plates, *Lett. Appl. Engng. Sci.* **2**, 329–339.

[30] Cosserat, E. and F. [1909] *Theórie des Corps Deformable*, Hermann, Paris.

[31] Cracium, I.A. [1995] Uniqueness Theorem in Linear Theory of Piezoelectric Micropolar Thermoelasticity, *Int. J. Engng. Sci.* **33**, 1027–1036.

[32] Dahler, J.S. and Scriven, L.E. [1963] Theory of Structured continua. I. General Considerations of Angular Momentum and Polarization, *Proc. Roy. Soc. London, Ser. A* **275**, 505–527.

[33] De Cicco, S. and Nappa, L. [1997] Torsion and Flexure of Micro-Stretch Elastic Circular Cylinders, *Int. J. Engng. Sci.* **35**, 573–583.

[34] De Gennes, P.G. and Prost, J. [1995] *Physics of Liquid Crystals, 2nd ed.*, Oxford Scientific, Oxford.

[35] De Groot, S.R. and Suttorp, L.G. [1972] *Foundations of Electrodynamics*, North-Holland, Amsterdam.

[36] Demiray, H. and Eringen, A.C. [1975] A Two-Dimensional Lattice Model for Composites Reinforced with Orthogonal Fibers, *Lett. Appl. Engng. Sci.* **3**, 295-309.

[37] Demiray, H. and Eringen, A.C. [1977] Wave Propagation in Viscoelastic Composites Reinforced by Orthogonal Fibers, *J. Sound and Vibration* **55** 509–519.

[38] Demiray, H. and Eringen, A.C. [1978] Wave Propagation in Orthogonally Reinforced Fiber Composites, *Letters in Appl. & Engng. Sci.* **1**, 435–485.

[39] DeWitt, R. [1973] Theory of Disclinations II Continuous and Discrete Disclinations in Anisotropic Elasticity, *J. Res. Natn. Bur. Stand.* 77A 49.

[40] Dhaliwal, R.S. [1973] The Axisymmetric Boussiness Problem for a Semi-Space in Couple Stress Theory, *Int. J. Engng. Sci.* **11**, 1161–1174.

[41] Dhaliwal, R.S. and Khan, S.M. [1976] The Axisymmetric Boussinesq, Problem for a Semi-Space in Micropolar Theory, *Int. J. Engng. Sci.* **14**, 769–788.

[42] Dragoş, L. [1984] Fundamental Solutions in Micropolar Elasticity, *Int. J. Engng. Sci.* **22** 265-275.

[43] Dresen, L., Kozak, J., Spicak, A., Waniek, L., and Teisseyre, R. [1984] Wave Propagation in Physical Models of Micromorphic Media, *Studia Geoph. et Geod.*, **28**, 272-285.

[44] Edelen, D.G.B. [1973] A Nonlinear Onsager Theory of Irreversibility, *Int. J. Engng. Sci.*, **10**, 481-490.

[45] Edelen, D.G.B. [1993] *The College Station Lectures on Thermodynamics*. Texas A & M University, College Station, Texas.

[46] Ellis, R.W. and Smith, C.W. [1967] A Thin-Plate Analysis and Experimental Evaluation of Couple-Stress Effects, *Exp. Mech.*, **7**, 372.

[47] Erbay, S. and Şuhubi, E.S. [1989] Nonlinear Wave Propagation in Micropolar Media—I and II, *Int. J. Engng. Sci.* **27** 895-920.

[48] Erbay, S., Erbay, H.A., and Dost S. [1991] Nonlinear Wave Modulation in Micropolar Elastic Media—I and II, *Int. J. Engng. Sci.* **29**, 845-868.

[49] Eringen, A.C. [1962] *Nonlinear Theory of Continuous Media*, McGraw-Hill, New York.

[50] Eringen, A.C. [1964a] Simple Micro-Fluids, *Int. J. Engng. Sci.* 205-217.

[51] Eringen, A.C. [1964b] Mechanics of Micromorphic Materials in *Proc. 11th Int. Congress of Appl. Mech.* (Görtler, H., ed.) Springer-Verlag, New York.

[52] Eringen, A.C. [1965] Theory of Micropolar Continua, in *Proceedings of the Ninth Midwestern Mechanics Conference*, Wisconsin, Aug. 16-18, 1965, Wiley, New York.

[53] Eringen, A.C. [1966a] Linear Theory of Micropolar Elasticity, *J. Math. Mech.* **15**, 909-923.

[54] Eringen, A.C. [1966b] Theory of Micropolar Fluids, *J. Math. Mech.* **16**, 1-18.

[55] Eringen, A.C. [1966c] A Unified Theory of Thermomechanical Materials, *Int. J. Engng. Sci.* **4**, 179-202.

[56] Eringen, A.C. [1967a] *Mechanics of Continua*, Wiley, New York. Revised edition 1980 Krieger, Melbourne, Florida.

[57] Eringen, A.C. [1967b] Linear Theory of Micropolar Viscoelasticity, *Int. J. Engng. Sci.* **5**, 191-204.

[58] Eringen, A.C. [1967c] Compatibility Conditions of Micromorphic Elastic Solids, NASA report (1967); *J. Math and Mech.* **19**, 473-481.

[59] Eringen, A.C. [1967d] Theory of Micropolar Plates, *J. Appl. Phys.* (ZAMP), **18**, 1, 12-30.

[60] Eringen, A.C. [1968a] *Theory of Micropolar Elasticity*, Chapter 7 of *Fracture*, Vol. II (Liebowitz, H., ed.), Academic Press, New York.

[61] Eringen, A.C. [1968b] *Mechanics of Micromorphic Continua*, *IUTAM Symposium, Mechanics of Generalized Continua*, pp.18-35 (Kröner, ed.) Springer-Verlag, New York.

[62] Eringen, A.C. [1969a] Micropolar Fluids with Stretch, *Int. J. Engng. Sci.* **7**, 115-127.

[63] Eringen, A.C. [1969b] Compatibility Conditions of the Theory of Micromorphic Elastic Solids, *J. Math Mech.* **19**, 6, 473-481.

[64] Eringen, A.C. [1970a] *Foundations of Micropolar Thermoelasticity: Intern. Cent. for Mech. Studies*, Course and Lectures, No. 23, Springer-Verlag, Wien.

[65] Eringen, A.C. [1970b] Balance Laws of Micromorphic Mechanics, *Int. J. Engng. Sci.* **8** 819-828.

[66] Eringen, A.C. [1971a] *Tensor Analysis Continuum Physics, Vol. I.*, Acadic. Press, New York.

[67] Eringen, A.C. [1971b] Micropolar Elastic Solids with Stretch, Professor Mustafa Inan Anisma, Ari Kitapevi Matbaasi, 1-18, Istanbul, Turkey.

[68] Eringen, A.C. [1971c] Micromagnetism and Superconductivity, *J. Math Phys.* **12**, 7, 1353-1358.

[69] Eringen, A.C. [1972a] Theory of micromorphic materials with memory, *Int. J. Engng. Sci.* **10**, 623-641.

[70] Eringen, A.C. [1972b] Theory of Thermomicrofluids, *J. Math. Anal. Appl.* **38**, 480-496.

[71] Eringen, A.C. [1973] Linear Theory of Nonlocal Microelasticity and Dispersion of Plane Waves, *Lett. Appl. Engng. Sci.* **1**, 11-17.

[72] Eringen, A.C. [1974] *Polar and Nonlocal Theories of Continua and Applications*, Bogaziçi University Publications, Istanbul, Turkey.

[73] Eringen, A.C. [1977] Screw Dislocation in Nonlocal Elasticity, *J. Phys. D: Appl. Phys.* **10**, 671-678.

[74] Eringen, A.C. [1978a] Nonlocal Continuum Mechanics and Some Applications, in *Nonlinear Equations in Physics and Mathematics*, pp. 271-318 (Barut, A.O.,ed.), Reidel, Dordrecht, Holland.

[75] Eringen, A.C. [1978b] Line Crack Subject to Shear, *Int. J. Fracture*, **14**, 367-379.

[76] Eringen, A.C. [1980] *Mechanics of Continua, 2nd ed.*, Krieger, Melbourne, Florida.

[77] Eringen, A.C. [1983] On Differential Equations of Nonlocal Elasticity and Solutions of Screw Dislocation and Surface Waves, *J. Appl. Phys.* **54**(9), 4703-4710.

[78] Eringen, A.C. [1984a] Nonlocal Stress Fields of Dislocations and Crack, in *Modelling Problems in Crack Tip Mechanics*, pp. 113–130 (Pindera, J.T., ed.), University of Waterloo, Canada, Martinus Nijhoff.

[79] Eringen, A.C. [1984b] On Continuous Distributions of Dislocations in Nonlocal Elasticity, *J. Appl. Phys.* **56**, 2675-2680.

[80] Eringen, A.C. [1984c] Theory of Nonlocal Piezoelectricity, *J. Math Phys.* **25**, 717-727.

[81] Eringen, A.C. [1984d] Electodynamics of Memory-Dependent Nonlocal Elastic Continua, *J. Math Phys.* **25** (11), 3235-3249.

[82] Eringen, A.C. [1985] Rigid Suspensions in Viscous Fluids, *Lett. Appl. Engng. Sci.* **22**, 491-495.

[83] Eringen, A.C. [1987] Theory of Nonlocal Elasticity and Some Applications, *Res. Mechanica* **21**, 313-342.

[84] Eringen, A.C. [1990] Theory of Thermo-Microstretch Elastic Solids, *Int. J. Engng. Sci.* **28**, 1291-1301.

[85] Eringen, A.C. [1991] Continuum Theory of Dense Rigid Suspensions, *Rheol. Acta.* **30**, 23-32.

[86] Eringen, A.C. [1992] Balance Laws of Micromorphic Continua Revisited, *Int. J. Engng. Sci.* **30**, 805-810.

[87] Eringen, A.C. [1994] Disclinations in Liquid Crystals, *Bull. Tech. Univ. Istanbul* **47**, 1-21.

[88] Eringen, A.C. and Claus, W.D. Jr. [1970] A Micromorphic Approach to Dislocation Theory and its Relation to Several Existing Theories, in *Fundamental Aspects of Dislocation Theory* (Simmons, J.A., DeWit, R., and Bullough, R., eds.), *Nat. Bur. of Stand. Publ.* **317**, 2: 1023-1040.

[89] Eringen, A.C. and Edelen, D.G.B. [1972] On Nonlocal Elasticity, *Int. J. Engng. Sci.* **10** 233-248.

[90] Eringen, A.C. and Ingram, J. [1965] A Continuum Theory of Chemically Reacting Media, *Int. J. Engng. Sci.* **3**, 197-212 and **5**, 289-322.

[91] Eringen, A.C. and Kafadar, C.B. [1970] Relativistic Theory of Micromagnetism, *J. Math Phys.* **11**, 5, 1984-1991.

[92] Eringen, A.C. and Kafadar, C.B. [1976] *Polar Field Theories, Continuum Physics, Vol. IV*, Academic Press, New York.

[93] Eringen, A.C. and Maugin, G.A. [1990] *Electrodynamics of Continua, I & II*, Springer-Verlag, New York.

[94] Eringen, A.C., Speziale, C.G., and Kim, B.S. [1977] Crack-Tip Problem in Nonlocal Elasticity, *J. Mech. Phys. Solids* **25**, 334-335.

[95] Eringen, A.C. and Şuhubi, E.S. [1964] Nonlinear Theory of Simple Microelastic Solids I & II, *Int. J. Engng. Sci.* **2** 189-203 and 389-404.

[96] Eringen, A.C. and Şuhubi, E.S. [1975] *Elastodynamics, Vol II*, Academic Press, New York.

[97] Ellis, R.W. and Smith, C.W. [1967] A Thin-Plate Analysis and Experimental Evaluation of Couple-Stress Effects, *Exp. Mech.* **7**, 372.

[98] Ewing, W.M., Jardetski, W.S., and Press, F. [1957] *Elastic Waves in Layered Media*, McGraw-Hill, New York.

[99] Gauthier, R.D. [1982] Experimental Investigations of Micropolar Media, in *Mechanics of Micro-Polar Media* (Brulin, O. and Hsieh, R.K.T., eds., World Scientific, Singapore, pp. 395-463.

[100] Gauthier, R.D. and Jahsman, W.E. [1975] A Quest for Micropolar Elastic Constants, *J. Appl. Mech.* **42**; *Trans. ASME*, Ser. E. **97**, 369.

[101] Grad, H. [1952] Statistical Mechanics, Thermodynamics and Fluid Dynamics of Systems with an Arbitrary Number of Integrals, *Comm. Pure Appl. Math.* **5**, 455-494.

[102] Green, A.E. and Adkins, J.E. [1960] *Large Elastic Deformations*, Chapter I, Clarendon Press, Oxford.

[103] Green, A.E. and Naghdi, P.M. [1967] Micropolar and Director Theories of Plates, *Quart. J. Mech. Appl. Math.* **20**, 183-199.

[104] Grot, R.A. [1969] Thermodynamics of a Continuum with Microstructure, *Int. J. Engng. Sci.* **7**, 801-814.

[105] Günther, W. [1958] Zur Statik und Kinematik des Cosseratschen Kontinuums, *Abh. Braunschweig. Wiss. Ges.* **10**, 195.

[106] Hirashima, Ken-ichi [1976] Fundamental Solutions for Stress Concentration Problems of a Two-Dimensional Cosserat Elastic Body, *Proc. 26th Japan National Congress for Applied Mechanics*, pp. 237–247. University of Tokyo Press, Tokyo.

[107] Hirth, J.P. and Lothe, J. [1968] *Theory of Dislocations*, McGraw-Hill, New York.

[108] Hjalmar, Inga Fischer- and Hjalmar, Stig. [1981] The Magnitude of Polar Effects Estimated from Simple Structural Micromodels, in *Continuum Models of Discrete Systems* (Brulin, O. and Hsieh, R.K.T. eds.), North-Holland, Amsterdam.

[109] Inga Fischer-Hjalmar [1981a] Micropolar Phenomena in Ordered Structures, in *Mechanics of Micropolar Media* (Brulin, O. and Hsieh, R.K.T., eds.), World Scientific, Singapore.

[110] Inga Fischer-Hjalmar [1981b] On the Validity of the Continuum Description of Molecular Crystals, *Int. J. Engng. Sci.* **19**, 1765-1773.

[111] Ieşan, D. [1967] Sur la Théorie de la Thermoelasticité Micropolaire Coupleé, *Comptes Rendus Acad. Sci., Paris*, **265**A, 271-274.

[112] Ieşan, D. [1969] On the Linear Theory of Micropolar Elasticity, *Int. J. Engng. Sci.* **7**, 1213-1220.

[113] Ieşan, D. [1971] Existence Theorems in Micropolar Elastostatics, *Int. J. Engng. Sci.* **9**, 59-78. For plane strain, see *Int. J. Engng. Sci.* **8**, 777-791, 1970.

[114] Ieşan, D. [1981] Some Applications of Micropolar Mechanics to Earthquake Problems, *Int. J. Engng. Sci.* **19**, 855-863.

[115] Ieşan, D. [1982] Saint-Venant's Problem, in *Micropolar Elasticity Mechanics of Micropolar Media* (Brulin, O. and Hsieh, R.K.T., eds.), World Scientific, Singapore, pp. 281-393.

[116] Ieşan, D. and Nappa, L. [1994] Saint-Venant's Problem for Microstretch Elastic Solids, *Int. J. Engng. Sci.* **32**, 229-236.

[117] Ieşan, D. and Nappa, L. [1995] Extension and Bending of Microstretch Elastic Circular Cylinders, *Int. J. Engng. Sci.* **33**, 1139-1152.

[118] Ieşan, D. and Quintanilla, R. [1994] Existence and Continuous Dependence Results in the Theory of Microstretch Elastic Bodies, *Int. J. Engng. Sci.* **32**, 991-1001.

[119] Ieşan, D. and Scalia, A. [1997]. On Saint-Venant's Principle for Microstretch Elastic Bodies, *Int. J. Engng. Sci.* **35**, 1277-1290.

[120] Ignaczak, J. [1963] A Completeness Problem for Stress Equations of Motion in the Linear Elasticity Theory, *Arch. Mech. Stas.* **15**, 225-234.

[121] Jahsman, W.E. and Gauthier, R.D. [1980] Dynamic Measurements of Micropolar Elastic Constants, in *Continuum Models of Discrete Systems 3*, (Kröner, E. and Anthony, K.H., eds.,) SM Study No. 15, University of Waterloo Press, Ontario, Canada, p.667.

[122] Jeffreys, H. and Jeffrey, B.S. [1950] *Methods of Mathematical Physics*, Cambridge University Press, Cambridge, p. 96.

[123] Kafadar, C.B. and Eringen, A.C. [1971] Polar Media—The Classical and Relativistic Theory, *Int. J. Engng. Sci.* **27**, 305 and 307-329.

[124] Kaloni, P.N. and Ariman, T. [1967] Stress Concentration Effect in Micropolar Elasticity, *J. Appl. Math Phys. (ZAMP)* **18**, 136-141.

[125] Kanatani, K.I. [1979] A Micropolar Continuum Model for Vibrating Grid Frameworks, *Int. J. Engng. Sci.* **17**, 409-418.

[126] Kessel, S. [1970] Spannungsfelder einer Schraubenversetzung und einer Stufenveersetzung im Cosserat Kontinuum, *Z. Angeus. Math. Mech.* **50**, 547.

[127] Khan, S. M. and Dhaliwal, R. S. [1977] Axisymmetric Problem for a Half-Space in Micropolar Theory of Elasticity, *J. Elasticity* **7**, 1, 13-32.

[128] Kim, B.S. and Eringen, A.C. [1973] Stress Distribution Around an Elliptic Hole in an Infinite Miropolar Plate, *Lett. Appl. Engng. Sci.* **1**, 381-390.

[129] Kim, Kil Soo and Piziali, R.L. [1987] Continuum Models of Materials with Beam-Microstructure, *Int. J. Solids Structures* **23**, 1563-1578.

[130] Kiral, E. and Eringen, A.C. [1990] *Constitutive Equations of Nonlinear Electromagnetic-Elastic Crystals*, Springer-Verlag, New York.

[131] Kossecka, E. and DeWitt, R. [1977] Disclination Dynamics, *Arch. Mech.* **29**, 749.

[132] Kirshma, R. G.V. and Venkatasubramanian, N.K. [1976] Saint-Venand's Problem for a Micropolar Elastic Circular Cylinder, *Int. J. Engng. Sci.* **14**, 1047.

[133] Kumar, R. and Chadha, T.K. [1985] Axisymmetric Problem in Micropolar Elastic Half-space with Stretch, *Int. J. Engng. Sci.* **23**, 1009-1029.

[134] Kunin, I.A. [1982, 1983] *Elastic Media with Microstructure I & II*, Springer-Verlag, Berlin.

[135] Lukaszewicz, G. [1999] *Micropolar Fluids: Theory and Application*, Birkhäuser, Boston.

[136] Magnus W. and Oberhettinger, F. [1949] *Formulas and Theorems for the Special Functions of Mathematical Physics*, Chelsea, New York.

[137] Maiti, N.C. [1987] Note on Problems of Spherical Cavity in Isotropic, Micropolar, Elastic Medium, *Int. J. Engng. Sci.* **25**, 127-132.

[138] Maradudin, A.A., Montroll, E.W., Weiss, G.E., and Ipatova, I.P. [1971] *Theory of Lattice Dynamics in the Harmonic Approximation, 2nd ed.*, Academic Press, New York.

[139] Maugin, M. [1970]. Un Principe Variationnel pour des Milieu Micromorphic Non Dissipative, *C. R. Acad. Sci., Paris*, **271**, 807-810.

[140] Maugin, G.A. and Miled, A. [1986] Solitary Waves in Micropolar Elastic Crystals, *Int. J. Engng. Sci.* **9**, 1477-1499.

[141] Miesowicz, M. [1935] *Nature* **17**, 261; *Bull. Acad. Polon. Sci.* **A**, 228 (1936); *Nature*, **158**, 27 (1946).

[142] Minegawa, S. [1977] Elastic Fields of Dislocations and Disclinations in an Isotropic Micropolar Continuum, *Lett. Appl. Engng. Sci.* **5**, 85-94.

[143] Minegawa, S. [1985] Dynamics of Dislocations and Disclinations in Anisotropic Elastic Micropolar Media, *Int. J. Engng. Sci. (Lett. Appl. Engng. Sci.)*, **23**, 251-255.

[144] Morse, P.M. and Feshbach, H. [1953] *Methods of Theoretical Physics, Vols. I & II*, McGraw-Hill, New York.

[145] Musimovici, M.A. [1967] *Proceedings of the Symposium. on Ferroelectricity*, (Weller, E.F. ed.), Elsevier, New York, pp. 123-160.

[146] Nabarro, F.R.N. [1967] *Theory of Crystal Dislocations*, Clarendon Press, Oxford.

[147] Nappa, L. [1995] Thermal Stresses in Micro-Stretch Elastic Cylinders, *J. Thermal Stresses* **18**, 537-550.

[148] Noether, E [1918] Invariante Variations probleme, *Nachr. König. Gesell. Wissen. Göttingen, Math.-Phys. Kl.*, 235–257, See also Olver, P.J. [1986], 277.

[149] Noor, A.K. and Nemeth, M.P. [1980a] Micropolar Beam Models for Lattice Grids and Rigid Joints, *Comput. Methods. Appl. Mech. Engrg.* **21**, 249-263.

[150] Noor, A.K. and Nemeth, M.P. [1980b] Analysis of Spatial Beam Like Lattices with Rigid Joints, *Comput. Methods. Appl. Mech. Engrg.* **24**, 35-59.

[151] Nowacki, J.P. [1977] Theory of Disclinations in Elastic Cosserat Media, *Arch. Mech.* **29**, 531-545.

[152] Nowacki, W. [1969] The Axially Symmetric Lamb's Problem in Semi-Infinite Micropolar Elastic Solid, *Proc. Vibration Problems (Warsaw)* **10**, 2, 97.

[153] Nowacki, W. [1970] *Theory of Micropolar Elasticity, Udline Course & Lectures No. 25*, Springer-Verlag, Wien.

[154] Nowacki, W. [1977] Theory of Disclinations Elastic Cosserat Media, *Arch. Mech.* **29**, 531.

[155] Oevel, W. and Schröter, J. [1981] Balance Equations for Micromorphic Materials, *J. Statist. Phys.* **25**, 4, 645-662.

[156] Olver, P.J. [1986] *Applications of Lie Groups to Differential Equations, Graduate Texts in Mathematics, Vol. 107,*, Springer-Verlag, New York.

[157] Parfitt, V.R. and Eringen, A.C. [1969] Reflection of Plane Waves from the Flat Boundary of a Micropolar Elastic Half-Space, *J. Acoust. Soc. Amer.* **45**, 5, 1258–1272.

[158] Paul, H.S. and Sridharan, K. [1980] The Penny-Shaped Crack Problem in Micropolar Elasticity, *Int. J. Engng. Sci.* **18**, 651–664. See also *Int. J. Engng. Sci.* **18**, 1431-1448.

[159] Perkins, R.W. and Thomson, D. [1973] Experimental Evidence of a Couple Stress Effect, *AIAA J.* **11**, 1053–1054.

[160] Petrosyan, L.G. [1984] Some Problems of Mechanics of Fluids with Asymmetric Stress Tensor, Erevan 1984 (in Russian).

[161] Pouget, J., Aşkar, A., and Maugin, G.A. [1986a] Lattice Model for Elastic Ferroelectric Crystals: Microscopic Approach, *Phys. Rev. B* **33**, 9, 6304–6319.

[162] Pouget, J., Aşkar, A., and Maugin, G.A. [1986b] Lattice Model for Elastic Ferroelectric Crystals: Continuum Approximation, *Phys. Rev. B* **33**, 9, 6320–6325.

[163] Pucci, E. and Saccomandi, G. [1990] Symmetries and Conservation Laws in Micropolar Elasticity, *Int. J. Engng. Sci.* **28**, 557–562.

[164] Puri, P. [1971] Axisymmetric Stress Concentrations Problems Cosserat Elasticity, *ZAMP*, **22**, 320.

[165] Rao, K.M. [1988] Longitudinal Wave Progation in a Micropolar Wave Guide, *Int. J. Engng. Sci.*, **26**, 135–141.

[166] Rao, K.R. and Chaplot, S.L. [1980] Lattice Dynamics of KNO_3, $LiNbO_3$, and $KNbO_3$, *Ferroelectrics*, **29**, 7–10.

[167] Rao, K.R. and Chaplot, S.L. [1980] Lattice Dynamics of KNO_3, $LiNbO_3$ and $KNbO_3$ *Ferroelectrics* **29**, 7.

[168] Rogula, D. (ed.) [1982] *Nonlocal Theory of Material Media*, Springer-Verlag, Berlin.

[169] Russo, R. [1985] On the Principle of Minimum Potential Energy in Asymmetric Elastostatics, *Int. J. Engng. Sci.* **23**, 659.

[170] Sandru, N. [1966] On Some Problems of the Linear Theory of Asymmetric Elasticity, *Int. J. Engng. Sci.* 81–96.

[171] Scalia, A. [1990] On Some Theorems in the Theory of Micropolar Thermoelasticity, *Int. J. Engng. Sci.* **28**, 181–189.

[172] Schijve, J. [1966] Note on Couple Stresses, *J. Mech. Phys. Solids* **14**, 113–120.

[173] Schaefer [1967] Das Cosserat-Kontinuum, *Z. Angew. Math. Mech.*, **47**, 34.

[174] Schiavone, P. [1989a] On Existence Theorems in the Theory of Extensional Motions of Thin Micropolar Plates, *Int. J. Engng. Sci.* **9**, 1129–1133.

[175] Schiavone, P. [1989b] A Generalized Fourier Approximation in Micropolar Elasticity, *J. Appl. Math. Physics* (*ZAMP*) **40**, 838–845.

[176] Schiavone, P. [1990] Generalized Fourier Series for Exterior Problems in Extensional Motions of Thin Micropolar Plates, *Int. J. Engng. Sci.* **28**. 1067–1072.

[177] Schiavone, P. [1991] Uniqueness in Dynamic Problems in Thin Micropolar Plates, *Appl. Math. Lett.* **4**, 81–83.

[178] Schiavone, P. and Constanda, C. [1989] Existence Theorems in the Theory of Bending of Micropolar Plates, *Int. J. Engng. Sci.* **27**, 463–468.

[179] Scott, J.F. [1974] Soft-Mode Spectroscopy: Experimental Studied of Structrual Phase Transitions, *Rev. Mod. Phys.*, **46**, 83–128.

[180] Singh, S.J. [1975] A Spherical Cavity in a Micropolar Medium and Related Problems, *Gerlands Beitr. Geophysik, Leipzig*, **84**, 55–66.

[181] Singh, B. and Kumar, R. [1998] Wave Propagation in a Generalized Thermo-Microstretch Elastic Solid, submitted to *Int. J. Engng. Sci.*.

[182] Smith, A.C. [1968] Inequalities Between the Constants of a Linear Microelastic Solid, *Int. J. Engng. Sci.* **6**, 65–74.

[183] Smith, A.C. [1970] Torsion and Vibrations of Cylinders of a Micropolar Elastic Solid, in *Recent Advances in Engineering Science*, and **5/11**, (Eringen, A.C., ed.), Gordon & Breach, London, p. 129.

[184] Sneddon, I. N. [1972] *The Use of Integral Transforms*, McGraw-Hill, New York.

[185] Soós, E. [1969] Uniqueness Theorems for Homogeneous, Isotropic, Simple Elastic and Thermoelastic Materials Having a Microstructure, *Int. J. Engng. Sci.* **7**, 257–268.

[186] Stevenson, R.W.H. (ed.) [1966] *Phonons*, Oliver & Boyd, Edinburgh.

[187] Stokes, V.K. [1984] *Theories of Fluids with Microstructure*, Springer-Verlag, Berlin.

[188] Şuhubi, E.S. and Eringen, A.C. [1964] Nonlinear Theory of Simple Micro-Elastic Solids II, *Int. J. Engng. Sci.* **2**, 389–404.

[189] Sun, C.T. and Yang, T.Y. [1973] A Continuum Approach Toward Dynamics of Grid Works, *J. Appl. Mech.* 186–192.

[190] Synge, J.L. [1960] *Handbuch der Physik III/1*, (Flugge, S., ed.), Springer-Verlag, Berlin.

[191] Tauchert, T.R. [1970] A Lattice Theory for Representation of Thermoelastic Composite Materials, in *Recent Advances in Engineering Science, Vol. 5, Part I*, (Eringen, A.C., ed.), Gordon & Breach, New York, pp. 325–345.

[192] Tauchert, T.R., Chang, T.Y., and Eringen, A.C. [1968] A Lattice Model for Composite Materials, *Gen. Tech. Corp. Rept.*, No. 8.8.

[193] Teisseyre, R. [1982] Some Seismic Phenomena in the Light of the Symmetric Micromorphic Theory, *J. Tech. Phys.* **38**, 1, 95–99.

[194] Teisseyre, R., Dresen, L., Kozák, J., and Waniek, L. [1985] Physical Properties of Micromorphic Medium: Theory and Experiment, *Acta Geophisica* **4** 341–356.

[195] Teodosiu, C. [1982] *Elastic Models of Crystal Defects*, Springer-Verlag, Berlin.

[196] Tomar, S.K., Kumar, R., and Kaushik, V.P. [1998] Wave Propagation of Micropolar Elastic Medium with Stretch, to appear in *Int. J. Engng. Sci.*.

[197] Twiss, J.T. and Eringen, A.C. [1971] Theory of Mixtures of Micromorphic Materials—Balance Laws, *Int. J. Engng. Sci.* **9** 1019–1044.

[198] Twiss, J.T. and Eringen, A.C. [1972] Theory of Mixtures of Micromorphic materials—II. Elastic Constitutive Equations, *Int. J. Engng. Sci.* **10**, 437–465.

[199] Usidus, C. and Sokolowski, M. [1973] Torsion of Cylindrical Shafts Made of Micropolar Material, *Bull. Acad. Polon. Sci. Sér. Sci. Tech.*, **21**, 19.

[200] Vallis, R.F. (ed.) [1965] *Lattice Dynamics*, Pergamon Press, Oxford, UK.

[201] Vertogen, G. and de Jeu, W.H. [1988] *Thermotropic Liquid Crystals, Fundamentals*, Springer-Verlag, Berlin.

[202] Voigt, W. [1887] Theoritiscke Studien über die Elastizitäts-verhältnisse der Krystalle, Braunschweig. *Abh. Wiss. Ges.* 34. 3–51.

[203] Vukobrat, M.D. [1989] Conservation Laws in Micropolar Elastodynamics and Path-Independent Integrals, *Int. J. Engng. Sci.* **27**, 1093–1106.

[204] Wagner, E.L. and Hornig, D.F. [1950] The Vibrational Spectra of Molecules and Complex Ions in Crystals III. Ammonium Chloride and Dextero-Ammonium Chloride, *J. Chem. Phys.* **18**, 296, 305.

[205] Wang, Fei-Yue [1990] On the Solutions of Eringen's Micropolar Plate Equations and Other Approximate Equations, *Int. J. Engng. Sci.* **28**, 919–925.

[206] Warren, J.L., Wenzel, R.G., and Yarnell, J.L. [1965] Inelastic Scattering of Neutrons, *IAEA*, Vienna.

[207] Weyl, H. [1939] *The Classical Groups*, Princeton University Press, Princeton, NJ.

[208] Woods, A.D.B., Cochran, W. and Brockhouse, B.N. [1960] Lattice Dynamics of Alkali Holide Crystals *Phys. Rev.* **119**, 2, 980–99.

[209] Wozniak, C. [1966a] Load-Carrying Structures of the Dense Lattice Type. The Plane Problem, *Arch. Mech. Stosow.* **18**, 581–597.

[210] Wozniak, C. [1966b] Bending and Stability Problems with Lattice Structures, *Arch. Mech. Stosow.* **18**, 781–796.

[211] Yang, J.F.C. and Lakes, R.S. [1982] Experimental Study of Micropolar and Couple Stress Elasticity in Compact Bone Bending, *J. Biomechanics* **15**, 2, 91–98.

[212] Zheng. Q.S. and Spencer, A.J.M. [1993] On the Canonical Representation for Kronecker Powers of Orthogonal Tensors with Applications to Material Symmetry Problems, *Int. J. Engng. Sci.* **31**, 617–635.

Index